SECURITY AND COOPERATION IN WIRELESS NETWORKS

Thwarting Malicious and Selfish Behavior in the Age of Ubiquitous Computing

As wireless networking becomes almost ubiquitous, it is important to anticipate potential malicious and selfish misdeeds. This self-contained text is the first to provide a scholarly description of security and non-cooperative behavior in wireless networks.

The major networking trends are analyzed and their implications explained in terms of security and cooperation. Key problems such as cheating with identities, illegitimate access to confidential data, attacks against privacy, and "stealing of bandwidth" are described along with the existing security techniques and putative methods of protection for the future. The fundamental questions of security: user and device identification; establishment of security associations; secure and cooperative routing in multi-hop networks; fair bandwidth distribution; privacy protection, and so on, are approached from a theoretical perspective and supported by real-world examples including ad hoc, mesh, vehicular, sensor, and RFID networks. The important relationships between trust, security, and cooperation are also discussed.

End of chapter homework problems test the reader and open new directions of thought; and two tutorials in the appendices, on cryptographic protocols and game theory, provide a review of the background material required to grasp the core concepts.

Ideal for senior undergraduates and graduate students of electrical engineering and computer science, this book will also be an invaluable resource on thwarting malicious and selfish behavior for researchers and practitioners in the wireless industry.

Supplementary material for this title, including lecture slides and instructor-only solutions, are available online at http://www.cambridge.org/9780521873710 and http://secowinet.epfl.ch.

LEVENTE BUTTYÁN is an Associate Professor in the Department of Telecommunications, Budapest University of Technology and Economics (BME), Hungary. JEAN-PIERRE HUBAUX is a Professor at the School of Computer and Communication Sciences, Ecole Polytechnique Fédérale de Lausanne (EPFL), Switzerland.

SECURITY AND COOPERATION IN WIRELESS NETWORKS

Thwarting Malicious and Selfish Behavior in the Age of Ubiquitous Computing

LEVENTE BUTTYÁN

Budapest University of Technology and Economics (BME), Hungary

JEAN-PIERRE HUBAUX

Ecole Polytechnique Fédérale de Lausanne (EPFL), Switzerland

CAMBRIDGE
UNIVERSITY PRESS

CAMBRIDGE
UNIVERSITY PRESS

University Printing House, Cambridge CB2 8BS, United Kingdom

One Liberty Plaza, 20th Floor, New York, NY 10006, USA

477 Williamstown Road, Port Melbourne, VIC 3207, Australia

314-321, 3rd Floor, Plot 3, Splendor Forum, Jasola District Centre, New Delhi - 110025, India

103 Penang Road, #05-06/07, Visioncrest Commercial, Singapore 238467

Cambridge University Press is part of the University of Cambridge.

It furthers the University's mission by disseminating knowledge in the pursuit of education, learning and research at the highest international levels of excellence.

www.cambridge.org
Information on this title: www.cambridge.org/9780521873710

© Cambridge University Press 2008

First published 2008

A catalogue record for this publication is available from the British Library

ISBN 978-0-521-87371-0 Hardback

To Zsombi, Benci, and Boti
To Catherine, Sylvie, Nathalie, and Emilie

Contents

Preface

We have entered the era of wireless networks. By now, the number of wireless phones has superseded that of wired ones. Wireless LANs are routinely used by millions of nomadic users. Wireless devices have become commonplace in offices, private homes, factories, and hospitals. And technologists promise us a world of ubiquitous computing, in which myriads of tiny, untethered sensors and actuators will communicate with each other, promptly taking care of our various needs and wishes.

In addition to this pervasiveness, we are witnessing a change of paradigm: initially, wireless devices had limited or no programmability and were managed (and secured) in a highly centralized fashion. Today, high-tier wireless end-systems are full-fledged personal computers and take an increasingly active role in the networking mechanisms. In the extreme case of multi-hop ad hoc networks, the end-systems *are* the network.

Unfortunately, this evolution is creating new vulnerabilities. Even existing wireless networks (and especially wireless LANs) exhibit a number of security weaknesses, some of which have been painstakingly fixed *a posteriori*. It is now clear that the security solutions devised for wired networks cannot be used as such to protect the wireless ones. An additional problem is that the frenzy to commercialize quickly new products and new services is in contradiction with the design of a well-thought (and possibly standardized) security architecture.

This textbook aims at preventing ubiquitous computing from becoming a pervasive nightmare. It contains a thorough description of existing and envisioned mechanisms devised to thwart misdeeds against wireless networks. Indeed, we believe that the protection of wireless networks now requires more attention and a more systematic *a priori* approach.

In addition to the usual security concerns of networking, we need to address selfish behavior. The reason is that each wireless communication makes use of a fraction of the spectrum that has been and will remain a scarce resource. Moreover,

xi

most wireless devices are battery-powered, and for them energy is scarce as well. Consequently, the behavior of a wireless device can affect the service enjoyed by a another, neighboring device. Likewise, the behavior of a wireless network can affect the performance of another wireless network, especially if both networks operate in the same frequency band. These are the reasons we mention "cooperation" in the title of this book; wherever appropriate, we will make use of game theory in order to formalize the problems.

We believe this textbook to be the first of its kind regarding the treatment of security and cooperation in wireless networks. Owing to the constant evolution of the field, one of the major challenges of writing such a book is ensuring that it will have a reasonably long shelf life (and that the material learned from this book has long lasting value). The strategy we have adopted is to focus on the principles and to keep examples as generic as possible.

What this book is not

This book covers a substantial amount of material, but it obviously does not aim at covering everything. In particular, it is not an introduction to security or cryptography, nor is it a tutorial on game theory (but we do provide an appendix on each of these topics for the convenience of the reader). It is not an introduction to wireless networks. It is not a book on wired networks security. It is not a handbook on jamming and anti-jamming techniques. It is also not a book on wireless security standards (the reader is referred to the numerous books recently published on this topic). Finally, the book is not about the computing aspects of security, such as the protection against viruses.

What this book is about

The book provides a thorough analysis of the major trends in wireless networks and explains the implications in terms of security and cooperation. It provides a detailed description of the problems and a precise explanation of mainstream solutions wherever they exist, and of potential solutions otherwise. The structure of the book is captured by the following figure.

The twelve chapters are organized in three parts. Part I is an **introduction**, providing some background information. Chapter 1 describes how existing wireless networks are secured. Chapter 2 contains a description of upcoming wireless networks, such as mesh, vehicular, sensor and RFID networks. It identifies general trends, such as increasing decentralization and growing programmability of the devices and discusses their implications in terms of security and cooperation. Chapter 3 is devoted to the difficult issue of trust in wireless networks; it explains the

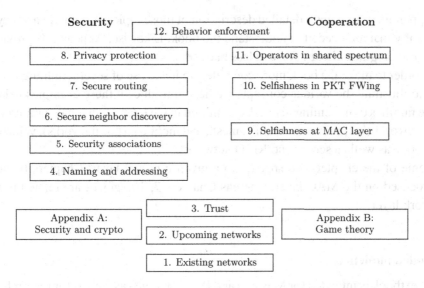

relationships between trust, security, and cooperation, and discusses the adversary model.

Part II describes the techniques aiming at thwarting **malicious behavior;**[1] as such, it makes use primarily of security techniques. Chapter 4 addresses the problem of naming and addressing; it explains how the Sybil and the replication attacks can be thwarted in such networks. Chapter 5 explains how security associations can be set up between wireless devices, notably by exploiting their physical proximity. Chapter 6 addresses secure neighbor discovery and explains the wormhole attack along with techniques to thwart it. Chapter 7 provides techniques to secure the fundamental operation of routing in wireless multi-hop networks. Finally, Chapter 8 addresses the crucial issue of privacy in upcoming wireless networks.

Part III focuses on the techniques intended to prevent **selfish behavior;**[2] therefore, it heavily relies on game theory. Chapter 9 focuses on the MAC layer. It first explains the techniques by which a WiFi selfish user can increase its share of the bandwidth, at the expense of well-behaved users; then it provides a detailed study of selfish behavior in pure ad hoc networks. Chapter 10 discusses the problem of selfishness in packet forwarding, and explains why incentives to cooperate are needed. Chapter 11 addresses the difficult question of the co-existence of operators in the same part of the spectrum. Finally, Chapter 12 describes examples of protocols that encourage selfish devices to adopt a *desirable* behavior.

---

[1] As we will see, malicious behavior encompasses many misdeeds, including the willingness to access unauthorized information or to deliberately affect the availability of the network for other users.

[2] Selfish behavior, as we will see, means the overuse of a common resource.

Appendix A contains a detailed description of those topics of **security** and **cryptography** that are needed to understand the book. Likewise, Appendix B provides a tutorial on **game theory** for wireless networks.

In order to make the book more concrete, we make use of several running examples to illustrate the various concepts we have introduced; these examples belong to the families of upcoming networks identified in Chapter 2: personal communication networks (including community, mesh, and mobile ad hoc networks), vehicular networks, as well as sensor and RFID networks.

Some of the chapters are specific to a given protocol layer: Chapters 6 and 9 are focused on the MAC layer, whereas Chapters 7, 10, and 12 are related to the network layer.

Intended audience

This textbook is intended for Master's and Ph.D. students as well as for researchers. It should also be of interest for the practitioners who want to get a broader view of the field.

Some familiarity with networking and security principles is useful for a proper understanding of this book.

About the title

The title of this book, *Security and Cooperation in Wireless Networks*, is well suited for the security aspects. But the word "cooperation" can be misleading, because it can be confused with the notion that wireless devices cooperate with each other at the physical layer (e.g., for beamforming). The usual term in networking is "non-cooperative behavior," but it is not particularly appropriate for the title of a book.

How to use this book

This book is designed to be covered in a one-semester course. If the students have little background on security, it is appropriate to start the course by covering Appendix A. Covering Part I should then be straightforward. At the end of Part I, the students could be encouraged to read the description of the security scheme of a wireless system not covered in the book (e.g., WiMAX) and check if they can understand it.

In Part II, each chapter can be addressed relatively independently, but the proposed order should make the understanding easier.

In current engineering and computer science curricula, game theory is usually not taught. Hence, with all likelihood, it will be necessary to first cover Appendix B before tackling Part III. Each of the four chapters of that part is fairly self-contained and can therefore be studied independently of the other. However, the beginning of the first of them (Chapter 9) is particularly intuitive because it addresses the concrete reality of WiFi systems. The last chapter (Chapter 12) is especially important as it combines the concepts of security and cooperation.

In case only a few hours per week are available, another approach consists in covering Part I and Part II in one semester, and then Part III in a follow-up (maybe optional) course in the following semester. Indeed, the two first parts of the book constitute a self-contained introduction to wireless security.

Additional material

The URL of the Web site of this book is http://secowinet.epfl.ch/ available directly or through www.cambridge.org/9780521873710. Additional material, such as slide shows (in pdf or PowerPoint[3] formats) is available there.

---

[3] Trademark of Microsoft Inc.

Acknowledgements

We would like to thank the students, former students, and post-docs in our research groups for their invaluable contributions. In particular, many thanks to Gergely Ács, whose research influenced Chapter 7; Naouel Ben Salem, whose research in mesh networks and in protocols for behavior enforcement was very helpful for Chapters 2 and 12; Srdjan Čapkun and Mario Čagalj, whose research shaped significantly the material of Chapter 5; Mario again, as his research was also a direct source of inspiration for Chapter 9, as was also the research carried out by Maxim Raya, Imad Aad, and Alaeddine El Fawal; Maxim again, along with Panos Papadimitratos, whose research in the security of vehicular networks proved to be very useful in several of the chapters, and in particular in Chapter 2; Mark Felegyhazi, whose research greatly influenced Chapter 10 and the second part of Chapter 11, and who contributed very significantly in clarifying the concepts in Appendix B, devoted to game theory; Julien Freudiger, for the questions of Chapter 12; Tamás Holczer and Péter Schaffer, whose research shaped the first part of Chapter 11; Hossein Manshaei for his contributions to the clarification of Bianchi's model and for some of the questions of Chapter 9; and Jun Luo, Jacques Panchard, and Marcin Poturalski who provided detailed feedback on many of the chapters.

Srdjan Čapkun deserves a specific acknowledgement for having been the first to teach a class based on a very early version of this book.

Several researchers provided very useful insights and a great support to this project; this was particularly important, considering the relative novelty of the topic. In particular, we would like to express our gratitude to Tansu Alpcan, David Basin, Jean Bolot, Daniel Figueiredo, Virgil Gligor, Matthias Grossglauser, Markus Jakobsson, Phil Janson, Frank Kargl, Edward Knightly, P. R. Kumar, Jean-Yves Le Boudec, Li Erran Li, Peter Marbach, James Massey, Cristina Nita-Rotaru, Charles Perkins, Adrian Perrig, Patrick Thiran, Don Towsley, David Tse, Nitin Vaidya, Jean Walrand, Dirk Westhoff, Heather Zheng, and Sheng Zhong.

We would like to thank the Swiss National Science Foundation for funding the National Competence Center in Research Mobile Information and Communication Systems (NCCR/MICS, sometimes nicknamed the "Terminodes project"); many of the ideas developed in this book have matured in the framework of this research program. We are indebted to those of our colleagues who have spent long hours to run the center, in particular Martin Vetterli, Thomas Gross, Karl Aberer, and Lothar Thiele. Martin deserves a special note: without his extreme dedication and visionary capabilities, the NCCR/MICS would not have taken off and, as a consequence, this book would have never existed.

We are indebted to Rafik Chaabouni who was instrumental in the formatting of the book and demonstrated a remarkable knowledge of LATEX idiosyncrasies, and to Thomas Thurnherr who edited several figures. We would like also to extend our gratitude to Holly Cogliati who provided us with many recommendations to improve our English expressions.

Anna Littlewood and Phil Meyler, both from Cambridge University Press, deserve our gratitude for having assisted the whole editorial process; Phil selected the cover figure and helped us formulating the title of the book in the most concise way.

LB is indebted to JPH for initiating the writing of this book. Without the enthusiasm and dedication of JPH, this book would not exist today. LB is also grateful to his colleagues, István Vajda and Boldizsár Bencsáth, and to his students, Gergely Ács, László Csik, László Dóra, Tamás Holczer, Péter Schaffer, and Ta Vinh Thong for taking care of the various ongoing projects of the CrySyS Lab while he was absent due to the writing of this book, and for reading and commenting on the manuscript at various stages of the writing. Special thanks go to István Vajda for the many useful discussions on provable security. Finally, LB is infinitely grateful to his wife Zita for an endless list of things.

JPH is grateful to his dean, Willy Zwaenepoel, for his comments (and warnings) about this project. He would also like to thank Victor Bahl for having suggested to him the idea of writing a textbook on wireless security a few years ago. Last but not least, JPH is also indebted to David Messerschmitt for hosting him as a visiting scholar at the EECS department of the University of California, Berkeley, several years ago. Many of the questions addressed in this book have their seed in the discussions with David as well as with Michael Katz and Sergio Verdù.

Part I
Introduction

1

The security of existing wireless networks

Before discussing wireless networks, it is necessary to take a broad look at networking in general and to see why malicious and selfish behavior is such a relevant issue. For this purpose, we will consider the Internet.

The Internet is probably the most impressive achievement ever in networking: A simple set of brilliant engineering rules has led to the deployment of the most pervasive network that, in spite of its size (or rather, thanks to it), supports a growing number of services and applications. At the core of these rules stands of course the principle of universal connectivity.

Unfortunately, the Internet is plagued by several major problems, fuelled by this very principle. Viruses and spam have become a daily issue for most users around the world, many people fall prey to phishing attacks, and denial-of-service (DoS) attacks are routinely perpetrated against the servers of major corporations. An additional problem is that some network providers tend to establish *walled gardens*, by which they offer specific capabilities exclusively to their customers. Finally, some providers are tempted to interconnect their network in a way that is beneficial to themselves, but can be detrimental to the rest of the community [209]. The situation is so critical that many prominent specialists, including some of the founding fathers of the Internet, call for a profound revamping of the network [102]. Ambitious research projects, funded notably by the NSF and by the European Commission, have adopted a *clean slate* approach to respond to this challenge.

All these problems have a common cause: they are the result of *human intention*, not technical failures. They also have common implications: they consume other users' time and nerves. They represent a formidable *tax* on the usage of the network, in terms of firewalls, filters, anti-spam software, anti-DoS systems, and the related workforce in charge of deploying and operating these tools.

It is clear that the problem is very complicated. One of the reasons is that most of the vulnerabilities we have mentioned do not revolve exclusively around the communication protocols: they can also be related to the operating system and (especially for viruses) to the programming techniques and they can depend on human factors. Yet, in this book we focus as much as possible on the issues primarily related to networking.

Another reason for this complexity is that it is extremely difficult to anticipate the kind of misbehavior that will affect a network while not yet deployed. In addition, competition encourages rapid deployment of new networking technologies and of new services, thus leaving little time to devise and implement (let alone standardize) protection mechanisms. Consequently, very often the protection mechanisms are designed *a posteriori* and constitute as many patches to the network. This leads to a growing complexity of the deployed systems (and complexity is often detrimental to security).

We believe that the widespread adoption of upcoming wireless networks creates even more formidable challenges in terms of misbehavior prevention. As malice and selfishness are the core problems addressed in this book, we make a distinction between these two kinds of misbehavior: malice aims at doing harm to known or unknown individuals or organizations, whereas selfishness consists of overusing the network resources (possibly at the expense of the other users). With this termi-nology, a virus designer is malicious, whereas a spammer is selfish. We will refine these concepts in Chapter 3.

Having discussed the lessons that can be drawn from the Internet, we will now see the peculiarities of wireless networks that are relevant to malice and selfishness. We will first discuss existing wireless networks, leaving the treatment of upcoming wireless networks to the next chapter.

1.1 Vulnerabilities of wireless networks

Existing wireless networks are primarily *personal* communication networks, mean-ing that the end systems are used by human beings to communicate either with other human beings or with servers. In the next chapter, we will see that some of the upcoming wireless networks have a different purpose, in the sense that com-munications, in a growing number of cases, will *not* involve human beings. As we will see, this has profound implications in terms of how these networks need to be protected.

The most obvious characteristic of wireless networks is that communication takes place over a wireless channel (which is usually a radio channel, but can also be an infrared channel). Such a channel suffers from a number of vulnerabilities, mentioned hereafter.

- The channel can be **eavesdropped**: by placing an antenna at an appropriate location, an attacker can overhear the information that the victim transmits or receives. Eavesdropping is often used to carry out attacks, notably *passive attacks*. Passive attacks consist in listening to the network and analyzing the captured data without interacting with the network. Such an attack can be illustrated with the weakness of WEP (described later in this chapter). Usually, the protection against such misdeeds is achieved by encrypting that information.

- The data can be **altered**: an attacker can try to modify the content of the message exchanged between (wireless) parties. These attacks are called *active attacks*. We will see later several cases of active attacks such as man-in-the-middle attacks perpetrated on GSM.
- The absence of wired link makes it easier to **cheat on identities**: being untethered, the attacker can more easily impersonate a legitimate user.
- The radio channel can be **overused**: the radio spectrum being a shared resource, there is a risk that a wireless operator or a user makes an excessive use of it. To solve the problem between cellular operators, the solution consists in allocating to each of them a licensed piece of the spectrum; but it can happen that several operators have to share the same spectrum, as it is the case today in WiFi. The problem of overuse by mobile users has not been an issue in cellular networks, because the bit rates were upper-bounded by the protocols, under the supervision of the base stations; but it can be an issue in WiFi because the stations can be programmed in a selfish way. We will come back to this problem in Chapter 9.
- The channel can be **jammed**, notably in order to perpetrate a DoS attack: by transmitting at the same time the victim transmits or receives data, an attacker can make it impossible for the victim to communicate. This problem has been studied in detail over the last decades. Typical solutions include spread spectrum and frequency hopping (and very often a combination of the two). We will not focus on anti-jamming techniques in this book, as they are more related to the physical layer; yet, in Chapter 9 we will see that the threat of jamming can actually *thwart* selfish behavior.

A second characteristic is that the users are usually **mobile**,[1] which has several implications.

- As the user roves with her mobile device, the device becomes a way to permanently trace her whereabouts, hence jeopardizing her **privacy**.[2] We will devote a full chapter to this crucial topic of privacy; in Section 1.3, we will see how this problem is (very partially) solved in existing wireless networks.

---

[1] The term "mobile" can designate a terminal that either communicates, moves, and then communicates again, or that communicates while moving (achieving the latter is of course technically more challenging). The precise meaning of this adjective will depend on the context in which it is used.
[2] The passive attacks mentioned above can be mounted against another component of privacy, namely the privacy of data.

- Mobility also means that a given device must be able to **roam** across wireless networks controlled by different operators. This requires that appropriate roaming agreements are made between operators, notably to define the pricing and billing policies.
- To be mobile the device must be small, meaning that it has **limited storage, computing power, and energy**. The last of these limitations is the most significant, as technological progress on batteries is much slower than on electronics. Usually, the problem is solved by minimizing the number of computational operations to be performed by the mobile station. This can, however, lead to poor engineering of the security protocols.
- A mobile station can easily be stolen, with the risk that it is **misused** or **reverse engineered** and that the data that it contains are accessed. The solution to this problem typically consists in encrypting the data it contains and embedding a tamper-resistant component in order to protect the cryptographic keys.

1.2 Security requirements

Based on the characteristics that we have just described, we are now in a position to discuss the requirements usually expected to be met by secure systems. This will help us to better understand how (and to what extent) they are fulfilled in existing wireless networks.

- The most obvious requirement is **authentication**: for example, an operator must be able to know who is trying to obtain connectivity through its network; likewise, the user wants to make sure that she is indeed connected to the wireless operator she chooses. Hence, authentication is a fundamental mechanism to support access control.
- **Access control** is the ability of an organization (e.g., a network operator) to grant appropriate access to resources (connectivity, data, . . .) based on the user's identity and the organization's policy.
- We have mentioned that the radio channel is particularly vulnerable to eavesdropping. Hence, **confidentiality** of the exchanged information is also an important requirement.
- As the radio channel is also highly vulnerable to active attacks, the **integrity** of data must be appropriately protected. The data to be protected are not only the users' data, but also the data related to the control of the network.
- Another requirement we have already mentioned is **privacy**. The network should not reveal the location of the user, nor the party with which she communicates (yet it is generally admitted that law enforcement agencies must have access to these two families of information, at least under some well-defined conditions).

- **Non-repudiation** is also an important requirement: for example, it should not be possible for a user, who has made use of a given service provided by a given operator, to pretend that she did not. In other words, it must be possible for an operator to prove that a given user really made use of the service that it provides, typically in case of a billing dispute.
- Last but not least, the network must provide a certain level of **availability**. This means in particular that it should provide higher priority to very important communications, such as an emergency call from a cellular phone; it should also guarantee a fair share of the radio resource to mobile users located in the same radio domain.

1.3 How existing wireless networks are secured

Let us now examine how the security requirements listed above are satisfied – or not – in existing wireless networks. The examples that we will consider here cover a wide range of network types beginning from wide area wireless networks and ending with personal area networks. More specifically, we briefly describe how security is provided in cellular networks, in WiFi LANs, and in Bluetooth. We do not intend to give a very detailed description of the security architectures of these systems; instead, and in line with the spirit of this book, we describe only the principles underlying those security architectures.

1.3.1 Cellular networks

Cellular networks have been deployed at a lively pace in the last decade, and are proliferating throughout the world. Today, cellular networks are so popular that in many countries, the number of mobile subscribers already exceeds the number of fixed telephone lines. Originally, cellular networks provided only voice communication services and they could also be used to send and receive short text messages. Today, the range of applications is much wider, including data communications, Internet access, multimedia applications (e.g., video telephony), and mobile payment services, just to name a few.

For political and historical reasons, cellular networks in different parts of the world are based on different standards. In this subsection, we focus on the European initiatives: GSM (Global System for Mobile Communications) and UMTS (Universal Mobile Telecommunications System). We note, however, that the principles are similar in other cellular networks (notably in the USA, in China, and in Japan).

Cellular networks are infrastructure-based networks. The infrastructure consists of base stations and a wired backbone network that connects the base stations together, as well as to the wired telephone system and to the Internet. Each base station serves only a limited physical area, called a *cell*, hence the name cellular. However, all the base stations of a given network operator together can cover a large area (typically a whole country in Europe). In addition, by connecting their backbones together and setting up appropriate roaming agreements, different network operators can jointly provide ubiquitous coverage and enable continent wide and even worldwide mobility for users.

The terminal equipment in cellular networks is typically a mobile phone. Mobile phones in a given cell are logically connected to the base station of the cell via wireless channels. They can initiate and receive calls to and from other mobile phones and fixed telephones via the base station (and the backbone infrastructure). In fact, the only wireless part in the system is the link between the mobile phone and the base station; the rest is a wired network.[3]

Setting up and running a cellular network is very expensive. A large share of the costs stems from the fact that cellular networks operate in licensed bands, meaning that the network operator must pay a licence fee for the use of the spectrum. The other part of the costs can be attributed to installing the base stations and deploying the backbone network, as well as to setting up the billing and the customer care infrastructure. At the end of the day, these costs are borne by the subscribers, who must pay for the services (including the access to the network) provided by the network operator.

GSM

GSM is a prominent example of cellular networks and we will now describe its security. From what we have just described, the main security requirement of GSM (at least from the operators' point of view) is subscriber authentication. Subscriber authentication is needed in order to support billing (i.e., to identify who must be charged for using the network).[4] In addition to subscriber authentication, GSM also provides some countermeasures for the inherent weaknesses of the wireless channel. More specifically, GSM provides confidentiality for voice communications and signalling over the wireless interface, and it protects the privacy of the subscribers by hiding their identity from eavesdroppers. Being a wide area system, GSM supports the roaming of subscribers across networks operated by different network operators. This means that the above mentioned GSM security services operate in a multi-party environment.

A fundamental assumption in the GSM security architecture is that there exists a long-term contractual relationship between a subscriber and a network operator;

---

[3] Base stations can also be connected to the backbone infrastructure via wireless links. However, those links are static and can be easily secured by the network operator.

[4] This guarantees only a weak form of non-repudiation, because a malicious operator could forge faked evidence of communications.

the latter is called the home network operator of the given subscriber. When setting up this relationship, the home network operator verifies the identity of the subscriber, and obtains further information about her, including the billing address. This contractual relationship is represented by a long-term secret key that is shared by the subscriber and the home network operator, and serves as the basis for the authentication of the subscriber.

In GSM, the secret key and other identity related information of the subscriber are not stored in the mobile phone, but in a separate security unit, called the SIM (Subscriber Identity Module). The SIM is implemented as a smart card with a small form factor, which can be inserted in and removed from the mobile phone. In effect, the key could have been stored in the non-volatile memory of the mobile phone itself, encrypted with a password. However, storing the key in a removable module has proved to be an excellent design choice, because it allows for the portability of the subscriber identity across different devices: the subscriber can remove the SIM from one mobile phone, insert it into another (e.g., when she buys a new device), and she still has the same phone number and receives a single bill.

Subscriber authentication in GSM is based on the so-called challenge–response principle. The subscriber receives an unpredictable random number as a challenge, and she must compute a correct response in order to be authenticated. The correct response is computed from the challenge and the long-term secret key of the subscriber. As the secret key is known exclusively to the subscriber and to the home network operator, no one else can compute the correct response. Thus, if the network operator receives the correct response, it believes that the response was produced by the subscriber; hence, she must be present. The unpredictability of the challenge ensures the freshness of the response: The network operator knows that the response must have been computed after it sent the challenge, because no one (not even the subscriber) could predict what the challenge would be. Clearly, the computations needed for authentication are not performed by the subscriber herself, but they are carried out by her mobile phone and the SIM without any user intervention.

We will now describe the steps of the GSM subscriber authentication protocol. For the sake of generality, we assume that the subscriber roams into a foreign network, usually referred to as the visited network. As the first step, the mobile phone reads the IMSI (International Mobile Subscriber Identity) from the SIM, and sends it to the visited network. Based on the IMSI, the visited network determines the identity of the home network of the subscriber. Then, the visited network forwards the IMSI to the home network via the backbone. The home network looks up the secret key K that corresponds to the subscriber identified by the IMSI. It then creates a triplet $(RAND, SRES, CK)$, where $RAND$ is an unpredictable random number used as the challenge, $SRES$ is the correct response to the challenge, and CK is a key to be used for encrypting communications over the wireless interface between the

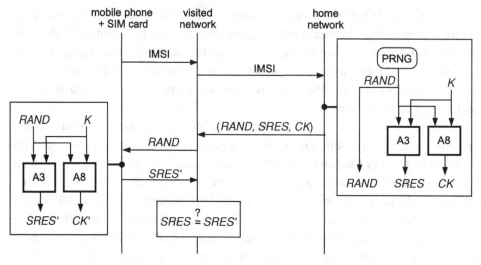

Figure 1.1. Illustration of the GSM authentication protocol.

mobile phone and the base station of the visited network. *RAND* is generated by a Pseudo-Random Number Generator (PRNG). *SRES* and *CK* are computed from *RAND* and *K* using the algorithms denoted by A3 and A8, respectively, in the GSM specifications. The triplet (*RAND*, *SRES*, *CK*) is sent to the visited network, which challenges the mobile phone with *RAND*. The mobile phone passes *RAND* to the SIM, which computes and outputs the response *SRES'* and the encryption key *CK'*. The mobile phone sends *SRES'* to the visited network, which compares it to *SRES*. If *SRES'* = *SRES*, then the subscriber is authenticated. In this case *CK'* = *CK* also holds. After the successful authentication of the subscriber, the communications between the mobile phone and the base station of the visited network are encrypted and decrypted with *CK* by using the stream cipher denoted by A5 in the GSM specifications. The steps of the protocol are summarized in Figure 1.1.

Note that the protocol ensures that the visited network can authenticate the subscriber without possessing the subscriber's long-term secret key. This is achieved with the help of the home network that provides a matching challenge–response pair to the visited network as part of the triplet. Similarly, the establishment of the encryption key between the mobile phone and the base station of the visited network is carried out with the help of the home network and the triplet mechanism. This requires some trust in the home network operator by the visited network operator, which is established by signing roaming agreements between the two operators. In practice, the home network can transfer several triplets to the visited network when the subscriber first authenticates herself (e.g., when she switches on her phone). In this way, there is no need to contact the home network every time the subscriber needs to be authenticated.

The identity of the subscriber is hidden from eavesdroppers on the wireless interface as follows. After each successful authentication, the subscriber receives a temporary identifier called TMSI (Temporary Mobile Subscriber Identifier) from the visited network. The TMSI is encrypted with the freshly established key CK, therefore, it cannot be eavesdropped. In the next authentication request, the mobile phone uses the TMSI, instead of the IMSI, to identify the subscriber. The TMSI is mapped to the IMSI by the visited network, and then the protocol proceeds as we described above.

When the subscriber moves into another visited network, the new network contacts the previous one and sends it the TMSI received from the mobile phone. The previous network looks up the data associated with the TMSI and transfers the IMSI of the subscriber and the remaining triplets (if any) to the new network, so that the new network can continue serving the subscriber. It can happen that the data associated with the TMSI are no longer available in the previous network (e.g., if the mobile phone has been switched off for a long time). In this case, the new network requests the mobile phone to send the IMSI in order to bootstrap the TMSI mechanism again.

To summarize, the GSM security architecture provides the following security services.

- **Subscriber authentication** is based on a challenge–response protocol and a long-term secret key shared by the subscriber and the home network operator. Data needed to authenticate the subscriber are transferred from the home network to the visited network in form of triplets, such that the long-term secret key is not revealed to the visited network.
- **Confidentiality of communications and signalling over the wireless interface** are ensured by encryption with a session key established between the subscriber's mobile phone and the base station of the visited network, during the subscriber authentication procedure, with the help of the home network operator.
- **Protection of the subscriber's identity from eavesdroppers on the wireless interface** is ensured by using short-term temporary identifiers instead of the real identifier of the subscriber during subscriber authentication. In some cases, the real identifier must be used; however, this happens rarely, and so it is difficult for eavesdroppers to track subscribers.

UMTS

The GSM security architecture provides a reasonable level of protection, but it has some deficiencies; hence the design of a new security architecture for UMTS, the Third Generation cellular network in Europe.

One main problem with the GSM security architecture is that it provides only unilateral authentication, where the subscriber is authenticated and the visited network operator is not. This means that someone can set up a fake base station and implement a man-in-the-middle attack. This probably seemed to be too far fetched in the 1980s when GSM was designed. But today, there are commercially available devices, called "IMSI catchers," that were originally intended for protocol testing purposes, but can also be used (or misused) to implement a fake base station attack.

The fake base station issue is further aggravated by the fact that GSM authentication triplets can be re-used indefinitely. Indeed, the subscriber cannot verify the freshness of the challenge that she receives in the subscriber authentication protocol. Thus, a fake base station can coerce the subscriber's mobile phone to re-establish an old, possibly compromised, encryption key with the fake base station.

Another problem is that the GSM security architecture does not provide integrity protection services for communications and signalling over the wireless interface. Although it is true that modifying messages on-the-fly in a wireless channel is quite challenging (if not impossible in practice), if the communication between the mobile phone and the visited network takes place through a fake base station, then the attacker does not need to carry out the modifications in the wireless channel, but it can implement the attack within the fake base station. In addition, as a stream cipher is used for encryption, the attacker can easily manipulate individual bits in encrypted messages without decrypting them. Of course, if the messages carry parts of a voice communication, then the attacker can only achieve some distortion, but it is very unlikely that it can alter the true content of the communication in an unnoticeable way. It can still, however, attack the signalling information. Moreover, besides voice communications, cellular networks are increasingly used for data communications, where flipping a single bit in a message can have devastating consequences.

Additional reasons for a new design include the short length of the encryption key (practically 54 bits only), and the weaknesses discovered in the commonly used implementation of the A3 and A8 algorithms, which, under specific conditions, allow an attacker to compromise the long-term secret key of the subscriber and clone her SIM card [67].

The UMTS security architecture addresses the weaknesses listed above. The design approach was to keep the general principles of the GSM security architecture, and to extend it with the necessary mechanisms for authenticating the network to the subscriber and providing integrity protection over the wireless interface. For this reason, the GSM triplets are replaced by authentication vectors that have five elements: ($RAND, XRES, CK, IK, AUTN$). As before, $RAND$ is an unpredictable random number, generated by a PRNG, and used as a challenge in the subscriber authentication protocol, $XRES$ is the expected response to $RAND$, and CK is an encryption key to be used between the mobile phone and the base station of the

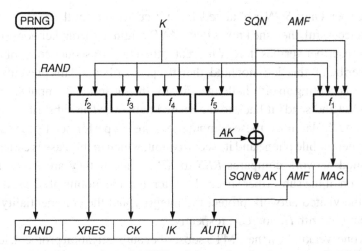

Figure 1.2. Construction of *AUTN* and the authentication vector in UMTS.

visited network. Both *XRES* and *CK* are computed from *RAND* and the long-term secret key *K* of the subscriber. In addition, *IK* is an integrity protection key and *AUTN* is a token that authenticates the home network to the subscriber and proves the freshness of *RAND*. *AUTN* consists of three fields: $AUTN = (SQN \oplus AK, AMF, MAC)$, where

- *SQN* is a sequence number maintained synchronously by both the subscriber and the home network;
- *AK* is called the anonymity key, and it is used to hide the value of *SQN* from eavesdroppers. *AK* is generated from *RAND* and *K*;
- *AMF* is an authentication and key management field used to pass parameters from the home network to the subscriber, but it is not fully specified in the UMTS standard;
- *MAC* is a message authentication code computed over *RAND*, *SQN*, and *AMF* using the long-term key *K*.

The construction of *AUTN* and the authentication vector is illustrated in Figure 1.2. Functions f_1, f_2, f_3, f_4, and f_5 are appropriate one-way functions defined in the UMTS standard.

The subscriber authentication protocol is modified in such a way that, upon request, the visited network receives an authentication vector from the home network and it passes not only the challenge *RAND* to the subscriber, but also the authentication token *AUTN*. The subscriber first generates the anonymity key *AK* and decodes the sequence number *SQN* received in *AUTN*. *SQN* is encoded with *AK* to protect the privacy of the subscriber. Otherwise, an eavesdropper could associate different executions of the authentication protocol with consecutive sequence numbers to the same subscriber. Once *SQN* is obtained, the subscriber verifies the *MAC*. If this verification is successful, then she knows that *RAND* originates from her home network. Then, the subscriber verifies if *SQN* is greater than the last sequence number stored by the subscriber. If this does not hold, then the protocol fails. This prevents the subscriber from accepting an old challenge. Finally, the subscriber computes a response *RES* to *RAND* and sends it back to the visited network. The subscriber also computes *CK* and *IK*. Naturally, these computations are not performed by the subscriber herself, but her mobile phone and its security unit, which in this case is called USIM.

The visited network compares *RES* to *XRES*, and if they are equal, then the authentication of the subscriber succeeds. After that, the mobile phone and the base station of the visited network protect the integrity and the confidentiality of their communications with *IK* and *CK*, respectively.

There is one weakness in the UMTS subscriber authentication protocol, identified in [396]: the visited network is not authenticated to the subscriber. Although the visited network can authenticate itself to the home network, the home network does not pass any confirmation regarding the identity of the visited network to the subscriber in the authentication token *AUTN*. This allows a malicious network operator *X* to masquerade as network *Y* to the subscriber. It would still authenticate itself as *X* to the home network, but the subscriber would not know this, and she would believe that she is served by *Y*. This can be a problem, as *X* and *Y* could use different tariffs, and the subscriber would learn that she actually used a more expensive network when she receives her bill at the end of the month. One solution to this problem is to include the identifier of the visited network in the *AMF* field of *AUTN*.

1.3.2 WiFi LANs

Security has always been considered an important issue in WiFi networks. Consequently, early versions of the IEEE 802.11 wireless LAN standard [187] already featured a security architecture, called WEP (Wired Equivalent Privacy). As its name indicates, the objective of WEP is to render wireless LANs at least as secure as wired LANs (without particular security extensions). For instance, if an attacker wants to connect to a wired Ethernet network, she needs physical access to the Ethernet hub. However, this is usually made difficult by placing the hub in a locked room. In case of an unprotected wireless LAN, the attacker has an easier job because she does not need to have physical access to any equipment in order to connect to the network. WEP is intended to transform this easy job into a difficult one. More precisely, WEP is intended to increase the level of difficulty of attacking wireless LANs such that it becomes comparable to the difficulty of attacking wired LANs (e.g., breaking into locked rooms).

Unfortunately, WEP did not make attacks as difficult as its designers hoped. This would not have been a problem if the weaknesses had been discovered in due time. But things happened differently: WEP was already deployed when cryptographers and security experts discovered its flaws. It became evident that WEP did not provide adequate protection. Soon after this discovery, tools that automate the cracking of WEP keys appeared on the Web.

In response to these developments, the IEEE came up with a new security architecture for wireless LANs, described in an extension to the 802.11 standard. This extension is called IEEE 802.11i. In this subsection, we discuss both WEP and IEEE 802.11i. The reason for discussing 802.11i is clear: this is the current approach to protect WiFi LANs. We discuss WEP because, despite its known weaknesses, many systems still support it (for backward compatibility), and thus probably many people and organizations still use it. Also, the design flaws in WEP illustrate many subtleties in security protocol design that are interesting in general.

WEP

There are two basic security problems in wireless LANs: first, owing to the broadcast nature of radio communications, wireless transmissions can be easily eavesdropped. Second, and more important, connecting to the network does not require physical access to the network Access Point (AP), thus any device can try to illegitimately use the services provided by the network. WEP attempts to solve the first problem by encrypting messages. The second problem is addressed by requiring the authentication of the mobile stations (STAs) before allowing their connection to the network.

The authentication of the STA is based on a simple challenge–response protocol, similar to that used in GSM systems. Once authenticated, the STA communicates with the AP by encrypted messages. The key used for encryption is the same as the one used for authentication. The encryption algorithm specified by WEP is based on the RC4 stream cipher (for the description of the operation of RC4 see, e.g., pages 397–398 of [339]). Stream ciphers produce a long pseudo-random byte sequence out of a short secret seed value; this pseudo-random sequence is then XORed to the clear message (byte by byte) in order to generate the encrypted message. WEP works in the same way. The sender (the STA or the AP) of a message M initializes the RC4 algorithm with the secret key and XORs the pseudo-random sequence K produced by RC4 to M. The receiver of the encrypted message $M \oplus K$ uses the same secret key to initialize the RC4 algorithm that will then produce the same pseudo-random sequence K. Then K is XORed to the encrypted message to obtain the clear message: $(M \oplus K) \oplus K = M$.

But the description above is not precise enough: there is one more thing that WEP does when encrypting messages. It is easy to see that if encryption worked as we described in the previous paragraph, then every message would be encrypted with the same pseudo-random sequence K, as RC4 is initialized with the same secret key before encrypting every message. This would be bad for several reasons. Let us assume, for instance, that an attacker eavesdrops two encrypted messages $M_1 \oplus K$ and $M_2 \oplus K$. By XORing these two messages together, she gets $(M_1 \oplus K) \oplus (M_2 \oplus K) = M_1 \oplus M_2$. This is equivalent to one message being encrypted with the other, but clear messages are far from being pseudo-random sequences. Thus, $M_1 \oplus M_2$ is a very weak encryption, and the attacker is likely to be able to break it using the statistical properties of the clear messages.[5]

In order to address this problem, WEP appends an IV (Initialization Vector) to the secret key before initializing the RC4 algorithm, where the IV changes for every message. This ensures that the RC4 algorithm produces a different pseudo-random sequence for every message. The receiver should also know the IV in order to be able to decrypt the messages received. For this reason, the IV is sent in clear together with the encrypted message. In principle, this is not a problem, as the knowledge of the IV is not enough to decrypt the message: the secret key is also needed for the proper initialization of the RC4 algorithm. As for the sizes, we note that the IV is 24 bits long and the secret key is usually 104 bits long,[6] although some vendors provide products that allow for longer keys. Figure 1.3 illustrates the WEP encryption and decryption procedure.

Figure 1.3 also shows that before encryption, the sender attaches an integrity check value (ICV) to the clear message. The purpose of this value is to enable the receiver to detect any malicious modifications of the message by an attacker. In the case of WEP, the ICV is a CRC value computed for the clear message. As a CRC value alone cannot enable the detection of malicious modifications (because the attacker can compute the new CRC value for the modified message), the CRC value is also encrypted in WEP. The rationale is that in order to modify the message in an unnoticeable way, now the attacker must encrypt the new CRC value, but she cannot do this without the knowledge of the secret key. This reasoning is not quite solid, as we will see below.

We must also mention how keys are handled in WEP. The standard states that each STA has its own key, known only to that STA and the AP. However, this makes key management on the AP's side complicated, since the AP must store a key for every STA. For this reason, most implementations do not actually support this option. The standard also specifies a default key, known to every STA and the AP. Originally, this key was intended to be used for the encryption of broadcast

---

[5] It is also possible that the attacker (partially) knows the content of one of the messages (e.g., the value of the header fields), in which case she can easily compute the (partial) content of the other message.

[6] In various marketing materials, this is interpreted as "128-bit security." This is of course misleading (as marketing materials in general), because out of 128 bits, 24 bits are transferred in clear, hence known by the attacker.

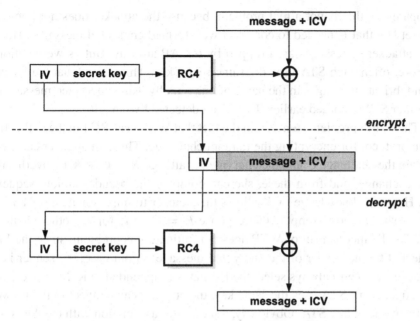

Figure 1.3. Encryption and decryption in WEP.

messages originated by the AP. But most WEP implementations support only this default key. Hence, in practice, in most wireless LANs there is a single common key. This key is installed in every mobile device and in the AP manually. Clearly, this solution can only be used to protect the communications from an outside attacker, but the devices that belong to the network can (in principle) decrypt each other's messages (and impersonate each other).

As will be clear from the brief overview below, WEP does not actually achieve any of its original design goals. The discovered flaws are instructive; they demonstrate the many pitfalls of security protocol design.

- **Authentication:** Authentication in WEP has several problems. First of all, authentication is not mutual, meaning that the AP does not authenticate itself to the STA. Second, the authentication and the encryption mechanism use the same secret key. This is not desirable, as an attacker can exploit the weaknesses of both the authentication and the encryption method to break the secret key. Having different keys for different functions is a better security engineering practice.

  The third problem is that the STA is authenticated only at the time when it tries to connect to the network. Once the STA is associated with the AP, anyone can send messages in the name of that STA by spoofing its MAC address.[7]

---

[7] When followed by "address," "protocol," or "layer," "MAC" means Medium Access Control, and not Message Authentication Code.

Apparently, this is not a real problem, because the attacker does not know the secret key that is needed to construct well-formed encrypted messages. Hence, the attacker's messages are dropped by the AP anyway. But as we mentioned before, often each STA uses the same secret key. This means that the attacker can fabricate messages in the name of one STA by using encrypted messages of another STA recorded earlier. This is not detected by the AP.

The fourth problem stems from the fact that WEP uses RC4 in the authentication protocol for encrypting the random challenge. Thus, an attacker can easily obtain the challenge C and the encrypted challenge $R = C \oplus K$ by overhearing the exchange, and from these, she can compute the pseudo-random sequence K. However, knowledge of K allows the attacker to impersonate the STA later on, as she can now compute the response $R' = C' \oplus K$ for any other challenge C'. The IV mechanism of WEP does not mitigate this problem, since the IV is selected by the sender of the encrypted message; in our case, the sender is the attacker, who will always select the IV that was appended to R. Moreover, as in practice, every STA uses the same key, the attacker can connect to the network in the name of any STA. Obviously, a successful association with the AP is only the first part of the attack; in order to send and receive messages in the name of a legitimate STA, the attacker needs to know the secret key. However, other flaws in WEP described below allow the attacker to retrieve the secret key.

- *Integrity protection:* The integrity protection of WEP messages is based on attaching an ICV to the message, where the ICV is a CRC value computed for the message and encrypted with the secret key. Formally, the encrypted message can be written as $(M || CRC(M)) \oplus K$, where M is the clear message, K is the pseudo-random sequence produced by the RC4 algorithm from the IV and the secret key, $CRC(.)$ denotes the CRC function, and $||$ denotes concatenation. It is well known that the CRC function is linear with respect to the XOR operation, which means that $CRC(X \oplus Y) = CRC(X) \oplus CRC(Y)$. Based on this observation, an attacker can manipulate protected WEP messages by flipping any of their bits unnoticeably, although she does not get access to the contents of the messages. Let us denote the changes that the attacker wants to make in the message by ΔM. Then the attacker wants to obtain $((M \oplus \Delta M) || CRC(M \oplus \Delta M)) \oplus K$ from the original protected message $(M || CRC(M)) \oplus K$ that she eavesdropped. For this purpose, it is sufficient to compute $CRC(\Delta M)$, and then to XOR $\Delta M || CRC(\Delta M)$ to the original protected message. The following derivation shows why this works:

$$((M || CRC(M)) \oplus K) \oplus (\Delta M || CRC(\Delta M))$$
$$= ((M \oplus \Delta M) || (CRC(M) \oplus CRC(\Delta M))) \oplus K$$
$$= ((M \oplus \Delta M) || CRC(M \oplus \Delta M)) \oplus K$$

where in the last step we used the linearity of the CRC function. Since $CRC(\Delta M)$ can be computed without the secret key, the attacker can succeed despite the encryption and the ICV mechanism.

Another related integrity requirement is the detection of replayed messages. Unfortunately, WEP does not use any replay detection mechanism, therefore, an attacker can replay any previously recorded message that will be accepted by the AP.

- **Confidentiality:** As we said before, when using a stream cipher, it is essential that each message is encrypted with a different pseudo-random sequence. In WEP, this is ensured by the IV mechanism, but this has some problems too. The origin of the problem is that the IV is only 24 bits long, which means that there are only approximately 17 million possible IV values. A WiFi device can transmit approximately 500 full-length frames in a second, thus, the whole IV space is used up in a few hours. Once all IVs have been used, they start to repeat, and repeating IVs mean repeating pseudo-random sequences used for encryption. The problem is aggravated by the fact that in many networks, there is a single secret key used by every device with potentially different IVs. Hence the IV space will be used up even faster. Another practical problem is that in many WEP implementations, the IV is initialized with 0 at startup, and then incremented by one after each message sent. This means that if there are several devices switched on nearly at the same time, then they all use the same sequence of IVs; if they use the same secret key too, then the pseudo-random sequences used for encryption will be the same. In this case, the attacker would not even need to wait, but it would get messages encrypted with the same pseudo-random sequence immediately.

The total collapse of WEP is caused by the inappropriate use of the RC4 cipher. It is known that there exist so-called weak RC4 keys [141]. A weak key is a seed value from which the RC4 algorithm produces an output that does not look random. More precisely, when a weak key is used to seed RC4, one can infer the bits of the seed from the first few bytes produced by the algorithm. For this reason, security experts suggest always throwing away the first 256 bytes of the RC4 output. This simple solution would have solved the problem of weak keys, but WEP did not adopt it. Also, because of the ever changing IV value (which is part of the seed), a weak key can be encountered sooner or later, and the attacker can easily know that a weak key is being used, because the IV is transmitted in clear. Based on these observations, some cryptographers constructed a method that breaks the full 104-bit secret key by eavesdropping on only a few hundred thousand messages. Compared with the previously described flaws, this one is the far most serious, because it allows the attacker to crack the secret key itself: and once she has the secret key, she can do everything. Moreover, the attack is not only powerful, but easy to automate, and thanks to some "helpful" people, automated attacking tools are readily available on the Web for public use (e.g., Aircrack, Weplab).

IEEE 802.11i

When the flaws in WEP became apparent, the IEEE began to develop a new security architecture for WiFi networks, described in the 802.11i specification [189]. The new concept is called RSN (Robust Security Network) in order to distinguish it from WEP. RSN was designed more carefully than WEP. It includes a new method for authentication and access control, which is based on the model defined in the 802.1X standard. The mechanisms for integrity protection and confidentiality are also changed, and they use the AES (Advanced Encryption Standard) [5] cipher instead of RC4.

However, it is not possible to switch from WEP to RSN overnight. The reason is that for efficiency reasons, many WiFi devices (mainly WLAN adapter cards) support the encryption algorithm in their hardware. Thus, old devices support RC4 and not AES. This problem cannot be solved by a simple firmware update; the hardware needs to be changed, which slows the deployment of RSN.

This has been recognized by the IEEE too, and they included an optional protocol in the 802.11i specification, which still uses the RC4 cipher but fixes the flaws in WEP. This protocol is called TKIP (Temporal Key Integrity Protocol).

Manufacturers immediately adopted TKIP, as it provides a solution to the problems of WEP, and it can be deployed immediately without changing the hardware. They did not wait until the 802.11i architecture was finalized by the lengthy standardization procedure, but they issued their own specification, called WPA (WiFi Protected Access), based on TKIP. In other words, WPA is a specification supported by WiFi manufacturers, and it contains a subset of RSN that can also run on old devices that support only the RC4 cipher. Authentication and access control, as well as key management, are the same in WPA and in RSN. The difference between the two concepts lies in the mechanisms used for integrity protection and confidentiality. We must also mention that RSN is also called WPA2 by many manufacturers.

Below, we first give an overview of the authentication, access control, and key management procedures of 802.11i. Then, we briefly summarize the operation of TKIP (used in WPA) and AES-CCMP (used in RSN).

Authentication and access control The model of authentication and access control in 802.11i was borrowed from the 802.1X standard [188]. IEEE 802.1X was originally intended for wired LANs, but it turned out that the same concepts can be used in wireless LANs too (with a few extensions).

The 802.1X model distinguishes three entities in the authentication procedure: the supplicant, the authenticator, and the authentication server. The supplicant wants to access the network, and for this reason, it wants to authenticate itself. The authenticator controls access to the network. In the model, this is represented by controlling the state of a port. The default state of the port is "closed," which means that data traffic is disabled. The authenticator can "open" the port if this is authorized by the authentication server. Actually, the supplicant authenticates itself to the authentication server, and if this authentication is successful, then the authentication server grants access to the network by instructing the authenticator to open the port.

In the case of WiFi networks, the supplicant is the mobile device and the authenticator is the AP. The authentication server is a process that can run on the AP in the case of smaller networks, or on a dedicated server machine in the case of larger networks. In WiFi, the port is not a physical connector, but a logical control implemented in software running on the AP.

In a wired LAN, a device authenticates itself once, when it is physically connected to the network. There is no need for further authentication (at least for network access control purposes), because the port used by the device cannot be used by someone else. This would require first disconnecting the device that currently uses the port, which would be detected by the hardware of the authenticator, and the port would be disabled. The situation is different in WiFi networks, because there is no physical connection between the STA and the AP. Hence, once the STA authenticates itself and associates with the AP, someone else can try to steal its session by spoofing its MAC address. For this reason, 802.11i extends 802.1X with the requirement of setting up a session key between the STA and the AP when the STA first requests access to the network; this session key can then be used to authenticate any further communications between the STA and the AP.

The authentication procedure in 802.11i uses EAP (Extensible Authentication Protocol) [9] to carry the messages that need to be exchanged between the STA and the authentication server (see Figure 1.4 for illustration). Note that EAP is only a carrier protocol: it does not provide authentication services itself, but it can carry the messages of any higher layer authentication protocol. That is why

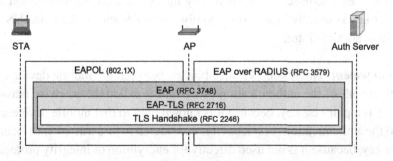

Figure 1.4. Authentication protocol architecture of 802.11i.

it is called "extensible." How the higher layer protocol messages are embedded into EAP messages must be specified for each and every higher layer protocol. Such specifications already exist for many widely used protocols such as the TLS (Transport Layer Security) Handshake and the GSM authentication protocols.

There are four message types in EAP: request, response, success, and failure. EAP request and response messages carry the messages of the embedded authentication protocol from the STA to the server, and from the server to the STA, respectively. The EAP success and failure messages are used to signal the result of the authentication to the supplicant.

As we indicated before, in 802.1X, the supplicant authenticates itself to the authentication server. This means that in WiFi networks, the EAP protocol and the embedded higher layer authentication protocol are executed by the mobile device requesting access and the authentication server. The AP relays messages without interpreting them. The AP understands only the EAP success and failure messages. When it sees an EAP success message passing, it enables the port and lets the mobile device connect to the network.

EAP messages between the mobile device and the AP are carried by the EAPOL (EAP over LAN) protocol defined in 802.1X. EAP messages between the AP and the authentication server can be carried by various protocols. WPA mandates the use of RADIUS [10] for this purpose, whereas RSN specifies RADIUS only as an option. In any case, RADIUS is already quite widely deployed, therefore it is expected to be often used in RSN, too.

As we mentioned before, the result of the authentication process in WiFi is not only the authorization for the mobile device to access the network but also a session key to protect further communications between the mobile device and the AP. However, as authentication takes place between the mobile device and the authentication server, the session key is also established between them, and it must be securely transferred to the AP. The RADIUS protocol makes this possible by means of the MS-MPPE-Recv-Key RADIUS attribute that has been specified for key transfer purposes. The session key is transferred in encrypted form, where the encryption uses a long-term key shared by the AP and the authentication server. This latter key is usually installed manually in the AP and in the RADIUS server by a system administrator.

Key management The session key established between the mobile device and the AP as the result of the authentication procedure is called the pairwise master key (PMK). It is a pairwise key, because it is known only to that mobile device and the AP (and the authentication server, but it is considered to be a trusted entity); and it is a master key, because it is not used directly for encryption or integrity protection of messages, but it is used to derive encryption and integrity keys. More precisely, both the mobile device and the AP derive four keys from the PMK: a data-encryption key, a data-integrity key, a key-encryption key, and a key-integrity key. These four keys together are called the pairwise transient key (PTK). We must note that AES-CCMP uses the same key for encryption and for integrity protection of data, therefore, in the case of AES-CCMP, the PTK consists of three keys only. Besides the PMK, the derivation of the PTK also uses as input the MAC addresses of the parties (the mobile device and the AP) and two random numbers generated by the parties.

The mobile device and the AP exchange their random numbers using the so-called four-way handshake protocol. This protocol also provides evidence to each party that the other party possesses the PMK. Messages of the four-way handshake protocol are carried by the EAPOL protocol in EAPOL messages of type Key. The contents of the messages and the operation of the four-way handshake protocol are described as follows.

(1) The AP sends its random number to the mobile device. When the random number is received by the mobile device, it has everything needed for the derivation of the PTK. Hence, the mobile device computes the PTK.
(2) The mobile device sends its random number to the AP. This message also carries a Message Integrity Code (MIC), computed by the mobile device using the key-integrity key just derived from the PMK. Upon reception of this message, the AP has everything needed for the derivation of the PTK. Hence, the AP computes the PTK and then uses the key-integrity key to verify the MIC. If the verification is successful, then the AP believes that the mobile device possesses the PMK.
(3) The AP sends a message that contains a MIC to the mobile device. The MIC is computed using the key-integrity key of the PTK. If the mobile device can successfully verify the MIC, then it believes that the AP possesses the PMK too. This message contains the starting value of a sequence number that will be used to number further data packets, and hence to detect replay attacks. In addition, this message signals to the mobile device that the AP has installed the keys and it is ready for encrypting all subsequent data packets.
(4) The mobile device acknowledges the reception of the third message. This acknowledgement also means that the mobile device is ready for encrypting all subsequent data packets.

Once the PTK is derived and the keys are installed, subsequent data packets between the mobile device and the AP are protected by the data-encryption and data-integrity keys. However, these keys cannot be used to protect broadcast messages sent by the AP. Those broadcast messages should be protected with keys that are known to all mobile devices and the AP. Therefore, the AP generates additional key material, called the group transient key (GTK). The GTK contains a group encryption key and a group integrity key and it is sent to each mobile device separately encrypted with the key-encryption key of the given mobile device.

TKIP and AES-CCMP Both TKIP (Temporal Key Integrity Protocol) and AES-CCMP (AES CTR Mode and CBC MAC Protocol) are based on the key hierarchy described in the previous paragraph. In particular, they use the data-encryption and data-integrity keys (of the PTK) to protect the confidentiality and the integrity of the data packets sent between the mobile device and the AP. However, they use different cryptographic algorithms. TKIP, just like WEP, uses RC4, but unlike WEP, provides more security. The advantage of TKIP is that it runs on old WEP hardware after some firmware upgrade. AES-CCMP needs new hardware that supports the AES algorithm, but it provides a clearer, more elegant and robust solution than TKIP does.

TKIP fixes the flaws in WEP as follows.

- *Integrity:* TKIP introduces a new integrity protection mechanism, called Michael. Michael operates at the Service Data Unit (SDU) level (i.e., it operates on data received by the MAC layer from higher layers before those data are fragmented). This makes it possible to implement Michael in the device driver, which in turn allows the introduction of Michael as a software upgrade.

  In order to detect replay attacks, TKIP uses the IV as a sequence number. Thus, the IV is initialized with some initial value and then incremented after the transmission of every message. The receiver keeps track of the IVs of the recently received messages. If the IV of a freshly received message is smaller than the smallest stored IV value, then the receiver drops the message; whereas if the IV is larger than the largest stored IV value, then it keeps the message and updates its stored IVs. If the IV of an incoming message falls between the smallest and the largest stored IV value, then the receiver checks if that IV is already stored; if so, then it drops the message, otherwise it keeps the message and stores the new IV.

- *Confidentiality:* Recall that the main problem with WEP encryption was that the IV size was too small and that the existence of RC4 weak keys was not taken into consideration. In order to overcome the first problem, in TKIP, the IV size is increased from 24 bits to 48 bits. This seems like an easy solution, but the difficulty is that the WEP hardware still expects a 128-bit long RC4 seed value. Thus, the 48-bit IV and the 104-bit key must somehow be compressed into 128 bits.

  As for the problem of weak keys, in TKIP, each message is encrypted with a different key. Thus, the attacker cannot observe a sufficient number of messages that are encrypted with the same (potentially weak) key. The message keys are generated from the data-encryption key of the PTK.

  TKIP's new IV mechanism and the generation of the message keys are illustrated in Figure 1.5. The 48-bit IV is divided into a 32-bit upper part and

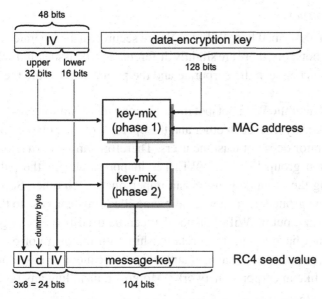

Figure 1.5. Generation of the RC4 seed value in TKIP.

a 16-bit lower part. The upper part of the IV is combined with the 128-bit data-encryption key of the PTK and the MAC address of the device. The result of this computation is then combined with the lower part of the IV in order to obtain the 104-bit message-key. The RC4 seed value for TKIP is obtained by concatenating the message-key to the lower part of the IV and a dummy byte (designed to avoid weak RC4 keys).

The designers of AES-CCMP had an easier job than the designers of TKIP, because they were not constrained by the peculiarities of the old WEP hardware. Thus, they simply replaced RC4 and based their design on the AES block cipher [5]. They defined a new mode for AES, called CCM, which is the combination of two previously known mechanisms: CTR (Counter) mode encryption and CBC–MAC (Cipher Block Chaining – Message Authentication Code) (see Appendix A for more details on these mechanisms). In the CCM mode, the sender of a message computes the CBC–MAC value of the message, attaches it to the message, and then encrypts the whole lot in CTR mode. The CBC–MAC computation covers the header of the message too, while the encryption is applied only to the message body. The CCM mode ensures both confidentiality and integrity of the message. Replay detection is ensured by sequence numbering the messages. The sequence number is integrated into the CBC-MAC value of the message by placing it in the initialization block of the CBC-MAC computation.

Public WiFi hotspots

So far, we have described how WiFi LANs are secured in a corporate environment. In public WiFi hotspots, there are slightly different security issues and solutions. The main differences between the corporate and the public settings are the following.

- The users of a public WiFi hotspot do not belong to a common group, hence they do not necessarily trust each other and the operator of the hotspot. Similarly, the hotspot operator does not trust the users. Therefore, any security solution based on a common group key (e.g., WEP) is inappropriate for the public setting. Firstly, using the group key, users can impersonate each other. Secondly, users can reveal the group key to anyone allowing illegitimate access to the hotspot.
- In addition, in a public WiFi hotspot that can be used free of charge, the users usually do not have a long-term relationship with the operator of the hotspot. This means, in particular, that user authentication cannot be based on long-term secret keys, like in corporate networks. Moreover, installing any type of key is a hassle for users.
- Finally, in the case of public access, the network behind the WiFi LAN can be insecure, unlike in a corporate environment, where the corporate intranet is considered to be secure. Hence, not only the wireless channel needs protection, but it could be more advantageous for users to use security services in a higher layer (e.g., in the transport layer).

The main concern of the public hotspot operator is to get paid for the services that it provides. But, as we saw above, the solution cannot be based on requiring the users to install keys. Hence, a much more pragmatic solution is adopted in most practical cases: password-based user authentication. The idea is the following: when a user buys a subscription, she gets a username and a password. The access points of the hotspot are run in open mode without any protection at the MAC layer. Thus anyone can connect to the hotspot, get an IP address, and begin sending IP packets to the Internet. However, the access points route every packet to a special gateway, called the *hotspot controller*, which blocks all IP traffic. In fact, the controller will let go through only those IP packets that carry an HTTP request to a special login page. Thus, the only action a user can take is to go to that login page and type her username and password. If this is done successfully, then the IP address of the user is inserted in a white list, and no more packets originating from that IP address are blocked by the hotspot controller. The hotspot controller can keep track of the connection time and the amount of traffic associated with each user, and it blocks the traffic again if the user exceeds her quota. To protect herself from other users of the hotspot, the user can use transport layer security solutions (e.g., TLS) or a solution based on IPsec (e.g., a VPN) at her own risk.

1.3.3 Bluetooth

Bluetooth is a wireless technology that uses short-range digital radio communications and offers fast and reliable transmission of both voice and data. The main objective of Bluetooth is to eliminate wires between nearby devices such as a mobile phone and a headset, a laptop and a mouse, or a computer and a printer. Unlike wireless LANs, where there are wireless stations and access points, in the case of Bluetooth, there are only wireless stations. However, the operation of Bluetooth networks (so called piconets) is based on the master–slave principle, where one of the stations takes the role of the master and the other stations (up to 7) become the slaves.

The Bluetooth specifications define a security architecture that aims at providing authentication and confidentiality services for communicating Bluetooth devices. Before presenting this security architecture, we should note that Bluetooth has some inherent characteristics that make the job of an attacker slightly more difficult than in the case of wireless LANs. First, Bluetooth devices use frequency hopping in order to avoid interference with other devices that operate in the same unlicensed ISM band. The frequency-hopping scheme uses 79 different channels and changes frequency 1600 times per second in a pseudo-random manner. This makes eavesdropping slightly more difficult, because the attacker must listen on practically all 79 channels in parallel. Second, as we mentioned above, Bluetooth is a short-range radio technology enabling communications over a few meters only. This means that an attacker must be physically close to the victims in order to eavesdrop on their communications, which further reduces the likelihood of attacks. Nevertheless, none of the inherent characteristics of the Bluetooth technology would stop a determined attacker, hence the need for security mechanisms in Bluetooth.

The Bluetooth security architecture is concerned with the establishment of a secured wireless link between two Bluetooth devices. This involves the authentication of the devices to each other and the setting up of a confidential channel between them. Both are based on a secret link key shared by the two devices. To generate the link key, a pairing procedure is used when the two devices communicate for the first time. We will first explain how the link key is established and then describe how it is used for the authentication of the devices and for the derivation of the encryption key. The presented mechanisms use the cryptographic functions E_1, E_{21}, E_{22}, and E_3, each of which is based on the SAFER+ block cipher [268].

There are two ways to establish a link key. The first method is used when one of the devices has memory limitations and can store only one key, otherwise the second method is used. However, both methods start by setting up a temporary initialization key K_{init}. This is illustrated in Figure 1.6 and explained as follows. First, one device selects a random number *IN_RAND* and sends it to the other device. Then, both

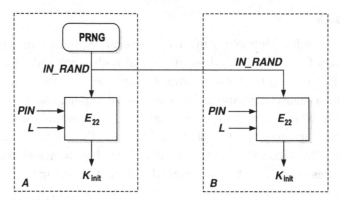

Figure 1.6. Setting up the temporary initialization key between two Bluetooth devices.

devices compute K_{init} as a function of *IN_RAND*, a shared *PIN*, and the length L of the *PIN*. The length of the *PIN* can vary between 1 and 16 bytes. Typically, the *PIN* is a 4-digit number with a default value of 0000. The *PIN* can be shared between the devices in several ways. If both devices have some input facility, then the user can choose a random *PIN* and enter it into both devices. If only one device has an input facility, then the user can enter the pre-configured *PIN* of the other device into the first device. Otherwise, pairing is not possible.

Let us now consider how the link key is established when one of the devices, say A, has memory limitations. In this case, A sends its long-term unit key K_A to the other device B encrypted with the initialization key K_{init} that they have just established. B obtains K_A by decrypting A's message, and K_A becomes the link key.

When none of the devices has memory limitations, the link key is established in the following way: both A and B choose a random number $RAND_A$ and $RAND_B$, respectively. A computes LK_K_A as a function of $RAND_A$ and its unique device address BD_ADDR_A. Similarly, B computes LK_K_B. Then, they exchange $RAND_A$ and $RAND_B$ encrypted with K_{init}. When A receives $RAND_B$, it can compute LK_K_B. Similarly, when B receives $RAND_A$, it can compute LK_K_A. Then, both can compute $LK\_K_A \oplus LK\_K_B$, which becomes the link key. The computation of the link key is illustrated in Figure 1.7.

When two devices share a link key (that they have just established or kept from a previous session), they authenticate each other using a simple challenge–response protocol, which is illustrated in Figure 1.8. One of the devices, referred to as the "verifier," generates a random number *AU_RAND* and sends it to the other device, called the "claimant." They both compute an authentication response from *AU_RAND*, the device address *BD_ADDR* of the claimant, and the link key K_{link}. The claimant then sends the obtained value $SRES'$ to the verifier, which then compares it to the value $SRES$ that it computed. If $SRES' = SRES$, then the authentication is successful.

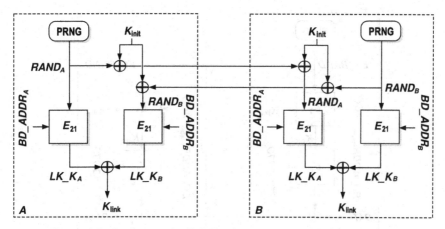

Figure 1.7. Setting up the link key between two Bluetooth devices.

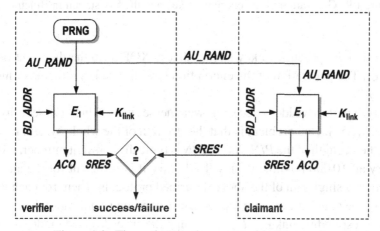

Figure 1.8. The authentication protocol in Bluetooth.

After that, the two devices run the same protocol with the roles swapped to achieve mutual authentication.

If the protocol above fails, then the verifier device will wait some time before a new attempt can be made. This waiting time increases exponentially with every failed attempt in order to make it impractical for an attacker to defeat authentication by trying different keys in rapid succession.

The encryption key K_{enc} is computed by both devices as a function of three elements: the link key K_{link}, the authenticated cipher offset ACO generated during the authentication protocol, and a random number EN_RAND generated by the master device. Encryption is performed with a stream cipher called E_0 in the Bluetooth specifications. Besides the encryption key K_{enc}, E_0 also inputs the unique address BD_ADDR_{master} of the master device, and the clock value $CLOCK_{master}$ of the master.

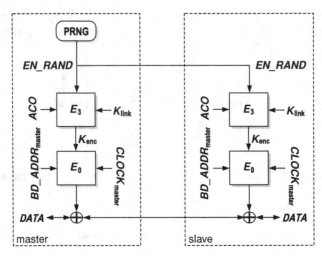

Figure 1.9. Generation of the encryption key and the key stream in Bluetooth.

The algorithm E_0 produces a key stream that is XORed to the data sent between the devices. The generation of the encryption key and the key stream is illustrated in Figure 1.9.

Security experts have identified some weaknesses in the Bluetooth security architecture [204, 351]. One problem is that the strength of the whole system is entirely based on the strength of the *PIN*. As the *PIN* is typically a 4-digit number, it is fairly easy to try all 10 000 possible values. To do this, it is sufficient for the attacker to eavesdrop on a single run of the above described protocols. Then, for each guessed value *PIN'*, the attacker can compute the corresponding initialization key K'_{init}, and then the corresponding link key K'_{link}, by using the eavesdropped random numbers. Each guessed link key K'_{link} can be tested using the challenge–response pair available from the eavesdropped execution of the authentication protocol. This means that the *PIN* can be cracked off-line, hence the mechanism of exponentially increasing the waiting times between failed authentication attempts is ineffective; there will be only one (successful) attempt once the *PIN* is cracked off-line. Moreover, many devices just use the default *PIN* (i.e., 0000).

Another problem is that for memory-constrained devices, the link key is the long-term unit key of the device. Hence, an attacker can easily obtain the unit key of a memory-constrained device *A* by establishing a link key with it. Once the unit key is obtained, the attacker can impersonate device *A*. The attacker can also decrypt the communication between *A* and any other device *B*, because the link key between *A* and *B* is also the unit key of *A*.

There is also a privacy problem that stems from the use of fixed and unique device addresses. As Bluetooth devices are often personal gadgets, a device address can be associated with a person. Then, the attacker can track the whereabouts of that person by tracking the use of the given device address.

Finally, cryptographers have discovered weaknesses in the E_0 stream cipher used in Bluetooth. Apparently, the encryption key can be broken with much less effort than the cost of a brute force attack (which is 2^{128}, as the encryption key is 128 bits long). The details of the attack are out of the scope of this overview; the interested reader is referred to [168] for more information.

1.4 Summary

In this first chapter, we were concerned with the security of existing wireless networks. First, we identified two important characteristics of wireless networks that have a strong effect on their security. The first characteristic is that communication takes place over wireless channels that are easy to eavesdrop on, jam, and overuse. The second characteristic is that users of wireless networks are usually mobile. This has some implications both in terms of security requirements and solutions. Besides the classical security requirements of authentication, confidentiality, integrity, and availability, we identified location privacy as a security requirement that is unique to mobile networks. We also argued that security architectures designed for wireless mobile networks must take into account the limited resources of portable mobile devices, and the lack of their physical protection. In addition, the security architecture should support the roaming of users across networks operated by different network operators.

In the second part of this chapter, we gave an overview of some existing wireless security architectures. More specifically, we described how cellular networks (GSM and UMTS), WiFi LANs, and Bluetooth are secured. These examples can serve as a reference to which security solutions developed in the rest of the book can be compared.

1.5 To probe further

The description of the security architectures in Section 1.3 was deliberately kept concise, because the focus of the book is not on how existing wireless networks are secured but rather on how upcoming wireless networks should be secured. The interested reader can find more information on the presented security architectures in many articles and books (see, e.g., [271] for GSM security, [290] for UMTS security, [124] for WiFi security, and [148] for Bluetooth security).

Details about the flaws in WEP can be found in [368, 61, 27]. At the time of this writing, the most prominent tools to crack WEP are Aircrack and Weplab[8]. These

---

[8] Both can be downloaded from htp://www.sourceforge.org/

tools are based on statistical analysis attacks originating from an unknown person nicknamed Korek. Rafik Chaabouni improved these attacks and found a new one during a semester project at EPFL (see [95] for the details).

Attacks against Bluetooth security are described in [204, 346], while privacy issues in Bluetooth are discussed in [204, 358].

Although the examples that we considered in Section 1.3 cover a broad spectrum, there are other examples of existing wireless security architectures. An early version of the WiMAX security architecture and a brief analysis of its weaknesses are presented in [211]. An updated version of the WiMAX security architecture is published as part of the IEEE 802.16e specification [190]. But the WiMAX standard was still not stable at the time of this writing therefore we did not address it.

Some security issues in underwater wireless networks are discussed in [109].

1.6 Questions

(1) What are the main vulnerabilities of wireless networks?
(2) What security services are provided by the GSM security architecture? What important security services does it not provide?
(3) Let us consider the authentication vector in UMTS. What is the purpose of the *AUTN* field? Does the *MAC* in *AUTN* authenticate the keys *CK* and *IK*? How is the freshness of *CK* and *IK* ensured?
(4) What are the main weaknesses of the WEP protocol?
(5) How does the authentication scheme of 802.11i differ from that of 802.1X?
(6) Why do you think the MAC address of the device is included in the computation of the message keys in TKIP (see Figure 1.5)?

2

Upcoming wireless networks and new challenges

2.1 Introduction

As we have seen in the previous chapter, the development of wireless networks has followed a centralized pattern: the network infrastructure (meaning all pieces of equipment except the terminal) has remained under the full supervision of the network operator, who traditionally used to be a large organization, very careful at respecting the legislation (and at nurturing the value of its own brand). As a result, the users generally tend to trust the operator, but do not generally trust the other users.

As we have also seen, current technology such as WiFi makes infrastructure equipment (and in particular access points) affordable to very small operators or even individuals, thus allowing the emergence of *community networks* and similar initiatives.

In this chapter, we will show that we are only at the beginning of this evolution, and that not only WiFi, but also other wireless technologies are about to dramatically transform the deployment and operation philosophy of wireless networks. As a consequence, the notions of authority and of trust need to be completely revisited, and this is exactly one of the reasons for writing this book: the novel organization of the wireless networks calls for a thorough study of the possible malicious and selfish behaviors, and of the techniques to thwart them.

In order to be as concrete as possible, we will first provide a certain number of examples of emerging wireless networks, spanning personal networks, vehicular networks, sensors, and RFID (Radio Frequency IDentification, described later in this chapter). Then, in spite of the substantial and deliberate diversity of these examples, we identify relevant trends common to all or to most of them. By the same token, from these trends we identify the most significant challenges that underpin Part II and Part III of this book.

2.2 Upcoming wireless networks

For each wireless network presented in this section, we provide a brief description and a set of security and cooperation challenges. If these challenges can be taken up by well-established techniques, we mention possible solutions right away. If, on the contrary, more sophisticated mechanisms are required, we refer the reader to the related chapters of Part II and Part III.

2.2.1 Personal communications

Personal communications have been and are likely to remain the most relevant and diverse kind of wireless communication systems. In the previous chapter, we have discussed their security requirements. In this subsection, we will describe the most relevant upcoming types of these networks; we start with those most similar to the existing cellular and WiFi networks and progressively relax the assumptions of one-hop radio connectivity between the mobile station and the base station and of a strong trust relationship between the user and the operator.

Small operators, operators in shared spectrum

The first type of networks resembles existing cellular and wireless data networks: a radio access device (a base station or a WiFi access point), installed and managed by an operator, provides mobile devices with one-hop access to the backbone. Yet, even this relatively classical type of networks will go through substantial modifications in the coming years, and this will have strong implications.

First of all, the increased programmability of the devices opens the door to selfish behavior with respect to the shared radio channel. In Chapter 9, we will see how a selfish user can modify the behavior of her wireless adapter to achieve this selfish goal, and how such a misdeed can be detected by the access point.

Moreover, the number of operators is likely to dramatically increase (especially in unlicensed frequency bands), because of the low cost of the access points. This means that the level of *trust* that can be associated with operators' brands can significantly decline. We will address this issue of trust in the next chapter.

Another important change is that the very notion of licensed band could be questioned. Indeed, the current practice consisting in allocating a different chunk of the spectrum to each operator is highly inefficient and can become an obsolete approach as soon as more sophisticated technology such as cognitive radios [138, 277] becomes available. The implications of such a change can be overwhelming: in this new setting, operators will have to cope with each other's presence, and program their base stations accordingly. In Chapter 11, we will show how to model this kind of situation. Note that this scenario already happens on a small scale, when several WiFi operators deploy their access points in the same area.

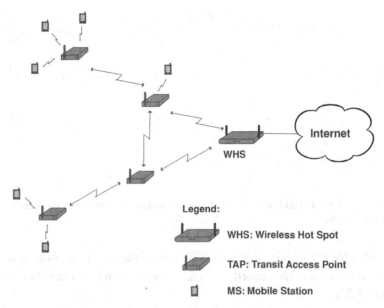

Figure 2.1. A Wireless Mesh Network: the (wireless) Transit Access Points (TAPs) relay the traffic between the Wireless Hot Spot and the mobile stations.

Wireless mesh networks

As shown in Figure 2.1, a typical mesh network is comprised of one Wireless Hot Spot (WHS), connected to the Internet, and of several Transit Access Points (TAPs) which relay traffic between the mobile stations and the WHS in a multi-hop fashion. A nice property of the mesh networks is that they are (in principle) relatively easy to deploy, as they require a single connection point to the Internet.

Wireless Mesh Networks (WMNs) represent a good solution to providing wireless Internet connectivity in a sizable geographic area; this new and promising paradigm allows for network deployment at a much lower cost than with classic WiFi networks.

WMNs are particularly interesting for us, because they contain some features (and vulnerabilities) typical of future networks such as wireless multi-hopping and are already in the standardization and early deployment phase. And, as we will see, they nicely illustrate the fact that performance (in this specific case fairness) and security are closely related. For these reasons, we will describe these networks in some detail.

WMNs, however, are not yet ready for wide-scale deployment for two main reasons. First of all, the communications being wireless and multi-hop (and therefore prone to interference), WMNs present severe capacity and delay constraints. Nevertheless, there are reasons to believe that technology will be able to overcome this problem, for example by using multi-radio and multi-channel TAPs. The

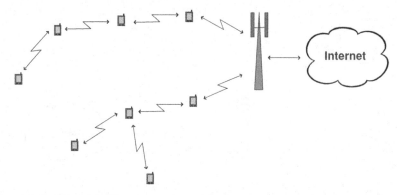

Figure 2.2. A hybrid ad hoc network: mobile wireless stations relay packets to and from the Internet.

second reason for the slow deployment of WMNs is the lack of security guarantees. The reader interested in the security peculiarities of mesh networks should refer to Subsection 2.2.6.

Hybrid ad hoc networks

In mesh networks, the relay stations (the TAPs) between the mobile station and the backbone are specialized devices under the control of one or several operators. A bold design decision consists in removing these relay stations and assigning the relaying task to other mobile stations. Such a network is usually called a "hybrid ad hoc network" or, in some cases, a "multi-hop cellular network."

The proper operation of these networks raises a number of formidable technical challenges and it is unclear, at the time of this writing, whether such networks will ever be implemented. These challenges include notably the problem of power management, as (by definition) *a priori* planning is not possible. With respect to the focus of this book, the routing protocol of such a network can be secured by making use of the protocols described in Chapter 7; the packet forwarding operation can benefit of stimulation mechanisms described in Chapter 10.

An example of a hybrid ad hoc network is provided in Figure 2.2.

Mobile ad hoc networks

A step further towards decentralization consists in removing completely the (on-line) infrastructure: the network then consists only of (mobile) nodes that relay each other's traffic (see Figure 2.3). These networks are usually called mobile ad hoc networks (often abbreviated as "MANET" or wireless ad hoc networks).[1]

---

[1] The first investigations and implementations of these networks took place in the seventies and were intended for military applications; at that time, these networks were known as "Packet Radio Networks."

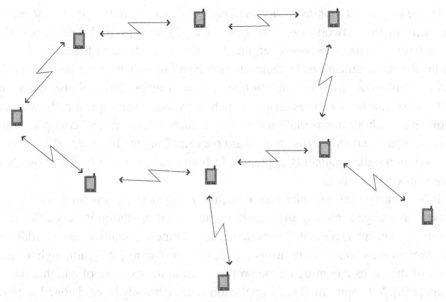

Figure 2.3. A mobile ad hoc network: mobile wireless stations relay packets mutually.

Such networks have been a strong stimulus to the research community, who have devoted to it hundreds of papers, essentially during the last ten years or so. It is very important to distinguish between the following two kinds of such networks.

a. Mobile ad hoc networks in hostile environments designate a set of mobile nodes that are expected to carry out a mission in an environment where the presence of a "strong" attacker is expected. This is typically the case of military networks (sometimes lamentably camouflaged as "rescue operation networks" in the literature). The need for security is of course particularly acute in this category. In this case, the authority would typically pre-load appropriate cryptographic keys in the devices, in compliance with the role of each of the users; these keys would then protect the communication between the devices during the unfolding of the mission. As long as a node is not compromised, it is reasonable to assume that it will have a highly cooperative behavior with respect to the other nodes of the network.

The security challenges typically encountered in this kind of networks include secure routing, prevention of traffic analysis,[2] and resistance of a captured device to reverse engineering and key retrieval. Secure routing is addressed in Chapter 7, and the other two are not addressed in this book; the interested reader can check the specialized literature, for example the proceedings of the IEEE Conference on Military Communications (MILCOM).

---

[2] Traffic analysis consists in establishing who is communicating with whom, typically in a network where all payloads are encrypted. It can also be used to locate a specific transmitter.

b. In self-organized mobile ad hoc networks, there is no authority whatsoever to take care of the network, *not even in the initialization phase.*[3] This means that the network is purely peer-to-peer, and that the nodes have to figure out how to secure the communications by themselves. We will see in Chapter 5 how two nodes can establish a security association between themselves. We will also show that selfishness can be a serious issue in such networks; more specifically, we will show that, without appropriate mechanisms, such a network can collapse, either because nodes selfishly refuse to forward packets (Chapter 10) or greedily overuse the common radio channel (Chapter 9). In both cases, we will explain how these problems can be solved.

It is unlikely that personal communication networks of this kind will be deployed on a large scale any time soon, because their operation is very difficult to ensure (power management, for example, is extremely complicated). In addition, all-wireless networks exhibit intrinsic scalability problems [157], although the mobility of the nodes can mitigate the problem, but at the expense of a higher packet delay [156]. Yet some small scale applications can certainly be envisioned: a group of people can get together, each equipped with a laptop or a PDA; they can be willing to establish a network between their devices, without having to rely on an infrastructure. Today's technology already allows doing this with laptops and PDAs, albeit somewhat painstakingly.

Other personal communication networks As mentioned in the previous chapter, there exist many other wireless personal communication networks, including Bluetooth and WiMAX. We do not discuss them here further, however, as their characteristics in terms of security and cooperation are already covered by the network types we have just described.

2.2.2 *Vehicular networks*

Initiatives to create safer and more efficient driving conditions have recently begun to draw strong support. Vehicular communications (VC) will play a central role in this effort, enabling a variety of applications for *safety, traffic efficiency, driver assistance*, and *infotainment*. For example, in order to improve safety, warnings for environmental hazards (e.g., ice on the pavement) or abrupt vehicle kinetic changes (e.g., emergency braking) will be provided by these systems.

Vehicular networking protocols will allow nodes, that is, vehicles or road-side infrastructure units, to communicate with each other over single or multiple hops. In other words, nodes will act both as end points and routers, with vehicular networks

---

[3] By self-organization, we refer in this book to the organization of security and not of other mechanisms such as routing.

emerging potentially as the largest instantiation of the *mobile ad hoc networking* technology. By their very nature, vehicular networks stand somewhere in between the two extreme cases of mobile ad hoc networks that we have just described: they cannot be fully self-organized, but they also cannot be placed under the strict control of a single authority.

The unique features of VC are a double-edged sword: a rich set of tools are offered to drivers and authorities, but a formidable set of abuses and attacks becomes possible. Hence, the security of vehicular networks is indispensable, because otherwise these systems could make anti-social and criminal behavior easier, in ways that would actually jeopardize the benefits of their deployment. What makes VC security hard to achieve is the tight coupling between applications, with rigid requirements, and the networking fabric, as well as the societal, legal, and economical considerations. Solutions to this problem involve the industry, governments, and the academia.

The reader interested in more details about the security of vehicular communications should refer to Subsection 2.2.7.

2.2.3 Sensor networks

Sensor networks are wireless networks that consist of a large number of sensor nodes and a few base stations or sinks (see Figure 2.4 for illustration). The sensor nodes are tiny devices that are equipped with sensing circuits that collect data about some physical phenomena, such as light, sound, vibration, humidity, temperature, etc. In addition, the sensor nodes have computing and wireless communication capabilities. The base stations are much more powerful than the sensor nodes, and their role is to collect the data gathered by the sensor nodes and to send those data to some application unit for further processing. For this reason, the base stations are often called sinks in the context of sensor networks.

The sensor nodes are usually battery powered, which has a profound effect on the design of sensor networks. Since recharging the batteries is often impractical, or even impossible in some deployment scenarios, the main design criteria for sensor networks is to reduce the energy consumption of the sensor nodes and increase network lifetime as much as possible. All networking mechanisms for these networks are designed with this requirement in mind.

In order to reduce their energy consumption, the sensor nodes cooperatively perform many functions. For instance, sensor nodes communicate with the base station using multi-hop wireless communications, where the nodes forward packets towards the base stations on behalf of other nodes. This reduces energy consumption in two ways: first, it is known that the energy needed for wireless transmission grows super-linearly with the distance of the transmission. Thus, the overall energy

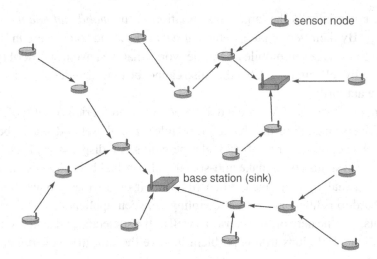

Figure 2.4. A sensor network: sensor nodes forward packets towards the base stations on behalf of other nodes in order to mitigate the overall energy consumption and interference.

consumption can be reduced, if packets are sent in several smaller hops to the base station instead of sending them directly in a single large hop. Second, communication via multiple smaller hops reduces the interference between the devices, which means that fewer re-transmissions are needed due to collisions.

In addition to packet forwarding, the sensor nodes can cooperate in the processing of the gathered data. The idea is that instead of relaying raw sensor readings to the base stations, the sensor nodes can perform in-network processing and aggregate data on their way to the base stations. This can greatly reduce the number of packets that need to be sent, and hence, reduce the energy consumption.

Sensor networks have many useful applications both in military and in civilian environments. In the military setting, they can be used in monitoring, surveillance, and reconnaissance applications. In the civilian setting, they can be used for environmental monitoring purposes (such as forest fire detection and earthquake prediction) and in health applications (such as telemonitoring of physiological data of elderly or chronically ill people, and drug administration). More civilian applications of sensor networks include building automation, smart environments, monitoring the status of structures, such as bridges, increasing the effectiveness of agricultural processes, water management, etc.

In terms of security requirements, sensor networks must ensure the integrity (and in some cases, the confidentiality) of the data delivered to the base stations. Similarly, the integrity (and the confidentiality) of control messages sent by the base stations to the sensors must be guaranteed. Availability is also an important security requirement, especially when the sensor network is used in life critical applications, such as earthquake prediction and telemonitoring of people's health conditions.

These are more or less standard security requirements that can also be found in traditional wired and wireless networks. However, the challenge is to satisfy these requirements under the special operating conditions of sensor networks. First of all, any security solution must take into account the requirement of reducing the energy consumption of the sensor nodes. In addition, as we mentioned before, sensor nodes are tiny devices, which means that their computing and storage capacity is severely limited. Thus, cryptographic algorithms and protocols that require intensive computation, communication, or storage cannot be used in sensor networks.

Furthermore, as sensor nodes are expected to be deployed in mass, they must be cheap, which makes their physical protection against tampering difficult. Moreover, sensor nodes are often deployed in areas where the access to them cannot be monitored. This means that an adversary can corrupt some of the sensor nodes. By doing that, the adversary can learn the content of the memory, including cryptographic secrets, of the corrupted nodes, and she can also modify the behavior of the corrupted nodes. These constraints greatly increase the difficulty of providing security for sensor networks.

As an illustrative example, in Chapter 5, we will elaborate on the problem of establishing shared keys in sensor nodes. Traditional approaches for key establishment are inefficient for sensor networks. Hence, we must develop new solutions that take into account the special characteristics described above. In addition, in Chapter 7, we study how routing in sensor networks can be secured.

2.2.4 RFID

RFID (Radio Frequency Identification) is a wireless technology that enables the identification of objects and people by computers. Humans usually solve the task of identification remarkably well by visual means. A huge amount of work in artificial intelligence and computer vision has been carried out by researchers to endow computers with similar capabilities. However, to a large extent, those efforts have failed: computers still cannot recognize objects and people visually in a reliable manner. RFID offers an alternative approach. The idea is to tag objects (and maybe even people) with smart labels (so called RFID tags) that emit identifying information in the form of a bit string, which can be easily interpreted by computers.

RFID systems have three types of components (see Figure 2.5 for illustration): RFID tags, RFID readers, and back-end databases. RFID tags store identifying information (typically a few hundred bits) about the objects or persons to which they are attached. RFID readers can read this identifying information out from nearby tags. The identifier obtained from a tag is used as an index into a back-end

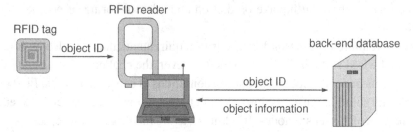

Figure 2.5. RFID systems have three types of components: RFID tags, RFID readers, and back-end databases. The reader looks up in the database the detailed object information, using the identifier obtained from the tag.

database where virtually unlimited information can be stored about the given object or person.

An RFID tag consists of a microchip and an antenna. The microchip stores the identifying information, and the antenna is used for communicating with the RFID reader. In addition, RFID tags can be active or passive. Active tags have their own batteries, while passive tags harvest energy from the reader's RF signal to power themselves up. Passive tags communicate with the reader by reflecting the reader's RF signal, and modulating the reflected signal with the identifying information stored in the microchip. This has the interesting side effect that the reader's signal is much stronger than that of the tag, and therefore, the reader can be eavesdropped from a larger distance. Passive tags cost only a few tens of cents, and therefore, they can be deployed in mass. Active tags are obviously more expensive, and hence, they are usually used to label valuable objects (e.g., an entire container of goods).

RFID today

RFID technology is already used today in many applications, including access control to buildings, toll-payment on highways, management of library books, and identification of pets, just to list a few.[4] Essentially, the operating principles in all these applications are the same.

In the case of access control to buildings, the users carry plastic cards that contain RFID tags. When a user draws her card near to the card reader at a door, her identifying information is transmitted from the card to the reader. The computer attached to the reader then looks up her access rights in a central database, and if she is authorized to enter through that door, then it is opened.

In the case of electronic toll-payment systems, the RFID tags are stuck to the windshields of the vehicles, and the RFID readers are installed at the toll gates.

---

[4] In addition, contactless credit cards and public transport cards also use RFID technology for the communication with payment terminals and ticket validating terminals, respectively.

When a user drives through a gate, her identifying information is read from the tag, and her account held in a central database is updated.

Many libraries use RFID to facilitate the management of books. In this application, the RFID tags are inserted into the books. When a book is checked out, the RFID reader at the check-out desk reads the identifying information from the tag, and updates the book's status in a central database. When the book is returned, the tag is read and the status information in the database is updated again.

The idea of tagging pets is to help the identification of lost animals, that allows them to be returned to their owners. Similarly, there exist human implantable RFID tags too. An intended application of those tags is to help finding the medical records of patients in a hospital, but they could equally be used for access control purposes.

RFID tomorrow

As we saw above, RFID is already used today in many applications, but an even more widespread use of this technology is expected in the future. There are plans to embed RFID tags in bank notes to make forgery more difficult and to combat against money laundering. Passports and ID cards will be based on RFID technology too in the near future.[5] Note that in contrast to access control cards and library books that can be carried by some people, but probably not everyone, virtually every person carries banknotes or some ID documents with her.

However, this is still not the end of the story. Many people believe that the "killer" application for RFID will be the replacement of optical bar codes printed on consumer products. Today, RFID tags are still too expensive to make this feasible, but in the near future their price could drop below the threshold that allows item level tagging. If this happens, virtually all objects will have RFID tags embedded in or attached to them. There is even a standardization effort that aims to prepare the grounds for this massive takeover. The organization behind this effort is called EPCglobal Inc., and it promotes the specifications for the so called EPC (Electronic Product Code) tag.

RFID has two main advantages compared with optical bar codes. First, optical bar codes usually indicate only the type of an object, whereas an RFID tag can store a *unique* identifier that identifies not only the type, but also distinguishes the object among many other objects of the same type. Second, reading of bar codes requires line-of-sight contact with the reader, whereas RFID tags can be read without line-of-sight contact and from a larger distance. This makes it possible to read RFID tags in large quantities rapidly and remotely.

---

[5] Indeed, at the time of this writing, the USA and some European countries have already started to issue electronic passports based on RFID technology.

The advantages of RFID for the manufacturers and for the merchants are clear: it enables automated, and hence, more efficient control and management of products throughout their whole life cycle, from the production line through the stock houses to the shelves in the stores. The advantages for the consumers, however, are less obvious. One advantage could be the possibility of fast check-out at point-of-sale (POS) terminals. The idea is that an RFID reader at the POS terminal can read all the tags of the goods in the shopping cart in a few seconds without the need to take the goods out from the cart. This could considerably speed up the check-out process and hence consumers would not need to stand in long queues when they do their shopping for the week-end. Another advantage would be that items could be returned by consumers without the need to keep the receipt received at the purchase, because based on the unique identifier in the tag of a given item, the merchant can look up in its database where and when that item has been purchased. Pilot experiments are already carried out in large retail shops.

Yet another possibility would be to pull together product databases and purchase records, and identify consumers (assuming that they are identifiable, for instance, they paid with their credit cards) that purchased a given product in a given period of time. This could be advantageous when a product turns out to be faulty or contaminated, and all the consumers that bought it must be quickly notified. At the same time, such an application would be dangerous, because it could be misused for profiling consumers.

A more futuristic application of RFID with some advantages for the consumers would be to make household appliances capable of interacting with items. One can imagine, for instance, that a smart washing machine automatically determines what program it should run by reading out the appropriate information from the RFID tags embedded in the clothes put into the machine. Or, one can image a smart refrigerator that would warn the user if some goods are about to expire or to run out; smarter ones can even order goods on-line on behalf of the user.

Although such smart appliances seem to be a far-fetched idea, interacting with objects is not a fantasy anymore. Large mobile phone manufacturers have started to integrate RFID readers in their handsets by means of a technology called Near Field Communications (NFC). Such an NFC enabled mobile phone allows its user to read identifying information out from RFID tags attached to nearby objects. In addition, the mobile phone is also able to immediately obtain (and display) related data from on-line databases through available GPRS or 3G data connections. One can imagine many useful applications of this technology. An example would be the following: let us assume that a user sees a movie poster on the street. Using her NFC enabled mobile phone, she can scan the RFID tag embedded into the poster, and immediately find more information about the movie on the Web (such as the trailer), including where and when it is played. Finally, she can even buy a ticket for the next performance using her mobile phone.

As we have seen above, RFID based automated identification enables many interesting applications. However, the widespread deployment of RFID technology can also lead to serious privacy problems. Imagine a world where virtually everything is tagged with RFID tags. In that world, the monitoring of the movement and the activities of people can be easily automated, meaning that tracking people could be cheap and continuous in space and time. Without privacy protecting measures, such a world could easily degenerate into the one described by Orwell in his book *1984* [295]. We believe that this is a very important problem, and we describe possible technical solutions to it in Chapter 8.

2.2.5 Mobility in the Internet

The need to cope with the emergence of networks such as the ones described above and with the growing mobility of hosts has led the Internet community to profoundly reconsider the overall organization of the network. We summarize here the efforts related to Mobile IPv6, insisting on the security challenges. The operating principles of Mobile IPv6 are described in RFC 3775 "Mobility Support in IPv6," of June 2004.[6] This discussion will help us establishing a link between the security concerns of the wireless and the wired parts of the network.

When a (wired or wireless) node changes location, in many cases it also changes links, thus affecting its address. The consequence of this address change can be very unpleasant to the user, as it can break all the existing connections of the mobile node that are using the address assigned when it was on the previous link. Mobile IP aims at solving this problem at the IP layer, thus making the mobility of the node completely transparent to upper layer protocols such as TCP.

Mobile IP is a flexible standard, supporting many modes of operation. We provide here only a brief description of the operating principles of Mobile IPv6, in order to introduce the subsequent discussion of security.

The various components are represented in Figure 2.6.

The *home link* is the link to which the Mobile Node (MN) is "usually" attached.

The *home address* is an address assigned to the mobile node when it is attached to the home link; the mobile node is always reachable through this address, regardless of its current real location. The *home agent* is a router on the home link permanently aware of the current location of the nodes that are away from home.

---

[6] The reader interested in the operating principles of Mobile IPv4 can refer to RFCs 3220 and 3344 (both from 2002) "IP Mobility Support for IPv4."

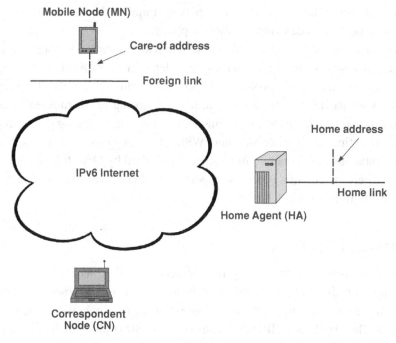

Figure 2.6. Components of Mobile IPv6.

The *foreign link* is a link different from the home agent's link, to which the mobile node is temporarily attached (either by wire or wireless). A *care-of address* is an address used by the mobile node while it is attached to a foreign link. The association of a care-of address with a home address for a mobile node is called a *binding*; correspondent nodes and home agents store bindings in a binding cache. A *correspondent node* is an IPv6 node communicating with a mobile node.

Two modes for mobility are supported by Mobile IPv6 for communication between a mobile node and a corresponding node.

a) *Bidirectional tunneling* In this mode, the mobile node tunnels the packets intended for the correspondent node through its home agent. Reciprocally, the home agent intercepts packets addressed to the mobile node's home agent and tunnels those packets to the mobile node via its care-of address.

b) *Route optimization* This mode uses the optimal route between the mobile node and the correspondent node. The mobile node registers its current address binding with the correspondent node. In this way, the correspondent node can send packets directly to the mobile node's care-of address. In addition to optimizing the path between nodes, this option also reduces the risk of congestion at the mobile node's home agent (as the latter is not involved in the packet forwarding process).

Route optimization is of course the most satisfactory solution in the long run and we will therefore focus on it in our brief discussion of security. The reader interested in more details can refer to RFC 4225 "Mobile IP Version 6 Route Optimization Security Design Background," December 2005.

Security principles and goal The security of Mobile IPv6 obeys two main principles. The first consists in complying with the end-to-end principle of Internet protocols. In this specific context, this means minimizing the involvement of the routers: in Mobile IPv6, only the home agent and the communicating nodes need to create state.

The second principle is related to trust:[7] it is assumed that the mobile node and the home agent know each other through a prior arrangement, whereas the mobile node and the correspondent node do not need to have any prior arrangement.

The security goal of Mobile IPv6 consists in being "as secure as the (non-mobile) IPv4 Internet." This means in particular that there is little protection against attackers that are able to attach themselves between a correspondent node and a home agent.

Attacks The target of an attack can be any node or network on the Internet (stationary or mobile). An attacker can either aim at diverting (stealing) the traffic destined to or sourced at the target node or cause a denial-of-service at the target node or network. It is important to notice that IPv6 uses the same class of IP addresses for both kinds of nodes (namely home and care-of addresses on one hand and stationary nodes on the other hand). This means that attacks that in principle would concern only mobile nodes are a threat to all IPv6 nodes.

Address stealing If binding updates were not authenticated, an attacker could send spoofed binding updates from anywhere in the Internet, and realize the attack illustrated in Figure 2.7.

The attacker might define the care-of address to be either its own current address, another address in its local network, or any other IP address. By selecting a care-of address allowing it to receive packets, the attacker would be able to send replies to the correspondent node, thus delaying the uncovering of the attack.

We have described only the basic address stealing attack. A number of attacks can be derived from it; in particular, as it breaks the communication paths, it can be used to mount denial-of-service attacks.

---

[7] The notion of trust will be discussed in detail in the next chapter.

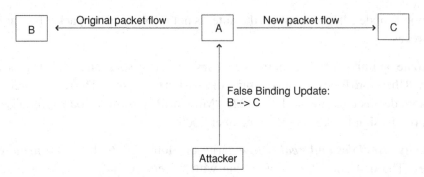

Figure 2.7. *Address Stealing* attack (RFC 4225): assume a packet flow from node A to node B. The attacker redirects the packets to a different address, C, by sending a *Binding Update* to A (hence node A believes that node B has moved to address C).

DoS attacks exploiting binding update protocols An attacker can try to exhaust the resources of a target (mobile node or correspondent) by sending spoofed IP packets that trigger a large number of binding update protocol instances.

Protection mechanisms A robust countermeasure against the address stealing and flooding attacks consists in a mutual authentication of the nodes involved in a binding update protocol, typically based on IPsec. However, to be usable between two *arbitrary* nodes, IPsec requires a global key management infrastructure (to be used typically by the Internet Key Exchange protocol, see RFC 2409), which does not exist, and is unlikely to come into existence any time soon.

Because of this major problem, a non-cryptographic solution was designed, which relies on the assumption of an uncorrupted routing infrastructure. The cornerstone of the solution is *Return Routability* (RR). The principles are illustrated in Figure 2.8.

It is intuitive that the presence of this test makes attacks much more difficult to carry out. Yet the detailed description and the security analysis of this protocol are beyond the scope of this book, and the interested reader is invited to refer to the related RFCs. Return routability can of course fall prey to a compromised routing infrastructure or to an attacker located between the verifier and the address to be verified.

Finally, we should mention that a possible protection against the mentioned DoS attacks exploiting binding update protocols can be realized by each node setting a limit on the amount of resources (processing time, memory, and communication bandwidth) that it devotes to processing binding updates. However, this can lead to a self-denying of some of the mobility mechanisms.

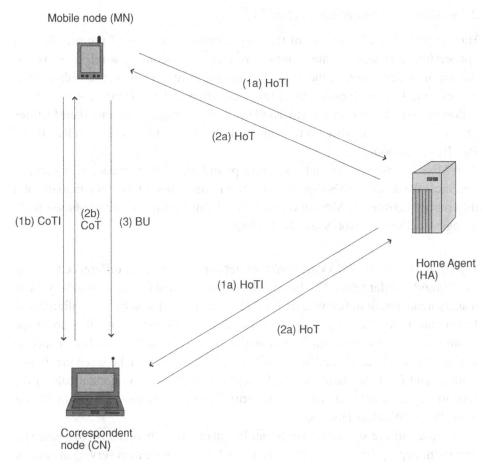

Figure 2.8. Return Routability (RFC 4225): the Mobile Node (MN) checks the routability to the Correspondent Node (CN) (a) via the Home Agent (HA) (message Home Test Init or HoTI) and (b) directly (message Care-of Test Init or CoTI); the correspondent node replies to both of them independently by sending a Home Test (HoT) in response to the Home Test Init and a Care-of Test (CoT) in response to the Care-of Test Init. It is only once the mobile node has received both Home Test and Care-of Test packets that it sends a Binding Update to the correspondent node. In addition, the bindings are short-lived, in order to mitigate the effect of a possible malicious binding update ("time shifting attack").

Privacy in Mobile IPv6 In this discussion of Mobile IPv6 security, we have considered only active attackers. However, it is clear that, simply by examining packets, eavesdroppers can track the movements of individual nodes and therefore of users. Mobile IPv6 is even more vulnerable to this kind of misdeed, as it adds potentially sensitive information into the packet, such as binding updates. The interested reader can refer to RFC 3041 "Privacy Extensions for Stateless Address Autoconfiguration in IPv6," of January 2001.

2.2.6 More on wireless mesh networks

Having provided an overview of the major trends as well as of the security and cooperation issues in the most significant types of upcoming wireless networks, we will now detail two of the examples that we have already mentioned: we will address wireless mesh networks in this subsection and vehicular networks in the following one. We have chosen to detail these two examples, because their features provide a very nice motivation of many of the mechanisms described in Part II and Part III of this book.

WMNs represent a new network concept and therefore introduce new security specifics. We describe these specifics by providing an overview of the fundamental differences between WMNs and two well-established infrastructure-based technologies: cellular networks and the Internet.

Difference between WMNs and cellular networks The major difference between WMNs and cellular networks – besides the use of different frequency bands (WMNs usually make use of unlicensed frequencies) – concerns the network configuration. In cellular networks, a given area is divided into cells and each cell is under the control of a base station. Each base station handles a certain number of mobile stations that are in its immediate vicinity (i.e., communication between the mobile stations and the base station is single-hop) and it plays an important role in the functioning of the cellular network; the entity that plays an equivalent role in WMNs is the WHS (Wireless Hot Spot).

Whereas all the security aspects can be successfully handled by the base station (with appropriate assistance from an on-line authentication server) in cellular networks, it is risky to rely only on the WHS to secure a WMN, given that the communications in WMNs are multi-hop. Indeed, centralizing all security operations at the WHS would delay attack detection and countermeasure and therefore would give the adversary an undeniable advantage. Furthermore, multi-hopping makes routing in WMNs a very important and necessary functionality of the network; and like all critical operations, an adversary can be tempted to attack it. The routing mechanism must thus be secured.

Multi-hopping has also an important effect on the network utilization and performance. Indeed, if the WMN is not well-designed, a TAP that is several hops away from the WHS would receive a much lower bandwidth share than a TAP that is next to it. This leads to severe unfairness problems, and even potentially to starvation.

Difference between WMNs and the Internet In WMNs, the wireless TAPs play the role that is played, in the classical (wired) Internet, by the routers. Given that wireless communications are vulnerable to passive attacks such as eavesdropping,

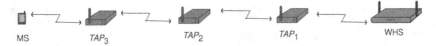

Figure 2.9. A typical communication in WMNs: the mobile station MS is within the transmission range of TAP_3 and relies on TAP_1 and TAP_2 to relay its traffic to and from WHS. Used with permission, from [48], © IEEE, 2006.

as well as to active attacks such as denial-of-service (DoS), WMNs are subject to all these attacks, whose effects are amplified by the multi-hop aspect of the communications.

Another fundamental difference between the Internet and WMNs is that, unlike Internet routers, the TAPs are not physically protected. Indeed, they are most often in locations that are accessible to potential adversaries, e.g., deployed on rooftops or attached to street lights. The absence of physical protection of the devices makes WMNs vulnerable to some serious attacks. Indeed, one very important requirement regarding the TAPs – for the concept of mesh networks to remain economically viable – is their low cost that excludes the possibility of strong hardware protection of the devices (e.g., detection of pressure, voltage, or temperature changes). Therefore, attacks such as tampering, capture or replication of TAPs are possible and even easy to perform.

This brief analysis of the characteristics of WMNs shows that, compared with other networking technologies, the new security challenges are mainly due to the multi-hop wireless communications and the fact that the TAPs are not physically protected. Multi-hopping delays the detection and treatment of the attacks, makes routing a critical network service and can lead to severe unfairness between the TAPs, and the physical exposure of the TAPs allows an adversary to capture, clone or tamper with these devices.

Security principles of WMNs Before further discussing the details of security in WMNs, let us consider a simple example: Figure 2.9 shows a branch of a WMN where a mobile station MS is within the transmission range of TAP_3 and therefore relies on it to get Internet connectivity; the data generated (and received) by the MS go through TAP_2, TAP_1 and WHS.

Let us consider an upstream message, i.e., a message generated by the MS and sent to the Internet. Before this message reaches the infrastructure, several verifications need to be performed.

- Given that Internet connectivity is a service that (usually) the MS has to pay for, the TAPs and the WHS have to authenticate the MS.

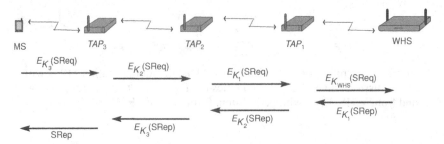

Figure 2.10. Session establishment: the mobile station MS generates a session request SReq and encrypts it using TAP_3's public key K_3. Then, TAP_3 decrypts SReq, encrypts it using TAP_2's public key and sends it to TAP_2, and so on until the message reaches the WHS. The session reply message SRep is then generated by the WHS and sent back to the MS. It is protected in the same way as SReq and it contains the information about the session key. Used with permission, from [48], © IEEE, 2006.

- The MS has to authenticate the TAPs to make sure that they belong to a valid operational WMN. It has at least to authenticate TAP_3, the TAP to which it is directly connected.
- The TAPs have to authenticate the other TAPs in the WMN to prevent TAP forgery and to detect intruders.
- Finally, the data sent or received by MS have to be protected (e.g., to ensure data integrity, non-repudiation and/or confidentiality).

Performing these verifications has to be efficient and lightweight, especially for the MS; we thus want to avoid, if possible, the use of asymmetric cryptographic operations by the MS. In fact, the MS being battery operated, the use of public key cryptography primitives is unsuitable as these primitives have a high computational overhead and are prone to DoS attacks. Indeed, if the authentication protocol requires the computation or the verification of a signature, this feature can be misused by an adversary that can continuously ask the MS to compute or verify signatures; this attack can drain MS's battery.

In Figure 2.10, we represent a simple way to perform the four aforementioned verifications in the WMN branch represented in Figure 2.9.

We assume, without loss of generality, that each node in the branch (i.e., TAP_1, TAP_2, TAP_3 and the WHS) has a public/private key-pair that is assigned to it by the network operator. These keys can be used to establish a session key k_S between MS and the WHS. This session key permits to secure the data sent and received by the MS while limiting the use of public key cryptography to the session establishment phase, which is much less frequent. Note that the session establishment is initiated by the MS, which reduces the risk of DoS attacks described above.

In Figure 2.10, we represent an example of such session establishment: first, the mobile station MS generates a symmetric key k and includes it in a session request message SReq, which it encrypts using TAP_3's public key K_3. Upon receipt of SReq, TAP_3 decrypts it using its private key K_3^{-1}, encrypts it using TAP_2's public key and sends it to TAP_2, and so on until the message reaches the WHS.

To exemplify, SReq can be computed as follows:

$$\text{SReq} = E_{K_{\text{WHS}}}(ReqID, roamingInfo, k, N)$$

where $ReqID$ represents the request identifier (to prevent replay attacks), $roamingInfo$ represents the information needed by WHS to authenticate MS, k is the key that WHS will use to encrypt the future session reply (SRep), and N is a nonce (used later to prevent replay attacks). SReq is encrypted using WHS's public key K_{WHS}.

The WHS uses $roamingInfo$ to authenticate the MS. This authentication can be done in different ways, depending on the content sent by the MS: for example, using a temporary billing account (e.g., credit card based authentication), a predefined shared secret (if the MS is a client of the operator managing the WMN), or a roaming system similar to the one used in cellular networks (if it is not a client of that operator); the latter has the advantage of preserving the anonymity of the MS with respect to the operator of the visited network.

Note that the fact for the WHS to receive a valid SReq message proves that all the TAPs in the route between the MS and the WHS are valid TAPs. Indeed, assume that an attacker replaced TAP_2 by a rogue device TAP_2'. When TAP_2' receives SReq, the message is encrypted using the public key of TAP_2 and therefore TAP_2' is not able to decrypt it correctly; the message TAP_2' sends to TAP_1 is thus corrupted. If TAP_1 is able to check the integrity of the message, it detects the attack and discards SReq, otherwise the attack will be detected by the WHS as the data in SReq would be meaningless; the WHS will then discard SReq.

If the session request SReq is valid, then the WHS generates a session reply message SRep and sends it back to the MS. SRep contains information that allows the MS (and, if needed, the TAPs in the route) to generate the session key k_S and is protected in the same way as SReq (i.e., encrypted and then decrypted successively using the public keys of the TAPs in the route). It is also protected against eavesdropping as the MS has to be the only mobile station that can interpret correctly the data in SRep; the WHS uses the key k generated by the MS to encrypt the data sent in SRep.

Once the session key k_S is set up, it is used to check the integrity of the exchanged messages, e.g., by computing Message Authentication Codes (MACs). The verification of the MACs can be done end-to-end (i.e., when the session key is known only to the WHS and the MS) or by each intermediate TAP (i.e., if the TAPs in the route also know k_S). The session key k_S can even be used to encrypt the exchanged messages if data confidentiality is a requirement. It is also possible to use MACs to authenticate the TAPs involved in the communication and to detect intruders during the session. Indeed, each two neighboring TAPs can establish (e.g., during session key establishment) or have a predefined symmetric key that they will use typically to compute Message Authentication Codes (MACs) on the exchanged messages and therefore to authenticate the nodes involved in the communication hop by hop.[8]

Three fundamental security operations Our study of WMNs' specifics has pinpointed three critical security operations: (i) securing the routing mechanism, (ii) enforcement of fairness, and (iii) detection of corrupt TAPs. These challenges are not the only ones: other network functionalities such as MAC protocols and node location information also need to be protected. In addition, WMNs are vulnerable to the same kind of selfish behavior as WiFi (see Chapter 9). Yet, we choose to focus on these three operations because they are, in our opinion, the most critical for WMNs.

Secure multi-hop routing By attacking the routing mechanism, an adversary can modify the network topology and therefore affect the proper functioning of the network. For example, the adversary can want to partition the network or to isolate a given TAP or a given geographic region, or to force the traffic through a specific TAP in the network (e.g., through a TAP that it has compromised) in order to monitor the traffic of a given mobile station or a region. Another example would be for the adversary to artificially lengthen the routes between the WHS and the TAPs, which would seriously affect the performance of the network.

To attack the routing mechanism, the adversary can tamper with the routing messages or perform DoS attacks.

(i) To prevent attacks against the routing messages, the operator can use one of the proposed secure routing protocols for wireless multi-hop networks, which we will describe in Chapter 7.
(ii) DoS attacks represent a simple and efficient way to attack routing. These attacks can be very harmful, are simple to perpetrate, and are very difficult to prevent.

---

[8] MACs are usually used to verify the integrity of a message, but they can also be used to authenticate the sender of the message. Indeed, assume that two parties A and B share a symmetric key k. A can generate a message m, use k to compute a MAC on it and then send both m and the corresponding MAC to B. Upon receipt of these data, B can use k to compute the MAC on m and compare it to the MAC it received; if the two MACs are identical, and given that A and B are the only two parties that know k, B can conclude that m was indeed generated by A. This authentication technique is weaker than the one using asymmetric key cryptography, but it is efficient.

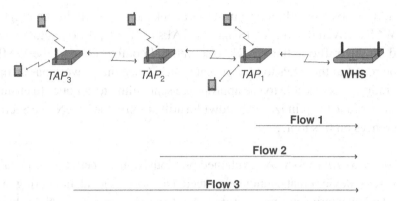

Figure 2.11. The fairness problem. In order to define the bandwidth sharing, it is important to take into consideration the number of mobile stations served by each of the TAPs. Flow 2 should thus have half as much as what flow 1 and flow 3 have, as TAP_2 is serving only one client, whereas TAP_1 and TAP_3 are serving two clients each. Used with permission, from [48], © IEEE, 2006.

The adversary can jam the communications between the TAPs in a given area and force the reconfiguration of the network. In order to solve this problem, the operator has to identify the source of the attack and, if possible, counter it; of course, thwarting this attack will generally require human involvement.

Fairness In WMNs, all the TAPs use the same WHS as a relay to and from the infrastructure and therefore the throughput obtained by the TAPs can vary significantly depending on their position in the WMN: the TAPs that are more than two hops away from WHS could starve (i.e., their clients are not able to send or receive significant traffic), which is highly unfair. The study conducted in [145] identifies the problem and proposes a solution that guarantees a TAP-fair share of the bandwidth. However, a TAP-based fairness is not necessarily the best solution for WMNs. Consider as an example the one-dimensional WMN presented in Figure 2.11: a per-TAP fairness policy leads to flows 1, 2 and 3 having each the same share of the bandwidth, without taking into consideration the number of clients that are served by each of these TAPs. The bandwidth sharing should be fair client-wise, because the purpose of a mesh network is to offer a service (typically Internet connectivity) to the mobile stations that are usually paying the same flat rate. That is why, in the example of Figure 2.11, flow 2 should have half as much as what flow 1 and flow 3 have, as TAP_2 is serving only one client, whereas TAP_1 and TAP_3 are serving two clients each (assuming all clients have equivalent needs).

The fairness issue is closely related to the number of hops between the TAPs and the WHS. This means that if the adversary manages to increase the number of hops between a given TAP and the WHS, it can decrease dramatically the bandwidth share of this TAP. A possible solution against this attack can be a periodic reconfiguration of the WMN. Given that the WHS and the TAPs are static, the operator can define – based on the traffic in the WMN – the optimal configuration of the WMN and force the routes at the TAPs to the optimal routes. Once the network has an optimal configuration, it is possible to use appropriate scheduling techniques to ensure per-client fairness and to optimize the bandwidth utilization in the WMN; see Section 2.5 for references on this topic.

Detection of corrupt TAPs   As explained previously, mesh networks typically employ low-cost devices that cannot be protected against removal, tampering or replication. An adversary can thus capture a TAP and tamper with it. Note that if the device can be remotely managed, the adversary does not even need to physically capture the TAP: a distant hacking into the device would work perfectly. The WHS plays a special role in the WMN and can handle or store critical cryptographic data (e.g., temporary symmetric keys shared with the mobile stations, long-term symmetric keys shared with the TAPs, etc.). Therefore, we assume that the WHS is physically protected.

We identify four main attacks that can be performed on a compromised TAP, depending on the goals the adversary wants to achieve.

The first attack consists in the simple **removal** or **replacement** of the TAP in order to modify the network topology to the benefit of the adversary. This attack can be detected by the WHS or by the neighboring TAPs when a sudden and permanent topology change is observed in the network.

The second attack consists in **accessing the internal state** of the compromised TAP without changing it. The detection of this attack is difficult, given that no state change is operated on the TAP. Disconnecting the device from the WMN might not be required for the adversary to successfully perform the attack; and even if a disconnection were required, the "absence" of the device might not be detected, as it can be assimilated to some congestion problem. If this attack is successful, it guarantees to the adversary the control of the corrupt TAP and a perfect analysis of the traffic going through it. This attack is more serious than simple eavesdropping on the radio channel because the adversary, by capturing the TAP, can retrieve its secret data (e.g., its public/private key-pair, the symmetric key shared with the neighboring TAPs or with the WHS, etc.) and can use these data to compromise, at least locally, the security of the WMN, especially data confidentiality and integrity, and client anonymity. Unfortunately, there is no obvious way to detect this attack. However, a possible solution that mitigates its effect is a periodic erasure and reprogramming of the TAPs; the adversary is then obliged to compromise the device again.

In the third attack, the adversary **modifies the internal state** of the TAP such as the configuration parameters, the secret data, etc. The purpose of this attack can be, for example, to modify the routing algorithm at the compromised node in order to change the network topology. This attack can be detected by the WHS using a software attestation mechanism, see Section 2.5.

Finally, the fourth attack consists in **cloning** a given TAP and installing the replicas at some strategically chosen locations in the mesh network, which allows the adversary to inject false data or to disconnect parts of the WMN. This attack can seriously disrupt the routing mechanism, but it can be detected using appropriate techniques for the identification of replicated nodes, see Chapter 4.

Two attack examples In order to illustrate the attacks described so far, we give two attack examples that an adversary can perpetrate against the WMN (see Figure 2.12a). In the first attack, the adversary corrupts TAP_2, whereas in the second attack, it performs a DoS attack – based on jamming – on the communication link between TAP_5 and TAP_6. Note that we assume the two attacks to be performed by the same adversary, which represents the worst case (as it gives more power to the adversary).

The goal of these attacks can be the following. First, by corrupting TAP_2, the adversary can retrieve its secret data and therefore can compromise the integrity and confidentiality of the data going through it, as well as the anonymity of the mobile stations attached to TAP_2, TAP_3, and TAP_4. Second, the DoS attack is a very simple and efficient way to partition the WMN and trigger a network reconfiguration, which will force more of the traffic to flow through the compromised TAP_2.

It is imperative to detect these attacks in order to react accordingly. A possible reaction to the corrupt TAP attack can be the replacement, by the network operator, of the compromised TAP_2 (see Figure 2.12b). The detection and disabling of the jamming station can be more delicate: finding the exact location of this station can be difficult and, even if it is found, the network operator might not have the authority to disable it (especially in the likely case where both the WMN and the jamming station are operating in unlicensed band); in this case, a network reconfiguration is required. This connectivity change affects the routing and can increase the number of hops from a given TAP to the WHS (for example, in Figure 2.12, TAP_6 was two-hops away from the WHS but after the network reconfiguration, it is seven-hops away), which, as shown previously, can dramatically affect the performance of the WMN. Note that the operator can decide to abandon a given TAP location if it is particularly exposed (the TAP located there is repeatedly corrupted), in which case it would be necessary to deploy additional devices to make up for the coverage gap.

Multi-operator WMNs So far, we have assumed the WMN to be managed by a single operator, but a mesh network can also designate a set of wireless devices

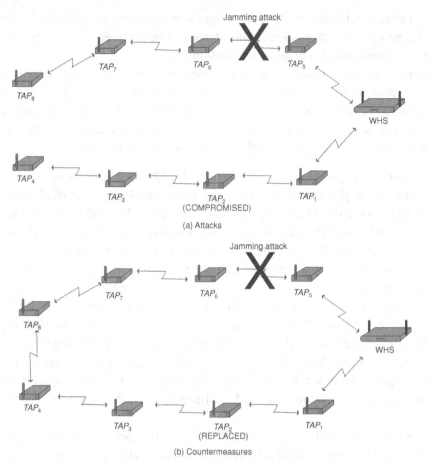

Figure 2.12. Two attacks and the related countermeasures. In (a), the adversary corrupted TAP_2 and placed a jamming station between TAP_5 and TAP_6. As shown in (b), the detection of these attacks leads to the reconfiguration of the WMN: the operator replaced the compromised TAP_2 by uncorrupted equipment and updated the routing. In this example, the reconfiguration leads to much longer routes for some TAPs (e.g., TAP_6 was two-hops away from the WHS and is now seven-hops away). Used with permission, from [48], © IEEE, 2006.

belonging to different networks and controlled by different operators. Ensuring security is more delicate in this case: in addition to the security challenges that we have already identified, one has to add challenges such as the mutual authentication of nodes belonging to different "operating domains" or the application of different charging policies for each of these domains (which can affect fairness).

Another important security challenge results from the utilization of the same spectrum by the different operators. If we assume that a mobile station can freely roam across TAPs that are managed by different operators and that it attaches to the neighboring TAP with the strongest signal, each operator can be tempted to configure its TAPs to always transmit at the maximum authorized level (and thus make sure that it is heard by the maximum number of mobile stations). This situation can lead to a bad performance of the WMN, but can be solved using Multi-Radio/Multi-Channel (MR-MC) TAPs in the WMN. Note that the use of MR-MC TAPs can also mitigate the effect of the DoS attack; instead of jamming a single channel, the adversary has to jam all the channels used by a given node to completely disable it.

2.2.7 More on vehicular networks

Having described mesh networks, we now will provide a description of the challenges of vehicular networks, and sketch a "reasonable" solution. The problem is much more involved than in the case of mesh networks and the progress towards the solution in the academic and industry communities has proved to be much slower.

Vulnerabilities

Any wireless-enabled device that runs a rogue version of the vehicular communication protocol stack poses a threat. We denote such rogue devices deviating from the defined protocols as *adversaries* or *attackers*.

The adoption of a variant of the widely deployed IEEE 802.11 protocol by the vehicle manufacturers makes the attacker's task easier.[9] And even possession of credentials cannot ensure alone the correct operation of the nodes. The nature of the attacker (internal or external, rational or malicious, independent or colluding, persistent or random) has an overwhelming influence on the amount of damage she can generate. Here, rather than analyzing specific protocols, we are after a general exploration of VC (vehicular communications) vulnerabilities.

Forgery Large portions of the vehicular network coverage area can be rapidly "contaminated" with false information by a single attacker who forges and transmits false hazard warnings (e.g., ice formation on the pavement) that are taken up by all vehicles in both traffic streams. This is illustrated in Figure 2.13.

In-transit traffic tampering Any node acting as a relay can disrupt communications of other nodes: it can *drop* or *corrupt* messages, or *meaningfully modify* messages. In this way, the reception of valuable or even critical traffic notifications or safety messages can be manipulated. Moreover, attackers can *replay* messages, e.g., to illegitimately obtain services such as traversing a toll check point. In fact,

---

[9] See http://grouper.ieee.org/groups/scc32/dsrc/

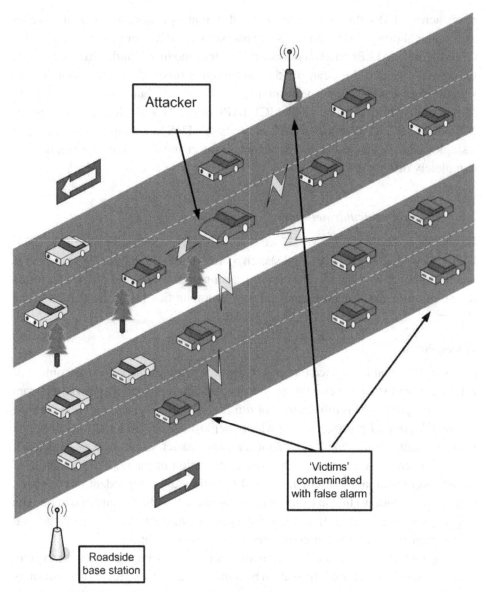

Figure 2.13. Message forgery: the attacker disseminates false alarms, e.g. in order to induce the drivers to brake abruptly. Used with permission, from [325], © IEEE, 2006.

tampering with in-transit messages can be simpler and more powerful than forgery attacks.

Impersonation Message fabrication, alteration, and replay can also be used towards impersonation. Arguably, the source of messages, identified at each layer of the protocol stack, could be of secondary importance. Often, it is not the source but the content (e.g., hazard warning) and the attributes of the message (freshness, locality, relevance to the receiver) that count the most. However, an impersonator can be a threat: consider, for example, an attacker masquerading as an emergency vehicle to mislead other vehicles to slow down and yield. Or, an adversary impersonating roadside units, spoofing service advertisements or safety messages.

Privacy violation With vehicular networks deployed, the collection of vehicle-specific information from overheard vehicular communications will become particularly easy. Then, inferences on a driver's personal data could be made, and thus violate her *privacy*. The vulnerability lies in the periodic or frequent messages generated by a vehicle: safety and traffic management messages, context-aware data access (e.g., maps, ferryboat schedules), transaction-based communications (e.g., automated payments, car diagnostics), or other control messages (e.g., over-the-air registration with local highway authorities). In all such occasions, messages will include, by default, information (e.g., time, location, vehicle identifier, technical description, trip details) that could precisely identify the originating node (vehicle) as well as the driver's actions and preferences (Figure 2.14).

On-board tampering Beyond abuse of the communication protocols, the attacker can select to tinker with data (e.g., velocity, location, status of vehicle parts) at their source, tampering with the on-board sensing and other hardware. In fact, it can be simpler to replace or by-pass the real-time clock or the wiring of a sensor, rather than modifying the binary code implementation of the data collection and communication protocols. Any VC security architecture should achieve a trade-off between robustness and cost due to tamper-proof hardware.

Jamming A jammer deliberately generates interfering transmissions that prevent communication within their reception range. As the network coverage area, e.g., along a highway, can be well-defined, at least locally, jamming is a low-effort exploit opportunity. As Figure 2.15 illustrates, an attacker can relatively easily, without compromising cryptographic mechanisms and with limited transmission power, partition the vehicular network.

Challenges

The operational conditions, the constraints, and the user requirements for VC systems make security a hard problem. We now discuss the most significant challenges specific to VC.

Network volatility The connectivity among nodes can often be highly transient and a one-time event. For example, two vehicles (nodes) passing by each other will remain, in general, only for a few seconds within their transceiver range. In other words, vehicular networks lack the relatively long-lived context and, possibly, the personal contact of the device users of a connection to a hot-spot or the recurrent connection to an on-line service across the Internet. Hence password-based

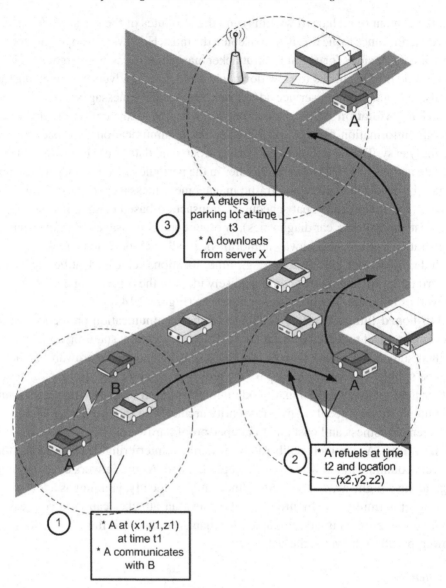

* A enters the
parking lot at time
t3
* A downloads
from server X

③

* A at (x1,y1,z1)
at time t1
* A communicates
with B

①

* A refuels at time
t2 and location
(x2,y2,z2)

②

Figure 2.14. Vehicle tracking: the attacker has deployed three rogue antennas and takes advantage of the messages transmitted by the victim vehicle (*A*) in order to track her. Used with permission, from [325], © IEEE, 2006.

establishment of secure channels, gradual development of trust by enlarging a circle of trusted acquaintances, or secure communication only with a handful of endpoints are impractical for securing VC.

Liability vs. privacy To make the problem harder, accountability, and eventually liability, of the vehicles and their drivers is required. Vehicular communication

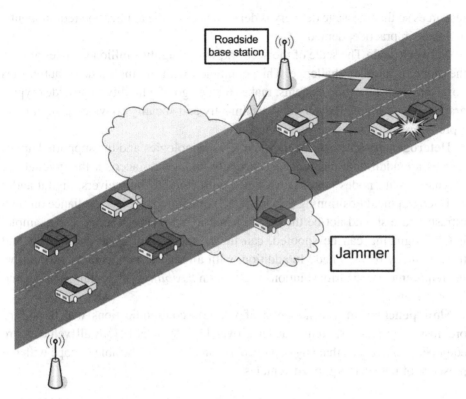

Figure 2.15. Spectrum jamming: communications are disrupted in the neighbor-hood of the jammer. Used with permission, from [325], © IEEE, 2006.

is envisioned as an excellent opportunity to obtain hard-to-refute data that can assist legal investigations (e.g., in the case of accidents). This implies that, to begin with, unambiguous identification of the vehicles as sources of messages should be possible. Moreover, context-specific information, such as coordinates, time intervals, and associated vehicles, should be possible to extract or reconstruct. But such requirements raise even stronger privacy concerns. This is even more so when drivers' biometrics are considered: biometrics, useful for enhancing vehicle access and control methods, are highly private and unique data cannot be reset or reassigned.

Delay-sensitive applications Many of the envisioned safety and driver-assistance applications pose strict deadlines for message delivery or are time-sensitive. Security mechanisms must take these constraints into consideration and impose low processing and messaging overhead. Not only must protocols be lightweight, they must also resist denial-of-service attacks. Otherwise, it would suffice for an adversary to generate a high volume of bogus messages and consume resources so that message delivery is delayed beyond the application requirements, and thus, in practice, denied.

Network scale The scale of the network, with roughly a billion vehicles around the globe, is another challenge. This, combined with the multitude of authorities governing transportation systems, makes the design of a facility to provide cryptographic keys a challenge *per se*. A technically and socially convincing solution is a prerequisite for any security architecture.

Heterogeneity The heterogeneity in VC technologies and the supported applications are additional challenges, especially taking into account the gradual deployment. With nodes possibly equipped with cellular transceivers, digital audio and Geographical Positioning Service (GPS) or Galileo receivers, reliance on such infrastructure should not be the weakest link in achieving security. For example, if GPS signaling can be spoofed, can the correctness of node coordinates and time accuracy be assumed? In addition, with a range of applications with differing requirements, security solutions must retain *flexibility*, yet, remain *efficient* and *interoperable*.

Slow penetration The adoption of wireless communications will be a very progressive process, spanning at least over two decades before all vehicles are equipped. This means that any deployed architecture must be able to cope with the presence of not (yet) equipped vehicles.

Security architecture

In this section, we present the components needed to protect VC against the wide range of threats that we have just discussed. This provides an AAA (authentication, authorization, accounting) framework for VC. We present in Figure 2.16 a "reasonable" possible architecture, the components of which are described next. As the field is still pretty immature at the time of this writing, this architecture should be considered as an "educated guess" (based on the many discussions that we had with several representatives of the automotive industry) rather than the ultimate solution.

Security hardware Among the vehicle onboard equipment, two logical blocks are needed for security, namely the *Event Data Recorder* (EDR) and the *Tamper-Proof Device* (TPD).[10]

The EDR will be responsible for recording the vehicle's critical data, such as position, speed, time, etc., during emergency events, similarly to an airplane's black box. These data will help in accident reconstruction and in attribution of liability. EDRs are already installed in many road vehicles, especially trucks. These can be extended to record also the safety messages received during critical events.

---

[10] In some proposals, both modules are implemented in the same hardware module.

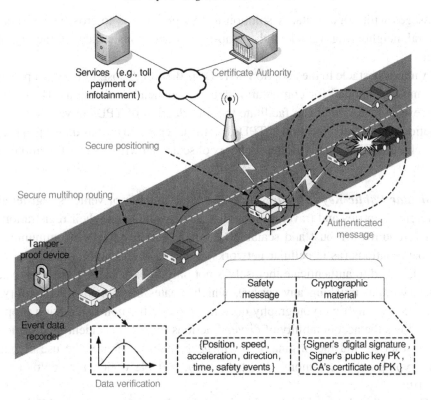

Figure 2.16. Overview of the security architecture. The represented scenario depicts an authenticated safety message generated from a vehicle involved in a collision. The message is relayed hop by hop in the direction opposite to the one of the vehicles, by a secured routing protocol. Each vehicle checks the correctness of the signature and the plausibility of the reported event. Secure positioning, also sketched in the figure, is an advanced (if not futuristic) feature by which each vehicle will be able to prove that it is really located at the position where it claims to be. Used with permission, from [325], © IEEE, 2006.

The car electronics, especially the data bus system, are easily accessible by the owner or by a mechanic. Hence the cryptographic keys of a vehicle need proper hardware protection, namely a TPD. The TPD will take care of storing all the cryptographic material and performing cryptographic operations, especially signing and verifying safety messages. By binding a set of cryptographic keys to a given vehicle, the TPD guarantees the accountability property as long as it remains inside the vehicle. The TPD has to be as independent as possible from its external environment, hence it should include its own clock and have a battery that is periodically recharged from the vehicle's electric circuits. Yet, despite all these features, the TPD will still suffer from the fact that it cannot control the correctness of the data it receives. This can result in the TPD signing messages with bogus data. A solution to this problem is to cross-check (among several neighboring vehicles) the plausibility and consistency of the reported data.

A major obstacle to the adoption of TPDs is their high cost. But current products are mainly intended for computation-hungry financial applications. Hence there are several factors that can facilitate the introduction of TPDs in vehicles: (i) the creation of a "lighter" version of TPDs, (ii) the leverage on the building-up expertise for vehicular EDRs, and (iii) the economy of scale that will drive costs significantly lower.

Vehicular Public Key Infrastructure The large number of vehicles registered in different countries and traveling long distances, well beyond their registration regions, requires a robust and scalable key management scheme. Communication via base stations (as in cellular networks) is not enough for VC, mainly because vehicles need to authenticate themselves not only to base stations but also to each other (without invoking any server), which creates a problem of scalability. In addition, symmetric cryptography does not provide the non-repudiation property that allows the accountability of drivers' actions (e.g., for accident reconstruction or in order to find the originators of *forgery* attacks). Hence, the use of public key cryptography is a more, if not the only, suitable option for deploying VC security.

This implies the need for a *Vehicular Public Key Infrastructure* (VPKI) where Certificate Authorities (CAs) will issue certified public/private key-pairs to vehicles. Similarly to current vehicle registration authorities, there will be several CAs, each corresponding to a given region (e.g., country, state, metropolitan area, etc.). Other candidates for taking the role of CAs are car manufacturers. In any of the two cases, the different CAs will have to be cross-certified so that vehicles from different regions or different manufacturers can authenticate each other. This will require each vehicle to store the public keys of all the CAs whose certificates it needs to verify. Alternatively, in the case where CAs are regional authorities, vehicles can request new public/private key-pairs delivered by the foreign region they enter.[11]

Authentication The fundamental security functions in VC will consist of authenticating the origin of a data packet. Authentication and the inherent integrity property counter the *in-transit traffic tampering* and *impersonation* vulnerabilities. In addition, authentication helps to control the authorization levels of vehicles.

---

[11] In this context, "foreign" means a region different from a vehicle's home region.

To authenticate each other, vehicles will sign each message with their private key and attach the corresponding certificate. Thus, when another vehicle receives this message, it verifies the key used to sign the message and once this is done correctly, it verifies the message. To reduce the security overhead, the common approach is to use ECC (Elliptic Curve Cryptography) – the most compact public key cryptosystem so far. But it is possible to reduce this overhead by signing only critical messages (e.g., with accident warnings) or one in every few messages (the frequency and redundancy of messages can allow this). In addition, given the frequency of safety message broadcasts (typically, every 100 ms according to the current draft standards), a vehicle can ignore redundant messages.

Privacy To address the *privacy* vulnerability, a reasonable solution consists in using a set of anonymous keys that change frequently (e.g., every couple of minutes) according to the driving speed. Each key can be used only once and expires after its usage; only one key can be used at a time. These keys are preloaded in the vehicle's TPD for a long duration, e.g., until the next yearly checkup; the TPD takes care of all the operations related to key management and usage. Each key is certified by the issuing CA and has a short lifetime (e.g., a specific week of the year). In addition, it can be tracked back to the real identity of the vehicle – the *Electronic License Plate* (ELP) – in case law enforcement necessitates this and only after obtaining a permission from a judge. This conditional anonymity will help determine the liability of drivers in the case of accidents. The downside of this approach is the necessity of storage space for all the keys for one year, but these can fit in only a few Mbytes [323].

Link with the chapters of Parts II and III

With respect to the mechanisms described in the rest of the book, the reader is invited to identify the similarities with other networks, but also the peculiarities of vehicular networks. The design of vehicular networks security can leverage, to some extent, on the techniques related to identity management, described in Chapter 5. Likewise, secure neighbor discovery and secure routing between vehicles can be based on the mechanisms provided in Chapters 6 and 7. The difficult problem of key revocation is described in Section 5.5. In addition, privacy can be designed according to some of the techniques described in Chapter 8. Finally, the cooperative features of vehicular networks (such as fine-grained traffic optimization) could be modeled by means of game theory (see Part III for some examples of applications to other kinds of networks), although this is still an open research topic.

2.3 Trends and security challenges in wireless networks

From the examples provided in the previous section, we can infer a certain number of trends. For this purpose, let us consider again the classical cellular network: an operator, alone in its licensed frequency band, operates a set of base stations; the mobile stations, each duly equipped by the operator with a secret key, communicate in one hop with those base stations. The analysis of the upcoming wireless networks that we have just described shows that all the characteristics of this model will be progressively relaxed.

Indeed, as we move from centralized networks to **distributed** or even **self-organized** networks, security (and in particular key management) must be re-designed. As we will see, even the apparently simple notions of naming and addressing require specific attention, and we will devote a full chapter to them. In a subsequent chapter, we will also show that the mobility of the nodes can be used to establish security associations between nodes.

Multi-hopping increases the "security distance" between the device under the control of the operator (base station, access point) and the mobile station. Consequently, appropriate measures must be taken to prevent malicious or greedy behavior from affecting the proper operation of the network.

The growing **programmability** of the devices provides the users with more flexibility, in the sense that they can for example easily install new applications on their devices. But at the same time, the devices can be misused to mount attacks of growing sophistication; likewise, greedy behavior becomes a serious threat.

In addition, wireless devices are particularly **vulnerable**, in the sense that they can be captured and potentially reverse-engineered. An attacker can clone a captured device, typically to mount a Sybil or a replication attack.

Another dimension is the **growing relevance** that wireless personal devices are taking: for example, mobile phones are used more and more to support payment operations, which of course renders them even more attractive for potential attackers.

In a growing number of cases, these personal devices are also expected to deal with **heterogenous networks**, for example by selecting the most advantageous connection offered in the neighborhood (e.g., WiFi vs. cellular). As the various networks are protected by mechanisms that can substantially differ from each other, this heterogeneity can be exploited by badly intentioned people.

Arguably, the most formidable change is the result of the emergence of wireless communications between **embedded devices** such as vehicles and sensors. The communication does not directly involve human beings anymore, but as we have seen, this fact does not (by any means) make security or cooperation easier.

The **miniaturization** of the devices means that they will always have to cope with limited computing power, transmission capabilities, and energy reserves.

Consequently, the security mechanisms will need to take these limitations into account; likewise, these limitations will fuel the temptation of selfish behavior.

As we have seen in the previous examples, many wireless devices are mobile, and their study requires making use of appropriate **mobility models**. The most popular one is certainly the random waypoint model, in which a given mobile chooses a random destination in the eligible space and moves to it in a straight line at a randomly chosen speed (up to an imposed maximum speed). Once it reaches that location, it stays there for an amount of time generated randomly (again upper bounded by a given value) and then it starts the process again.

Another important evolution to take into account is the fact that software radios are now becoming **cognitive** or **smart radios**. Radios of this kind are able to sense their environment in order, for example, to switch to a less congested frequency. This can pave the way to *underlay systems*, in which a chunk of the spectrum is reserved to a *primary operator* (e.g., a television broadcast operator), but can also be used by *secondary users*, provided that the latter bring minimal interference to the former. Software-defined radios will tremendously facilitate this evolution.

Another possible underlay can be based on **Ultra Wide Band** (UWB) radios, which transmit at very low power over an extremely wide band (in the order of several GHz). Their proper operation is technically challenging and requires, in particular, a very tight synchronization. An interesting additional feature of UWB technology is that it can be used for distance estimation.

As is well known, the pervasiveness of the wireless technology raises delicate trade-offs between usability and **privacy**. This latter topic is so important that we will devote a full chapter to it (Chapter 8).

2.4 Summary

In this chapter, we have seen that the evolution of wireless networks leads to a large number of challenges in terms of security and cooperation. In Figure 2.17, we provide the relationship between the types of networks that we have just described and the mechanisms of security and cooperation that we will define and discuss in the chapters of Parts II and III.

2.5 To probe further

Gambiroza, Sadeghi, and Knightly study the characteristics of **wireless mesh networks** in terms of capacity and delay constraints [145]. Ben Salem and Hubaux [47] show how to ensure per-client fairness in such networks and how to optimize the bandwidth utilization. Kodialam and Nandagopal [224] explain how WMNs' capacity can be increased by multi-radio and multi-channel TAPs.

Network families \ Mechanisms	Naming and addr.	Security associations	Securing neighbor discovery	Secure routing	Privacy protection	Enforcing fair MAC	Enforcing PKT FWing	Cooperation among operators	Sec. prot. for beh. enforcement
Small operators, comm. NWs	x	x			x		x	x	x
Cellular op. in shared spectrum	x	x			x		x	x	x
Mesh networks	x	x	x		x	x	x	x	?
Hybrid ad hoc networks	x	x	x	x	x	x	x	x	x
Mobile a. h. NWs in hostile env.	x	x	**x**	**x**	x	**x**			
Self organized mobile a. h. NWs	**x**	**x**	**x**	x	**x**	x	**x**		x
Vehicular networks	x	x	**x**	x	x	?	x	?	?
Sensor networks	x	**x**	x	x	x		x	?	?
RFID	x	x			**x**			?	?

Figure 2.17. Upcoming networks vs. mechanisms. Each row corresponds to an upcoming network that we have described in this chapter (but we refrained from including "Mobility in the Internet", because it is not a wireless network as such). Each column describes a possibly relevant mechanism, where an "x" means that the mechanism is particularly crucial for that network, an "x" means that the mechanism is needed for that network, a "?" means that it is unclear whether that mechanism is needed, and a blank means that that mechanism is (probably) not needed. The first five columns correspond to the chapters of Part II of this book, whereas the last four correspond to the chapters of Part III.

As for mobile ad hoc networks, the routing protocols are discussed in [311]. An outlook for self-organized mobile ad hoc networks can be found in [185]. The reader interested in references on the security aspects of routing in mobile ad hoc networks is referred to Chapter 7.

The research on **vehicular communications** security is just beginning, with few pioneer papers so far. In [58], Blum and Eskandarian describe a security architecture for VC intended mainly to counter the so-called "intelligent collisions" (meaning that they are intentionally caused). But this is only one type of attacks and building the security architecture requires awareness of as many potential threats as possible. They propose the use of a PKI and a virtual infrastructure where cluster-heads are responsible for reliably disseminating messages (by a sequential unicast instead of broadcast) after digitally signing them. Gerlach [149] describes the security concepts for vehicular networks. Hubaux, Capkun, and Luo [186] take a different perspective of VC security and focus on privacy and secure positioning issues. They point out the importance of the trade-off between liability and anonymity and introduce Electronic License Plates (ELP), unique electronic identities for vehicles. Parno and Perrig [307] discuss the challenges, adversary types and some attacks; they also describe several security mechanisms that can be useful in securing these networks. Raya and Hubaux [323] describe a security and privacy architecture for VANETs with first evaluations of the security overhead; along with Papadimitratos, they further refine the issue, including revocation aspects in [325]. El Zarki *et al.* [392] describe an infrastructure for VC and briefly mention some related security issues and possible solutions.

The reader interested in the privacy aspects of vehicular (and other) networks is referred to Chapter 8.

The **IEEE P1609.2** standard [197] is part of the DSRC standards for vehicular communications supported by the US Vehicle Safety Communication Consortium (VSCC). It proposes using asymmetric cryptography to sign safety messages with frequently changing keys so that anonymity is preserved. There is no mechanism proposed for certificate revocation. Instead, certificates have short lifetimes and are periodically requested by vehicles through roadside base stations, implying the need for a pervasive infrastructure.

In Europe, vehicular communications security is partially considered within the projects **NoW** (Network on Wheels, http://www.network-on-wheels.de/) and **GST** (Global System for Telematics, http://www.gstproject.org/) as well as by the Car2Car Communication Consortium (**C2C-CC**, http://www.car-to-car.org/). It is being fully addressed by the European project **SEVECOM** (SEcure VEhicular COMmunications, http://www.sevecom.org) that focuses on providing a full definition and implementation of security requirements for vehicular communications.

Finally, a solution to provide privacy in vehicular networks is provided by the CARAVAN scheme [335].

In terms of **mobility models**, the random waypoint model is described in the work by Johnson and Maltz [210]. Some limitations and recommended precautions have been elaborated by Yoon, Liu, and Noble [388]. As for vehicular communications, Choffnes and Bustamante [101] have explored the integration of road mobility and traffic models, and stressed the substantial difference between vehicular mobility and the random waypoint model. Finally, the reader interested in more advanced aspects of mobility models can refer to the work by Le Boudec and Vojnovic [64], which studies the properties of a broad family of mobility models.

Cognitive radios have attracted a tremendous amount of research interest over the last years. For a survey, refer to Akyildiz *et al.* [17].

The detection of pressure, voltage, or temperature changes can be realized by appropriate techniques, but there is no absolute guarantee of perfect **tamper-proofness**, as mentioned by Anderson and Kuhn [26]. Seshadri *et al.* [343] propose **software-based attestation techniques** which can be used, as we have seen, to detect compromised TAPs. Finally, the technique proposed by Parno, Perrig, and Gligor [308] can be used for the distributed detection of node replication attacks.

Note: **intrusion detection** techniques are now routinely used in (wired) networks. We do not discuss them here, because they do not seem to exhibit significant peculiarities in the case of wireless networks (beyond the protection techniques presented in Part II). Another aspect that we do not develop in this book is the recent area of **secure positioning**; the interested reader can refer to [89].

A comprehensive survey on sensor networks and applications can be found in [18] written by Akyildiz *et al.* In addition, several projects have explored the research field of sensor network security. A recent example from Europe is the UbiSec&Sens Project (http://www.ist-ubisecsens.org). The reader interested in the standardization efforts on wireless sensor networks and their security should refer to ZigBee (http://www.zigbee.org).

Many details about the fundamentals of RFID technology can be found in [140]. A historical overview of the evolution of RFID security from World War II until today is presented in [329]. A comprehensive survey on the security and the privacy issues in RFID systems, including an outlook to potential research problems in the field, can be found in [213] written by Juels.

2.6 Questions

(1) In your opinion, what are the three principal reasons why hybrid ad hoc networks are so difficult to implement?

(2) Why is cooperation between nodes a non-issue in the case of mobile ad hoc networks in hostile environments?

(3) Assume a group of ten persons, each equipped with a laptop containing an IEEE 802.11 adaptor. Is that enough to set up a self-organized mobile ad hoc network? If yes, give a simple solution by which they could make their communications confidential to non-members of the group. What if they want peer-to-peer confidentiality within the group? What happens if a new member joins the group? What happens if a member leaves the group?

(4) Why is it not possible to rely exclusively on symmetric cryptography in order to secure vehicular networks?

(5) Why is privacy an issue in vehicular networks, considering that today's vehicles have license plates?

(6) Would it be possible to get rid of the certification authority, and let each vehicle generate its own signatures? Why?

3

Trust assumptions and adversary models

Before diving into the mechanisms of malice and selfishness prevention, which will be the topic of the rest of this book, we will now focus on the notion of trust and we will refine the definition of the adversary model.

3.1 About trust

As we have already hinted in the previous chapter, building and maintaining trust will be much more difficult in upcoming wireless networks than in existing ones. Yet, trust is absolutely fundamental for the future of (wireless) communications. Once computing has become ubiquitous, it will probably be de facto *mandatory*, as are already today mobile phones and personal computers; but what if it is not trustworthy? What if it is as unsafe as today's Internet? Moreover, no business is possible without trust, and wireless networks are essentially driven by business considerations.

Trust can be defined as the belief that another party (a person, an organization, but also a device) will behave according to a set of well-established rules and will thus meet one's expectations. This notion is fundamental in all human societies (and also in many animal groups); generally, a breach of trust is considered to be a major offense.

But trust is a fuzzy notion, be it considered across persons or across areas of competence: no matter how close they are to each other, different people may trust very different things, even in front of the same evidence. Likewise, a person A may trust a person B for the accomplishment of a certain task, but not another: most people trust their mother in general, but rarely for piloting a helicopter; similarly, a subscriber trusts a cellular operator to provide her with connectivity over a given territory, but not necessarily for striking the most advantageous roaming deals (from the subscriber's point of view) with other operators. To make things worse, even in a given area of competence, trust is neither symmetric nor transitive.

It is important at this stage to position trust with respect to security and to cooperation.

Trust *pre-exists* security. As mentioned, trust is a "natural" phenomenon, and it has existed for millennia, before any concept of security was invented. Security is simply a technique to *infer* trust: if I trust something, security can help me trusting something else. For example, if I trust that my personal computer is not compromised, that the security protocol I use is not flawed, and that the cryptographic algorithm running on both sides is not (yet) broken, then I can trust that what I see on my screen is indeed a Web page corresponding to my bank and I can carry out my e-banking transactions with the legitimate belief that I will not be defrauded. It should be clear from this simple example that any security mechanism requires some level of trust in its underlying components.

Cooperation *reinforces* trust. In the definition that we have provided, trust is about the ability to *predict* the behavior of another party. People being what they are, a reasonable assumption is to assume selfishness of the other parties. Therefore, if a system is designed in such a way that the socially *desirable behavior* coincides with a party's vested interest, then it is likely that that party will indeed behave as desired.[1] Hence the possible emergence of a virtuous cycle: I observe the other party's cooperative behavior. This lets me believe that she will continue to be cooperative in the future, and hence my trust in her. It also encourages me to be cooperative, which will reinforce the trust that she has in me, etc.

Because of the complex characteristics of trust, and as it is very deeply rooted in our human nature, trust is difficult to quantify and to model, in the same way as the "quality of service" of a communication application is difficult to assess in a fully objective way. It is in fact easier to describe the *reasons* to trust someone or something, which are the following.

Moral values As mentioned, any society has its rules, and in many cases we will consider that other parties obey these rules, typically because of their education or because they fear bad publicity, should their misbehavior be disclosed. So for example, we trust a large cellular operator to protect our privacy as long as there is no strong reason (e.g., a legal enquiry) to depart from that attitude.

Experience about a given party Previous interactions are of course revealing about the trustworthiness of a given party; these interactions can be either first hand or be reported by other parties, meaning that *reputation* is a fundamental component of trust. Of course, the frequency of the interactions as well as the

---

[1] As explained in Appendix B, this situation corresponds to the case in which Pareto-optimality coincides with a Nash equilibrium.

durability of the other parties (and of their identifiers) are very important to make experience relevant.

Rule enforcement organization If the stakes are high (e.g., the risk of accident when driving a car), the obedience to the rules is further "encouraged" by a specialized agency. For example, the way cellular operators use the radio spectrum is usually regulated by a governmental agency; the way mobile users make use of the radio spectrum is usually controlled by the operator.

Rule enforcement mechanism As it is not possible to "put a cop behind each wireless device," technical mechanisms must be deployed to either make attacks more difficult or to encourage the desired behavior.[2]

As an example of the former case, it is much more efficient to encrypt radio communications rather than to deploy police force everywhere to check that no one is eavesdropping. Several examples of the latter case are described in Part III of this book.

Usual behavior Although malicious behavior refers to poorly understood psychological mechanisms, it is possible to consider that one behavior is much more frequent than another. For example, usually a driver chooses an itinerary to reach her destination by taking into account exclusively her own benefit and not the implications of this decision on the other drivers; but it is (fortunately) very unusual that a driver throws a box of nails on a highway, just for the dubious pleasure of generating an accident. Likewise, network users will often keep trying to set up a communication in spite of the fact that the network is congested; but very few will make the effort to jam a given area simply to "enjoy" complicating other people's life.

3.2 Trust in the era of ubiquitous computing

In the previous chapter, we have explained the major characteristics of upcoming wireless networks. Our discussion of trust building will now be useful to explain why this evolution has profound implications in terms of trust.

We have seen that the number and diversity of operators will increase, that the wireless communication chain between the end device and the operated devices will become longer, that the mobility of the devices will increase, and that the overall number of devices will explode. Consequently, the two first items of the previous list (moral values and experience about a given party) will lose relevance: the compliance to the first becomes more difficult to observe and the increasing mobility of the devices and the shorter lifetime of organizations make the second

---

[2] This encouragement can be realized by either providing rewards in case of good behavior (e.g., by means of micropayments) or by punishing misbehavior (e.g., by reducing the provided quality of service).

more difficult. Rule enforcement organizations will have to evolve, because some of the techniques they use are not scalable (this is the case for example when sending engineers in various parts of the country to make measurements about the power used by base stations). Hence these organizations will have to rely more and more on rule enforcement mechanisms.

Rule enforcement mechanisms are indeed the way of the future. Whenever necessary, they will take into account the knowledge about usual behavior. These mechanisms can be classified in two categories. The first category aims at *preventing* bad things from happening and is typically based on security and cryptographic techniques. The second category aims at *encouraging* desirable behavior (or discouraging undesirable behavior). It usually quantifies the benefit to the user and leverages on game theory and mechanism design. Both categories can be complemented by anomaly detection mechanisms.

3.3 Adversary

Considering the diversity of upcoming wireless networks, it would be foolish to try to define a common adversary model: a threat on a vehicular network is not the same as one on a sensor network, for example. In the previous chapters, we have already described some possible misdeeds (hence giving some information about the attacker); in each of the following chapters of this book, we will define what the specific adversary is. Yet at this stage we will make several comments of general interest.

3.3.1 Malice and selfishness

As mentioned in the first chapter, an intuitive distinction between malice and selfishness consists in stating that the former refers to the willingness to do harm (which includes the access to personal data), whereas the second corresponds to the overuse of common resources such as a network or a radio spectrum.

In the classical security view, only the former is considered: for some reason there is an attacker, and it is willing to perpetrate its attack no matter what. This makes a lot of sense in the original application area, namely warfare: "we" are right, and we must make all possible efforts to fight our enemy and defeat it (breaking its cryptographic codes can be tremendously helpful to achieve that goal, as History has shown). But as we move from military to commercial settings, the motivation to deploy security mechanisms becomes weaker, leading to the unpleasant situation of today's Internet, because (i) the attacker is much more difficult to identify, (ii) those who deploy the security mechanisms are not necessarily those who benefit from them (we will come back to this issue shortly), and (iii) the attempts to overuse the network resources (as is the case with spam) can be very difficult to thwart.

This shows that **malice and selfishness must be considered jointly**, if we want to seriously protect the wireless networks of the future. For this reason, we believe that the specialists in charge of these tasks must have an appropriate understanding of *both* security and game theory. Indeed, security techniques are useful to thwart malice, whereas game theory can help modeling (and therefore preventing) selfishness. But this segregation in two camps is a bit artificial, as we will see towards the end of this book. Yet the distinction between malice and selfishness is useful, and we will make use of the following definitions.

Definition 3.1 A *misbehavior* is the action of a party or group of parties consisting in deliberately departing from the standardized or otherwise prescribed behavior in order to reach a specific goal.

It is thus assumed that the standardized or prescribed behavior is of public knowledge.

Definition 3.2 A misbehavior is *selfish* (or *greedy*, or *strategic*) if it aims at obtaining an advantage that can be quantitatively expressed in the units (bitrate, joules, or coverage) of wireless networking or in a related incentive system (e.g., micropayments); any other misbehavior is considered to be *malicious*.

From this last definition, we see that a technique aiming at increasing one's share of the bandwidth (in general at the expense of other users) is selfish. Likewise, an operator who increases the power of its base stations (thus leading to an overall degradation of the communication quality of the mobile users connected to the base stations of other operators) is selfish as well. A denial-of-service attack is malicious, but it can obviously rely on techniques borrowed from selfish attacks. Finally, an attack aiming at obtaining information about or from another user of the network (hence an attack against privacy) is malicious.

The distinction between Part II and Part III of this book is based on this definition; as we will see, Part II corresponds to what are usually considered to be *security* concerns, whereas Part III focuses on *cooperation* issues. The last chapter of Part III shows how mechanisms to enforce cooperation can be designed based on security techniques.

An additional reason to consider both security and cooperation is that one of the explanations for the lack of deployment of security mechanisms is the lack of incentives to do so, especially when the failure to deploy a security mechanism falls on other people. This topic is considered to be important enough to have triggered the creation of a workshop devoted to it: the Workshop on the Economics of Information Security (WEIS).

Yet another reason why malice and selfishness should be jointly studied is that, in a number of cases, the techniques to thwart them can (and in some instances, should) be combined. Here are a few examples.

- A mechanism aiming at enforcing a given behavior (designed for example with the help of game theory) needs to be secured in order to be effective. For example, reputation-based systems make sense exclusively if the involved parties can verify each other's identities.
- A security mechanism can be modelled and studied as a game: the attacker is modelled as being one of the players while the other players represent the defendants; applications to intrusion detection in wired and in ad hoc networks can be found in [246] and in [261], respectively. In another example, players are peers running a protocol in which they progressively unveil information; see [76] for an application to the modelling of a rational exchange protocol.
- More generally, there is always a trade-off between security and usability, meaning that security should be properly calibrated with respect to the objective threat. Game theory, as it allows expression of preferences of the various parties, offers the perspective of substantial progress on that front.

3.3.2 Adversary models

A popular adversary model used in security is defined by Dolev and Yao [119]. This model notably assumes that the attacker can (i) be a legitimate party (e.g., a registered network user), (ii) send and receive messages to any party in the network, and (iii) be a potential "man-in-the-middle" everywhere in the network (meaning that she is able to read, modify, block, replay, or insert any message anywhere in the network). Finally, the model assumes that the cryptographic primitives are unbreakable.

Nevertheless, in order to properly protect upcoming wireless networks, we need to modify this model.

- First, we need to include **selfish opponents**, as we have just explained.
- The Dolev–Yao attacker model may be **too strong** for our purpose, in the sense that the attacker of a wireless network does not necessarily have access to *all* communication links between all devices: for example, the attacker's pervasiveness is a reasonable assumption against a specific mesh network, but not against a continent-wide vehicular network.
- The notion of **physical location** of the (wireless) parties becomes very important, as we will see in several of the following chapters.
- Likewise, the topology and the **communication primitives** of the network become very relevant. For example, as we will see, an attacker can try to disrupt the

communication between legitimate parties by jamming a communication link or by fiddling with the route establishment protocols.

- The risk of **capture** and **cloning** must be taken into account, as we have already seen for the case of mesh networks.
- The huge number of parties (e.g., several thousand sensors per human being; a total of one billion road vehicles) makes **key management** a challenge *per se*.
- Finally, specific attention must be devoted to the assumption of unbreakability of the cryptographic primitives: no matter how much progress is made in technology, there will always be business opportunities for low tier devices, whose computing and communication capabilities will be very limited, thus calling for the design of **ad hoc cryptographic primitives**; in this case, the system model must take this issue into account.

Considering all these peculiarities as well as the diversity of the wireless networks that we have described in the previous chapter, it is clear that any attempt to define a single attacker model in wireless networks is doomed to fail. Consequently, in the following chapters we will describe the attacker's model that we assume for each considered problem.

Note: It would be naive to believe that, just because the opponent needs to be in power range of the victim to perpetrate an attack, these attacks will be less frequent or less harmful than against wired networks. Indeed, the wireless attack can be carried out over the Internet, from a compromised device; or the attack can be perpetrated by devices that the opponent has previously installed in a given area of interest, and which she can monitor from a remote distance. Progress in technology will make this easier and easier to accomplish, unfortunately.

3.4 Summary

In this chapter, we have seen that some level of trust is needed for the proper functioning of a wireless communication system. We have also explained that the current trends in wireless networks require a thorough re-examination of how trust can be built and maintained in those networks. We have explained that malice and selfishness must be considered jointly, and that this can lead to solutions based on security and game theory considerations. Finally, we have refined the notion of adversary model in a wireless setting.

3.5 To probe further

Trust has been investigated by a number of computer scientists. When checking the literature, it is very important to bear in mind that different authors may have different definitions or interpretations of the notion of trust.

Trust for inter-realm authentication in large distributed systems is discussed in the contribution by Gligor, Luan, and Pato [152]. Blaze, Feigenbaum, and Lacy have proposed "PolicyMaker" [55], a decentralized trust management language and system supporting the specification of trusted actions and trust relationships. This work inspired a subsequent trust management system called KeyNote, described in an IETF RFC [54]. These ideas were further explored by Yu, Winslett, and Seamons [389], notably for automated trust negotiation. Kohlas and Maurer provide a solution for confidence valuation in a public key infrastructure based on uncertain evidence [225]. Trust in decentralized systems can be based on reputation. A general reflection on reputation in future communication systems can be found in the work by Mundinger and Le Boudec [282].

Only a few researchers have tackled the issue of trust in (wireless) networks. Stajano and Anderson [356, 357] explore the role of physical contact between devices in order to establish trust; we will further develop this issue in Chapter 5. Anderson, Chan, and Perrig [25] argue that in order to bootstrap trust between sensors, it makes sense for them to whisper their key *in clear text* to their neighbors, thus departing from traditional key establishment protocols; we will come back to this issue, again in Chapter 5. A reputation-based system, aiming to reinforce mutual trust, is described in the work by Buchegger and Le Boudec [69].

A trust evaluation framework and its application to mobile ad hoc networks are described in a contribution by Sun *et al.* [361]. Alternative solutions are provided by Theodorakopoulos and Baras [363], Jiang and Baras [206], and Zouridaki *et al.* [405].

Finally, Eschenauer, Gligor and Baras explain the peculiarities of trust establishment in mobile ad hoc networks by making use of a military example [130].

In the area of computing, the most remarkable industrial effort so far is probably the notion of Trusted Platform, developed by the Trusted Computing Group.[3] Yet this endeavor has been criticized, because it tremendously reinforces the power of the hardware vendors.

3.6 Questions

(1) Why is trust a problem of growing relevance for wireless networks?
(2) In today's cellular networks, on what issues does a roaming user need to trust her home network? Same question for the visited network.
(3) For each of the upcoming networks mentioned in Chapter 2, mention whom (and with respect to which operation) a user needs to trust.

---

[3] https://www.trustedcomputinggroup.org

(4) Check the Web site of the Trusted Computing Group. Explain why trusted computing reinforces the power of hardware vendors.
(5) Consider the examples of attacks mentioned in Chapter 2. Try to model them as a game by identifying the players, their possible strategies, and their potential payoffs. (Read Appendix B if needed.)

Part II

Thwarting malicious behavior

This second part of this book is about malicious behavior in wireless networks. Each of the chapters describes a fundamental aspect, by first introducing some possible attacks, and then detailing the corresponding countermeasures.

Chapter 4 describes the question of how to designate an end station in a network. It shows that the question is far from being solved, even in the Internet. It then describes the related attacks, namely the Sybil and the replication attacks. Finally, it explains how they can be thwarted.

Chapter 5 is about bootstrapping security between wireless devices located in radio range of each other. An attacker can try to fool one of the parties by establishing a security association with the attacker (herself) rather than with another intended party. The described countermeasures take advantage of physical vicinity or of the mobility of the nodes.

Chapter 6 focuses on the notion of (radio) neighbor. With the wormhole attack, it is possible to let a given node believe that another node is in its radio range, when in reality it is not. This chapter explains why this attack is dangerous, and details the several techniques to thwart it.

Chapter 7 addresses the problem of secure routing in multi-hop wireless networks. It explains that, if unprotected, routing is vulnerable to a vast collection of devastating attacks. It then explains the basic mechanisms to prevent them.

Finally, Chapter 8 details the formidable challenge of privacy raised by wireless networks. The problem being particularly difficult to quantify and to comprehend, the chapter is based on three highly complementary examples: RFID, vehicular networks, and routing in ad hoc networks.

It is important to mention that most of the protection mechanisms described in these chapters are not (yet) implemented in operational products, as they refer to upcoming networks. Yet we strongly believe that a thorough understanding of these networks is crucial to being able to properly design and implement protocols in this complex field.

As this part heavily relies on security and cryptographic mechanisms, the reader unfamiliar with these concepts is strongly encouraged to refer to Appendix A.

4

Naming and addressing

In any network, nodes need to be addressable, notably in order for the routing protocol to be able to convey traffic to them. As the node addresses usually have arcane formats, it is common practice to also make use of names, which are easier to manipulate by human beings; there are specific servers (such as the Domain Name System (DNS) in the case of the Internet) that convert names into addresses.

In static networks, it is common practice to relate the address of a node to its location in the network; in this way, routing can typically be organized in a hierarchical fashion. This principle becomes problematic, however, as soon as some nodes start moving.

Naming and addressing strategies have been heavily debated within the Internet community, essentially *because* of mobility and security, as we have seen in Subsection 2.2.5; it is very difficult to predict how naming and addressing will evolve in the coming years. But this topic is crucial for us, because naming and addressing mechanisms are vulnerable to a number of attacks.

In this chapter, we will first describe an ambitious naming and addressing architecture envisioned for the Internet. We will then focus on the network layer and describe specific attacks related to the mobility and to the intrinsic vulnerability of the nodes. Finally, we will describe the corresponding protection techniques.

4.1 The future of naming and addressing in the Internet

The Internet has two global namespaces, the DNS (Domain Name System) names and the IP addresses. Both are tied to pre-existing structures (administrative domains and network topology, respectively). Unfortunately, this organization is not really fit for mobility or for the addressing of myriads of tiny (wireless) devices. For example, if a node moves, it will then be attached somewhere to the network; it is still the same node, but its address has changed. Likewise, with such a solution it is impossible to designate an object (e.g., a Web page) without having to relate to the domain or machine on which it is located.

85

This analysis has led a number of researchers to propose different approaches for naming and addressing. At the time of this writing, there is no real consensus on how this should be done in the future generations of the Internet, but several leading ideas are emerging. In the following, we will discuss a proposal made by Balakrishnan *et al.* [38], which is inspired by a number of ongoing research efforts, as pointed out by the authors. The reason we address this proposal is that it contains extremely useful concepts for our discussion. We will of course focus on the aspects of highest relevance to ourselves, namely mobility and security. Balakrishnan *et al.* insist on the tremendous difficulty of modifying the core of the Internet (its routers); consequently, their solution works with the existing IPv4 addressing scheme and is IPv6 capable. In order to ensure faithfulness to the original ideas of the authors, we have included (with the kind agreement of the authors) many of the sentences *verbatim* from the original paper.

Their proposal is built on four principles.

Principle 1 "Names should bind protocols only to the relevant aspects of the underlying structure; binding protocols to irrelevant details unnecessarily limits flexibility and functionality."

This apparently trivial principle is in fact frequently violated in today's Internet. Consider for example a search of the Web site of this book. A search engine asked to retrieve "Security and Cooperation in Wireless Networks" will typically return a URL such as http://secowinet.epfl.ch/, which includes a domain name. In addition, that information will be converted into an IP address visible to the web browser, instead of being confined in a lower-layer software procedure.

Avoiding this kind of violation requires the definition of two new identification layers. Principle 1 means that the applications must be able to refer to services with persistent names, that are independent of the machine hosting the service. This capability is supported by two new identification layers, the first of which relies on *service identifiers* (SIDs). These SIDs are typically the output of mapping services that take as input names called *user-level descriptors* (ULDs). User-level descriptors correspond to strings of characters understandable to humans, such as email addresses and search queries.

The second new identification layer is based on the following fact: transport protocols exchange data between two endpoints, and the network locations of the endpoints are irrelevant to the transport layer mechanisms. Yet, in today's Internet, hosts name TCP connections by a quadruple that includes two IP addresses. The unfortunate consequence is that a TCP connection breaks when the IP address of an endpoint changes. Admittedly, there exist solutions to work around this problem (such as Mobile IP, as we have seen in Chapter 2), but none of them addresses the architectural problem. Hence the second new naming layer contains topologically independent *endpoint identifiers* (EIDs).

These two new identification layers require two additional name resolution mechanisms: from SIDs to EIDs and from EIDs to IP addresses.

To illustrate this layering principle, consider again the case of Web browsing from a client. A user types a ULD (in this case a search query) in a search engine running on the client. As we have seen, the search engine returns an SID. The application then resolves that SID, thus receiving one or more EIDs that identify the end-hosts that run the service. The client will then establish one or more connections (e.g., TCP) with the service EIDs. The transport layer then resolves the EID to the current set of IP addresses to which the EID is attached.

The second principle focuses on the independence between the identifiers and the underlying networks.

Principle 2 "Names, if they are to be persistent, should not impose arbitrary restrictions on the elements to which they refer."

There exist different techniques to implement this principle. The most radical one consists in making use of a completely *flat* namespace able to represent all present and future identifiers. In a flat namespace, identifiers have no structure, which guarantees compliance with Principle 2. It is the approach adopted here for both SIDs and EIDs.

The two principles mentioned so far focused on the role of names, identifiers, and addresses. As we have seen, a ULD leads to an SID, which resolves into an EID, which in turn can be converted into an IP address. Yet, in many cases (and we will see some examples in this book), more flexibility is needed in the resolution process, which is expressed by the following principle.

Principle 3 "A network entity should be able to direct resolutions of its name not only to its own location, but also to the locations or names of chosen delegates."

This principle allows a destination which is unwilling to handle a request directly to direct the request to a chosen *delegate*. This principle can also provide some protection against DoS attacks, as we will explain shortly.

As we will see in Chapter 7, some routing protocols in ad hoc networks allow *source routing*: in the header of each packet to be sent, the source includes the whole list of node identifiers through which the considered packet is expected to travel. An extension of this mechanism is *loose source routing*, in which only a few nodes along the route are imposed.

A similar mechanism can be highly desirable in the namespaces that we have introduced: a source should be able to indicate that its packets should traverse a series of endpoints (specified by a series of EIDs), or that their communications traverse a series of services (specified by a series of SIDs). This leads to the fourth and last principle.

Principle 4 "Destinations, as specified by sources and also by the resolution of SIDs and EIDs, should be generalizable to sequences of destinations."

The described identifier layers are represented in Figure 4.1.

As mentioned, EIDs and SIDs can typically be organized as flat namespaces. This is of course a major difference with existing naming techniques based on DNS, the scalability of which is based on a hierarchical organization. Relatively recent research on peer-to-peer systems has paved the way to scalable and highly distributed flat namespaces, organized around the notion of distributed hash tables (DHTs) [8]. DHTs are a vibrant research theme, and we refer the interested reader to the related literature. Here we just briefly discuss the challenges raised by the application of DHTs to the name resolution process.

DHTs have emerged in the framework of peer-to-peer (P2P) systems, but it is clear that a self-organized and untrusted P2P system would be inappropriate for a crucial Internet mechanism. What has to be envisioned is a set of machines providing the name resolution service using a flat namespace resolution algorithm such as DHTs. Recent advances in DHT research have shown that they can guarantee global uniqueness of the names, as well as an acceptably low resolution time.

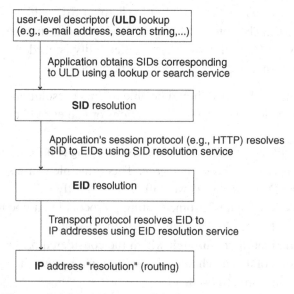

Figure 4.1. The naming layers in a possible future organization of the Internet. Used with permission, from [38], © ACM, 2004.

A difficulty of this approach, as compared with DNS, is the incentive for properly running and managing the endpoints participating in the DHT mechanism; a related question is why the end users would trust the infrastructure. A possible solution could be based on *Resolution Service Providers* (RSPs), which would form a competitive yet cooperative commercial market like current ISPs. The various ISPs would have peer relationships to exchange updates, in a way similar to how the tier-1 ISPs interconnect today with each other.

4.1.1 Resistance to attacks of the described architecture

The architecture that we have just sketched can help to resist several attacks. This is particularly important for mobile devices which, as they rove, are much more exposed to attacks than when they are attached to their home network.

At the SID level, the delegation mechanism that we have seen allows the owners of services and data items to invoke application-level proxies. For example, consider a mobile user who wants to receive email from an SMTP mail server after having it filtered for **viruses** and **spam** at a third-party site offering this kind of service. The indirection by means of the SIDs makes it easy to specify that all messages addressed to that specific mobile need to be first sent to the third party.

Phishing attacks are also relevant here, as they consist in luring a victim to access a Web page controlled by the attacker. A URL such as http://www.ieee.org gives some information about the organization running a given Web server (although this information can be very misleading in certain cases). On the contrary, a name from a flat namespace provides the human user with no clue of this kind. A solution to this naming opacity is to let specific third-parties offer directory services mapping SIDs to human-readable names; this is not an easy task, however.

A last issue to be mentioned here is **denial-of-service** resistance: a given system (a mobile node, typically) can protect itself from attackers by placing a forwarding intermediary between itself and untrusted correspondents and by installing traffic filters at the forwarding intermediary.

4.1.2 Naming and addressing in the running examples

The proposal that we have described is very attractive, because it provides mechanisms able to support mobility of personal communication devices. Hence it is interesting to see whether the running examples that we have introduced in Chapter 1 could leverage on this approach.

Obviously, personal communications would benefit tremendously from the flexibility brought by these principles; in particular, the emergence of numerous wireless operators as well as the progressive deployment of mesh networks could leverage on the fact that services and endpoint identifiers are decoupled from user-level descriptors on the one hand and from IP addresses on the other hand.

The situation is somewhat different in vehicular networks because, as we have seen, these networks have very specific requirements. First, the identifiers of the vehicles must be renewed at a high pace, in order to provide an appropriate level of privacy. This means that in this case the identifier EID should be renewed at an equivalent pace, in such a way that the vehicle is permanently addressable. But this can make rather complicated the resolution from SID to EID and from EID to an IP address, and it is not clear who would be in charge of the proper unfolding of this operation. Second, safety operations are of course strongly related to the geographic location of the vehicles. This means that it must be possible to address the vehicles located in a given area, hence geocasting is likely to be a frequent operation. Ironically, this brings back the temptation to address nodes by their topological location in the network; however, the topology of the network evolves very fast in this case. Finally, real-time constraints (especially for safety-related operations) are very stringent in these networks, and this is of course at odds with any address resolution mechanism (which generally involves access to remote servers).

Likewise, the architecture that we have described does not easily fulfill the peculiarities of wireless sensor networks. One reason is that the resolution operations that we have described involve additional communication overhead, which is of course undesirable for energy-limited devices. Another reason is that, as is the case with vehicular networks, the geographic location is extremely relevant (much more than the identifier of a specific sensor). This means that naming and addressing in sensor networks require very specific solutions. Consequently, the appropriate solution probably consists in the use of proxy gateways, able to cope on the one hand with the rules of the Internet and on the other hand with the constraints of sensor networks.

To conclude, the architecture that we have described is extremely appealing for personal communications; but specific, additional mechanisms are needed to fit the constraints and peculiarities of wireless embedded systems.

The purpose of this discussion was to show that the fundamental operations of naming and addressing, which may seem straightforward at first sight, raise a number of formidable challenges. It is very important to keep these challenges in mind when studying the issues of security (including privacy, of course) and cooperation in wireless networks.

We will now make our purpose more specific by focusing on the network and MAC layers. We will first identify relevant attacks and then describe appropriate countermeasures.

4.2 Attacks against naming and addressing

Numerous attacks can be envisioned against naming and addressing, and we refrain here from giving an exhaustive list; in particular, we do not describe attacks against DNS, which are well-known and somewhat remote from our scope. We rather focus our discussion on attacks that are directly related to wireless networking.

4.2.1 Neighborhood attacks

A first family of attacks, we will denominate "neighborhood attacks," consists of exploiting vulnerabilities of the *neighbor discovery protocols*. Indeed, in many networks, nodes have to discover their local environment and to advertise their own presence. The details vary of course substantially from standard to standard, and, in compliance with the philosophy of this book, we will refrain from diving into this level of details. We will therefore consider a generic model, loosely inspired from IPv6, and explain the threats. In Chapter 6, we will address a related problem: how an attacker can pretend to be in radio range, when in reality she is not.

The considered model is depicted in Figure 4.2: several nodes are located in radio range of each other (we will not discuss here the case in which some nodes are hidden to some others, but this case can certainly open additional opportunities to an attacker). One or several of these nodes can be routers, in which case they provide connectivity to the backbone (the rest of the Internet, in practice); in this case, they will (generally) also have a wireline interface.

As shown in the figure, we focus on the coexistence of two addressing schemes: IP addresses, which as we have seen are of global relevance, and MAC addresses, which are used at the local level. The protocol operates according to the following principles: a node advertises its own presence by a Neighbor Advertisement message and can request information about its neighbors by a Neighbor Solicitation. Likewise, a router advertises its presence by a Router Advertisement message and information about routers can be obtained by a Router Solicitation message.

The IP address can be either allocated by the local authority (typically by DHCP), or by the node itself (this case is often called "address self-configuration" or "stateless address allocation"). In this latter case, a mechanism called Duplicate Address Detection is generally used to make sure that no two nodes are using the same address.[1]

Assume an attacking node is present in the local network (or that an attacker has compromised a legitimate node of the local network). This attacker can then

---

[1] In IPv6, the heavy weight bits of the IP address correspond to the "network prefix," which is dependent on the networking location of the considered subnet.

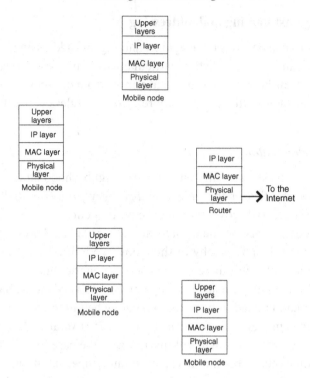

Figure 4.2. Scenario related to the attacks against neighbor discovery protocols.

perpetrate redirect attacks (diverting the traffic from where it should go to another destination) or DoS attacks (inhibiting communication). It can also try to combine these two attacks to mount a flooding DoS attack: redirect as many traffic flows as possible towards a given victim node, in such a way that the latter is overwhelmed.[2]

There are several ways by which an attacker can mount these attacks. Here are a few examples.

An attacking node can *spoof* a Neighbor Advertisement or Router Advertisement message. In this way, it can cause packets for legitimate nodes, both hosts and routers, to be sent to some other link-layer address. The countermeasure consists in having the mentioned messages be secured in some way. This can be based on security associations between all nodes (if they exist), but this approach can require much manual (hence undesirable) configuration. If such security associations do not exist, a solution can consist in making use of Cryptographically Generated Addresses, which we will describe shortly.

Another attack consists in *disrupting the Duplicate Address Detection* protocol: the attacking node responds to every DAD attempt made by an entering node; in

---

[2] These attacks exhibit some similarity with the ones described in Section 2.2.5. However, here we focus on the case in which the nodes are in power range of each other.

this way, the entering node is unable to obtain an address. The countermeasures are similar to those mentioned for the spoof attack.

4.2.2 Sybil and node replication attacks

The proper operation of any naming or addressing scheme usually requires that each participant be assigned a unique name or address, in order to avoid ambiguities. This property is often enforced by a central authority who assigns the names and addresses and by authentication mechanisms that make it possible to verify the ownership of names and addresses (in order to make the approach scalable, these mechanisms are in fact usually supported by a hierarchy of authorities).

Yet, the trend is towards more decentralization, to such an extent, as we will see later in this chapter, that in some cases nodes are expected to generate not only their own address but also their own public/private key-pair; in particular, this is of course the case in self-organized mobile ad hoc networks.

An attacker can try to break the fundamental principle of address ownership and uniqueness by mounting a so-called *Sybil attack*. This attack, initially described in the framework of peer-to-peer systems, consists of creating an arbitrary number of identities associated with the same entity. In a distributed system, the only reliable way to thwart this attack is to have a central, trusted authority to vouch for a one-to-one correspondence between entity and identity (techniques consisting in challenging a set of entities, for example about their computational resources, are in practice ineffective to detect Sybil attacks).

In the case of wireless networks, Sybil attacks are a major concern as well and they call for appropriate countermeasures. For example, vehicular networks have strong requirements in terms of liability. It is therefore mandatory to make sure that each vehicle has a single identity, which means that it needs to be assigned and certified by a trusted authority (the protection of privacy requires this identity to never be sent over the communication channel).

A fundamental difference between peer-to-peer systems and wireless networks is that in the latter, two entities (two nodes) can be in the *vicinity* of each other. As we will see in Chapter 5, this property can be used in order to support the establishment of security associations, even if the identities of the nodes are not certified by any central authority.

Another fundamental difference is that in wireless networks, an attacker can *capture* a node. By doing so, it can notably perpetrate a *replication attack*, which is the "dual" of a Sybil attack: instead of assigning different identities to the same entity, the result of a replication attack is that several nodes share the same identity. By doing so, an attacker can perpetrate a number of misdeeds, such as impersonating legitimate parties, leading astray routing protocols, and breaking schemes based on shared secrets.

Thwarting replication attacks is difficult even in the presence of a central author-ity. In the next section, we will describe a countermeasure specific to the case of sensor networks.

4.3 Protection techniques

4.3.1 Centralized solutions

As we have seen in Chapter 1, the most traditional protection technique (used typically in cellular networks) consists of having the network operator manually distribute the identity along with a symmetric key to the subscriber; we have also described the authentication protocols to be used when a subscriber roams in a foreign network.

In the Internet, a protocol called the Internet Key Exchange (IKE) [163] offers a centralized solution. However, IKE requires the involved parties to be able to verify each other's certificates, a solution which would require a global key management infrastructure in the general case. This latter requirement is usually considered to be too demanding; for this reason, we will focus on the description of a distributed solution.

4.3.2 Distributed solution: Cryptographically Generated Addresses

As we have seen, it is often the case that nodes generate their own address, for example, this must be the case (by definition) in self-organized mobile ad hoc networks, and as another example, the mobile nodes roving around mesh networks could also benefit from this mechanism. But of course this flexibility opens the door to a number of possible attacks, as an attacker can forge addresses.

A recently devised technique called "Cryptographically Generated Addresses" (CGAs) [31] consists of binding the IP address of a given node to its public key. We will first explain the principle, and then explain how to overcome specific lim-itations. To simplify things, we will base our discussion on IPv6, but this principle can be applied to any network in which a high number of bits (say at least 50) representing the node address can take an arbitrary value. In IPv6, this part of the address is called the "interface identifier". In contrast, the other main field (the sub-net prefix) contains information related to the overall organization of and routing in the global Internet.

As shown in Figure 4.3, the node first has to generate its public key (along, of course, with the corresponding private key). Then, by application of an appropriate one-way hash function, the node generates the arbitrary sequence of bits to be inserted in the interface identifier.

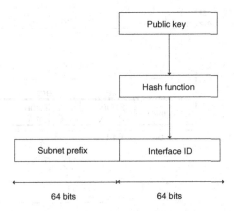

Figure 4.3. Simplified principle of Cryptographically Generated Addresses.

Therefore, it is possible to ask a given node to *prove* that it is the legitimate owner of a given address: only the node that has generated the address knows the private key and is thus able to properly respond to the challenge (assuming of course second pre-image resistance of the hash function, see Appendix A).

It is important to notice that this mechanism does not prevent an attacker from generating "legitimate" addresses, and therefore cannot prevent Sybil attacks. Consequently, CGA-based authentication does not prove that a node with the authenticated address *exists*. But it does prevent an attacker from stealing or spoofing an address already chosen by another node. In this way, CGAs constitute a useful building block to thwart the neighborhood attacks that we have previously described.

A vulnerability of this principle is that it can fall prey to a brute-force attack, because of the limited number of bits of the address that can contain an arbitrary string of bits (64, in our example). Indeed, this number is too small to guarantee second pre-image resistance: an attacker could pre-compute a large database of interface identifiers from public keys generated by himself, and use this database to find matches to victims' addresses.

It would of course be highly impractical to increase the size of the address field. The solution consists of increasing the cost of both address generation and brute-force attack by the same factor, while keeping constant the cost of address usage and verification; this technique is known as *hash extension*. The operating principles are depicted in Figure 4.4. The generation of a new CGA is based on three inputs: the 64-bits subnet prefix, the public key of the address owner, and a parameter called "Sec." This last parameter is an unsigned 3-bit integer, which qualifies the level of expected security (0 being the weakest, and 7 being the strongest). As we will see, increasing Sec by one adds 16 bits to the length of the hash that the attacker must break.

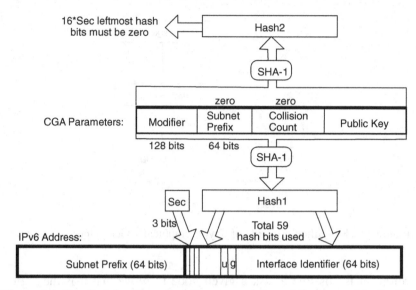

Figure 4.4. Detailed data flows in address generation of Cryptographically Generated Addresses (used with permission, from [31]).

As this last parameter must be conveyed in the address, the idea is to devote 3 bits out of the 64 bits of the identifier address to it; as an additional 2 bits are reserved,[3] we are left with 59 bits.

An important field is the Modifier: the address generator has to iteratively attempt to modify it in order to fulfill the requirement of the hash extension. The idea is that the attacker will be forced to do the same amount of work if it wants to spoof a legitimate address.[4]

Both the Modifier and the Collision Count are public.

A CGA address is typically generated as follows.

(1) Set the modifier mechanism to a random 128-bit value.
(2) Concatenate the modifier, 64+8 zero bits, and the encoded public key. Execute the SHA-1 algorithm on the concatenation. The leftmost 112 bits of the result are Hash2.
(3) Compare the 16*Sec leftmost bits of Hash2 with zero. If they are all zero (or if Sec = 0), continue with Step (4). Otherwise, increment the modifier and go back to Step (2).
(4) Set the collision count value to zero.

---

[3] They appear in Figure 4.4 and are called u and g. Their meaning is of no relevance to our discussion.

[4] A note for the cryptographic-inclined reader: this technique is similar to "salting" encrypted passwords in order to make them resistant to dictionary attacks. The complexity of the salt is controlled by the Sec parameter.

(5) Concatenate the modifier, subnet prefix, collision count and encoded public key values. Execute the SHA-1 algorithm on the concatenation. The leftmost 64 bits of the result are Hash1.
(6) Form an interface identifier by setting the two reserved bits in Hash1 both to 1 and the three leftmost bits to the value Sec.
(7) Concatenate the subnet prefix and interface identifier to form a 128-bit IPv6 address.
(8) If an address collision with another node within the same subnet is detected, increment the collision count and go back to Step (5). However, after three collisions, stop and report the error.

The value of the security parameter Sec determines the cost of generating a new address. As mentioned, the weakest level of security is when Sec = 0, in which case the hash extension is not used. This is appropriate for nodes with modest security concerns, or for nodes that frequently change addresses.

For security parameter values greater than zero, the brute-force search in Steps (2)–(3) takes, on the average, $O(2^{16*Sec})$. The idea is of course that an attacker would have to make an equivalent effort, meaning that the ratio between the cost of a brute-force attack and the cost of address generation remains at the constant value 2^{59}. The system is engineered in such a way that the address generation process can be fully carried out on a server (and not on the potentially anemic mobile device that will use it).

The parameter collision count is used to modify the input to Hash1 if there is an address collision. In practice, it is recommended to not allow collision counts higher than 2, because it is extremely unlikely for three collisions to occur. Hence, the collision must be caused by a configuration or an implementation error (or that a DoS attack is taking place, in case the CGA-based proof of ownership is not used).

The verification of the address ownership is realized by the execution of the following steps.

(1) Check that the collision count value is 0, 1 or 2, and that the subnet prefix value is equal to the subnet prefix (i.e. leftmost 64 bits) of the address. The CGA verification fails if either check fails.
(2) Concatenate the modifier, subnet prefix, collision count and the public key. Execute the SHA-1 algorithm on the concatenation. The 64 leftmost bits of the result are Hash1.
(3) Compare Hash1 with the interface identifier (i.e. the rightmost 64 bits) of the address. Differences in the two reserved bits and in the three leftmost bits are ignored. If the 64-bit values differ (other than in the five ignored bits), the CGA verification fails.

(4) Read the security parameter Sec from the three leftmost bits of the interface identifier of the address.
(5) Concatenate the modifier, 64+8 zero bits, and the public key. Execute the SHA-1 algorithm on the concatenation. The leftmost 112 bits of the result are Hash2.
(6) Compare the 16*Sec leftmost bits of Hash2 with zero. If any one of these is nonzero, CGA verification fails. Otherwise, the verification succeeds. If Sec = 0, verification never fails at this step.

As mentioned, CGA is a building block that can be used to thwart some of the attacks against neighbor discovery protocols. As mentioned by the author of this proposal, it can also be used to secure Mobile IPv6, and to create IPsec security associations.

4.3.3 Thwarting Sybil and node replication attacks

The protection against Sybil attacks very much depends on the deployment scenario. As we have mentioned, if the system contains a central, trusted authority, the solution is relatively straightforward. Otherwise, it is possible to take advantage of some physical aspects related to the radio communication. Examples include radio fingerprinting [159] or geographic location.

Preventing the node replication attack can be based on the physical protection of nodes (to avoid them being captured, or to make them very difficult to replicate in case they get captured). Again, these countermeasures are very much scenario-dependent. In order to provide a concrete example, here we will describe how node replication attacks can be detected in sensor networks [308]. The reason for this choice is that sensor nodes typically employ low-cost commodity hardware components with very limited (if any) protection against tampering. Even if an adversary compromises a single node, she can replicate it indefinitely, spreading her influence throughout the network.

An intuitive solution to this problem would consist in requiring all the nodes to transfer a list of their neighbors' claimed locations to a central base station that can identify potential conflicting location claims. The drawback of this approach is that it creates a single point of failure: the adversary can compromise the base station, or systematically interfere with its communications. In addition, some networks do not even have a powerful, central base station. Hence the intention is to aim for a decentralized solution.

The solution must take into account the limitations of sensor nodes; in particular, it should minimize the amount of communications between nodes and require only minimal memory storage at each of the nodes. The solution assumes that the adversary cannot deploy nodes with arbitrary IDs; hence, the adversary needs to capture at least one node. We will also assume that the nodes know their own geographic position, for example by means of GPS receivers or appropriate connectivity information. Likewise, we assume the nodes to be static and the network to utilize an identity-based public key system,[5] such that each node α is deployed with a private key, K_α^{-1}, and that any other node can calculate α's public key using α's ID, namely $K_\alpha = f(\alpha)$. A last assumption is that any cloned node has at least one legitimate node as a neighbor.

The basic idea is that each node α transmits its location to all its neighbors, thus forwarding this information to a subset of nodes, called its *witnesses*. In the case of a replication attack on node α, one of the witnesses can receive two different location claims corresponding to the same ID of α. This information will provide evidence of a replication attack on node α; consequently, the witness will broadcast this information to the whole network and node α is revoked.

The selection of a set of neighbors is a crucial decision. If it is too small, the probability of detection will be too low, and if it is too large, the communication overhead could be unacceptable. In addition, this set must be *unpredictable*, otherwise the attacker might compromise or jam all of the witnesses of the node on which she is perpetrating the replication attack.

Let us now provide a precise description of the protocol. As a first step, each node α broadcasts its location claim, along with a signature authenticating the claim. The location claim has the format $ID_\alpha, l_\alpha, H_{K_\alpha^{-1}}(ID_\alpha, l_\alpha)$, where l_α represents α's location and H is a hash function. Each of the node's neighbors verifies α's signature and the plausibility of the claimed location l_α. Then, with probability p, each neighbor selects g random locations in the scope of the network and uses geographic routing to forward α's claim to the nodes closest to the chosen locations.

When a witness receives a location claim, it first verifies the signature. Then it checks the ID against all location claims it has already received. If it receives two location claims for the same ID α, it blacklists α from further communication by immediately flooding the network with the pair of conflicting locations. Every node receiving this pair can independently verify the signature and agree with the revocation decision.

The probability of replication detection can be computed in the following way. Call d the average number of neighbors of a node (d is thus the average degree in the connectivity graph). The number of witnesses receiving a given location claim will be[6] $E[N_{\text{Receive}}] \approx p \cdot d \cdot g$.

---

[5] This system could be replaced by a more traditional PKI; this would require, however, transmitting the certificates of the public keys of the nodes, which would lead to a substantial communication overhead.

[6] This equality is only an approximation because the neighbors are assumed to choose the witnesses independently from each other, which can lead to some redundant selection of witnesses.

Assume that the attacker inserts L replicas of α. We want to determine the probability that two conflicting location reports collide at at least one of the witness nodes. Following the usual derivation of the birthday paradox, the probability P_{nc_1} that $p \cdot d \cdot g$ recipients of claim l_1 do not receive any of the $p \cdot d \cdot g$ copies of claim l_2 is:

$$P_{nc_1} = \left(1 - \frac{p \cdot d \cdot g}{n}\right)^{p \cdot d \cdot g},$$

where n is the total number of nodes of the network. Likewise, the probability P_{nc_2} that the $2 \cdot p \cdot d \cdot g$ recipients of claims l_1 and l_2 do not receive any of the $p \cdot d \cdot g$ copies of claim l_3 is given by:

$$P_{nc_2} = \left(1 - \frac{2 \cdot p \cdot d \cdot g}{n}\right)^{p \cdot d \cdot g}.$$

In the same way, the probability P_{nc} of no collision at all is:

$$P_{nc} = \prod_{i=1}^{L-1} \left(1 - \frac{i \cdot p \cdot d \cdot g}{n}\right)^{p \cdot d \cdot g}.$$

The approximation that $(1 + x) \leq e^x$ allows us to simplify numeric computations:

$$P_{nc} \leq \prod_{i=1}^{L-1} e^{\frac{-i \cdot p^2 \cdot d^2 \cdot g^2}{n}}$$

$$\leq e^{\frac{-p^2 \cdot d^2 \cdot g^2}{n} \sum_{i=1}^{L-1} i}$$

$$\leq e^{\frac{-p^2 \cdot d^2 \cdot g^2}{n} \frac{L(L-1)}{2}}.$$

The probability of collision can be lower-bounded as:

$$P_c \geq 1 - e^{\frac{-p^2 \cdot d^2 \cdot g^2}{n} \frac{L(L-1)}{2}}.$$

P_c is actually the probability of detecting an attack consisting of inserting L replicas of the same node.

This means for example that in a network of $n = 10\,000$ nodes, if each node has $g = 100$ witnesses, and an average of $d = 20$ neighbors that forward the request with a probability of $p = 0.05$, the system will detect a single replication of α with a probability greater than 63%; if α is replicated twice it will be detected with a probability greater than 95%.

Consequently, this mechanism provides a robust protection against replication attacks. However, it is easy to see that it is quite demanding in terms of communications and memory. Each node generates $p \cdot d \cdot g$ messages that must be evenly spread throughout the network. If the network has a "reasonable" shape (such as a circle or a square), the average distance between any two randomly chosen nodes is $O(\sqrt{n})$. Moreover, as it aims at meeting the conditions of the birthday paradox, the value $p \cdot d \cdot g$ must also be $O(\sqrt{n})$. Assume that the nodes employ a duplicate suppression algorithm in which each node only broadcasts a given message once. Then the two values obtained so far must further be multiplied by n, resulting in an overall communication cost of $O(n^2)$. As for the storage, even if the size of each claim can be reduced to the payload of a packet (around 36 bytes), the network mentioned previously ($n = 10\,000$, $g = 100$, $d = 20$, and $p = 0.05$) would require, on average, each node to store 3700 bytes, a high demand at least for low-tier sensors.

A technique to reduce the communication overhead, proposed by the same authors, consists in observing that the described solution does not take advantage of the fact that the relaying nodes involved in the multicast are not involved in the security process, in spite of the fact that they are provided with the location claims. Hence, an alternative solution, called "Line-selected multicast," consists of defining the set of witnesses as the set of nodes located along an appropriately chosen segment. The analysis of the solution then boils down to the computation of the probability of intersection between segments.

To conclude, we should stress that an adversary able to perpetrate Sybil and replication attacks is a powerful one; consequently, in most cases the related countermeasures will be *less than perfect* solutions. Hence, before any deployment, an appropriate *risk analysis* must be carried out, in order to predict the implications of undetected attacks.

4.4 Summary

In this chapter, we described an ambitious naming architecture for the Internet, and discussed the extent to which it would fulfill the requirements of upcoming wireless networks. We then described the attacks specific to naming and addressing, namely the neighborhood, the Sybil, and the node replication attacks. We also described several countermeasures, emphasizing those that are distributed; in particular, we introduced Cryptographically Generated Addresses and detailed a technique to thwart replication attacks in sensor networks.

4.5 To probe further

As mentioned, the layered naming architecture that we have presented is based on a proposal by Balakrishnan *et al.* [38]. In this field, many investigations and proposals have been made since the seminal work of Saltzer in 1982 [334]. The interested reader may in particular look at the Host Identity Protocol (HIP) [281].

Cryptographically Generated Addresses are described in [31] and in the subsequent RFC [32]. This work relies on the idea of binding the network address to the public key of the host, and was investigated by Nikander [291], by O'Shea and Roe [297], and by Montenegro and Castelluccia [279]. CGAs are used in particular in the Secure Neighbor Discovery protocol (SEND) [28]. The discussion on the neighborhood attacks was also inspired by an RFC devoted to trust models and threats [292].

The Sybil attack against peer-to-peer systems is described by Douceur [120]. Techniques to prevent the Sybil attack in sensor networks have been proposed by Newsome *et al.* [289]; some of these techniques are based on key pre-distribution schemes, which we will address in Chapter 5. The solution to thwart replication attacks against sensor networks is directly derived from the contribution by Parno, Perrig, and Gligor [308].

4.6 Questions

(1) Consider a vehicular network. In practice, what does it involve for an attacker to mount a Sybil attack? How dangerous would it be? Referring to Section 2.2.7, how do you think such an attack could be prevented?
(2) Same questions for a replication attack against a vehicular network.
(3) Is CGA useful to thwart the replication attack against a mesh network described in Section 2.2.6?
(4) In CGA (Figure 4.4), why are the 64 bits of the subnet prefix not involved in the operation?
(5) In CGA (see Figure 4.3), why is it necessary that the hash function has the property of second pre-image resistance?

5

Establishment of security associations

In the previous chapter, we have seen how a given device can be properly and unambiguously designated by a name or an address. In this chapter, we will explain how two wireless devices can securely identify each other and get ready to communicate securely with each other; in other words, we will see how they can perform authentication and key establishment.

Authentication and key establishment are strongly related to each other, because of their mutual dependency: once two (or more) entities have authenticated each other, they can usually establish a key, in order to secure their future communications; conversely, an already established key *can* be very useful to perform future authentication.[1]

These two operations are considered to be among the most fundamental (if not *the* two most fundamental) mechanisms of network security. As a result, a huge number of protocols have been proposed (and a sizeable number of them have already been standardized and implemented) in order to support authentication and key establishment in (wired) networks [65]. The choice of a protocol depends notably on the role of the trusted server (if any), on whether the key is established by one of the principals (and then transported to the other(s)) or agreed among the principals, and on the underlying cryptographic mechanisms (symmetric or asymmetric).

In Part I, we have already stressed that wireless links are particularly vulnerable to eavesdropping, and that mobile devices can be captured (and the secrets they contain can be compromised); an additional problem we have mentioned is the fact that, in many upcoming wireless networks, nodes cannot rely on the presence of an online trusted server (whereas most standardized authentication and key establishment protocols do rely on such a server). For example,

---

[1] It is important to keep in mind that, rather often, the purpose of a key establishment protocol is to establish a *session key*, which by definition is short lived.

two vehicles can be in power range of each other, but not of a roadside unit; likewise, at a given point in time, a group of sensors may be able to communicate with each other and yet have no connectivity with the sink; finally, by definition, a self-organized mobile ad hoc network does not rely on any trusted server.

A possible solution consists in using public key cryptography: each device carries a certificate, delivered by the trusted authority. This of course requires assuming that some organization accepts the burden of delivering the certificates, which is not always realistic. But there is a more general problem: as the trusted server is off-line, the verifying device cannot ask whether the certificate is still valid or whether it corresponds to a key that has been compromised. A solution then consists in the trusted server delivering short-lived keys and certificates. But this requires a frequent interaction between each device and the server, which of course is not desirable, as it would require the server to become every time available on-line. Another solution is that the server periodically broadcasts certificate revocation lists, but again, this leads to a strong assumption about the connectivity between the devices and the server.

Fortunately, as we will see in this chapter, we can take advantage of the physical proximity of the devices to solve the problem. This physical proximity can be either of long duration (as is the case in a static sensor network) or sporadic, if nodes are mobile. We will also see that the mobility of the nodes can be exploited in order to disseminate cryptographic material, and in the next chapter, we will see how a node can check whether another node claiming to be a neighbor is indeed located in power range.

We will first address key establishment in sensor networks, then describe authentication and key establishment in peer-to-peer personal communication networks, and finally present key revocation in vehicular networks. The first section, devoted to sensor networks, is meant to be a didactic overview, whereas the subsequent sections focus on more detailed examples and are thus a bit more technical.

5.1 Key establishment in sensor networks

Securing the operation of sensor networks requires the cryptographic protection of messages exchanged between the nodes. Typically, every message needs origin authentication and integrity protection, and some messages (e.g., control packets containing sensitive information) may also need confidentiality services. Due to the very limited resources of the sensor nodes both in terms of CPU power and available energy, it is preferable to base the protection on symmetric key cryptographic primitives.[2] This, however, raises the problem of how to establish the necessary symmetric keys in the network.

5.1.1 Requirements

In fact, there are different types of keys that are needed in a sensor network. The requirements are mainly determined by the typical communication patterns, which are the following: *unicast* (i.e., addressing a message to a single node), *local broadcast* (i.e., addressing a message to all the nodes in the local neighborhood), and *global broadcast* (i.e., addressing a message to all the nodes in the entire network). Unicast messages are typically used when the sensor nodes send their sensor readings to the base station or to another sensor node performing some aggregation task, and when the base station sends control information to a specific sensor node. Local broadcast is often needed by networking mechanisms such as routing. Global broadcast messages are typically originated by the base station and they are used to distribute control information that concerns all the nodes in the network.

In addition, there is a requirement to support *in-network processing*. In-network processing means that some sensor nodes aggregate the data received from their downstream nodes into a more compact report before relaying it further towards the base station. This reduces the amount of bits transmitted, and therefore, increases the efficiency and the lifetime of the network. This kind of aggregation must be supported by enabling the aggregating node to access the content of the messages sent by the downstream nodes. This typically requires hop-by-hop protection of messages instead of end-to-end protection between the sensor nodes and the base station.

Another form of in-network processing is *passive participation*, which means that a node can take actions based on overheard messages. For instance, a node may decide not to report a sensed event if it overhears a neighboring node reporting the same event. Passive participation requires that the nodes can access the content of overheard messages even if those messages are cryptographically protected and not destined for them.

5.1.2 Key types

In order to support the different communication patterns and in-network processing, the following types of keys are useful in sensor networks.

---

[2] This seems to be commonly accepted today (at the time of this writing). We note, however, that sensor nodes could be manufactured with custom hardware to support asymmetric key cryptography (e.g., circuitry for modular arithmetics). If such nodes are produced in mass, then the manufacturing cost would not be prohibitive; smart cards are a good example of mass produced, low-cost devices capable of performing asymmetric key operations.

- *Node keys:* A node key is a key that is shared by a sensor node and the base station. It is used to protect unicast messages exchanged between the sensor node and the base station that do not need in-network processing.
- *Link keys:* A link key is a key shared by two neighboring nodes (i.e., two sensor nodes or a sensor node and the base station). Link keys provide protection for unicast messages exchanged between neighboring nodes. They can be used for encryption, message authentication, and integrity protection. They allow for in-network processing by hop-by-hop protection of data packets sent from the sensor nodes to the base station. They can also be used to set up other keys between neighboring nodes, such as cluster keys.
- *Cluster keys:* A cluster key is a key shared by a node and all of its neighbors. This key is used to encrypt (and decrypt) local broadcast messages. In addition, hop-by-hop encryption of a data packet with local broadcast keys makes passive participation possible, as it ensures that the neighbors of the transmitting nodes can learn the content of the packet.
- *Network key:* The network key is a key that is shared by all the nodes in the network. It is used to encrypt (and decrypt) global broadcast messages.

Note that cluster keys and the network key are broadcast keys, and they cannot be used for message authentication. The reason is that a node receiving a message with a message authentication code computed with a broadcast key cannot be sure who the originator of the message is; indeed, any node that possesses the broadcast key may have sent the message.

Broadcast authentication based on symmetric-key cryptography can be realized with the TESLA protocol [314]. The operation of TESLA is described in Appendix A, therefore, we do not present it here. We note, however, that TESLA requires the distribution of the root element of a TESLA key chain in an authenticated manner to the potential receivers of the broadcast messages. In the case of global broadcast, the root element of the TESLA key chain of the base station can be pre-loaded in every sensor node before its deployment. In the case of local broadcast, the root element of the TESLA key chain of any node can be sent to each of its neighbors in a unicast message authenticated with the link key shared with that neighbor. In addition, in both cases, new root elements can be distributed and authenticated using TESLA itself when the current key chain is about to run out of elements.

5.1.3 Setting up node keys, cluster keys, and the network key

Setting up a node key is easy, as the key can be pre-loaded in the sensor node before its deployment. Also, setting up a cluster key with the help of link keys already shared by a node and its neighbors is easy: the node can generate a cluster key and encrypt it for each of its neighbors using the link key that it shares with that neighbor. The link keys can also be used to authenticate the cluster key. Since the number of neighbors is typically quite limited, sending the cluster key in separate unicast messages to each neighbor results in an acceptable overhead.

The network key can also be pre-loaded into the sensor nodes before their deployment. However, sensor nodes may be compromised and the network key may be leaked. Moreover, unlike the leakage of a node key, a link key, or a cluster key that has only a localized effect, the leakage of the network key affects the entire network. For this reason, when node compromise is detected, there is a need to revoke the compromised node and to update the network key.

The compromised node can be revoked by instructing its neighbors to delete the keys that they share with the compromised node and to update their cluster keys. The updated cluster keys are not distributed to the compromised node. Hence, the compromised node is practically excluded from the network, as it will not be able to send and receive encrypted messages to and from its neighbors.

Once the compromised node is revoked, the network key can be updated with the help of the cluster keys in the following iterative way. The base station generates a new network key and sends it to its immediate neighbors encrypted with the base station's cluster key. The neighbors of the base station decrypt the message, re-encrypt the network key with their own cluster keys, and re-broadcast the encrypted network key. This process is repeated until each node in the network receives the updated network key. Note that the compromised node will not be able to obtain the new network key as it is encrypted with the updated cluster keys of its neighbors that the compromised node does not possess. Note also that the authenticity of the network key can be ensured by a broadcast authentication mechanism such as TESLA.

What remains to solve is the problem of establishing link keys between neighboring sensor nodes. One may think of pre-loading these keys as well into the sensor nodes before deployment, but there are some problems with this approach. First, in many applications, the post-deployment layout of the network may not be known *a priori* (e.g., sensors are thrown out from airplanes), and therefore, it is not known which nodes will be neighbors and need a link key. In addition, sensor nodes can later be added to an already deployed network, for instance, in order to replace depleted or faulty nodes. It is difficult to anticipate at the time of network deployment where these new nodes will be added later, and thus, which nodes need to be pre-loaded with additional keying material to be able to interact with the newcomers.

In the next two subsections, we present two approaches to solve the link key establishment problem between neighboring nodes. The first approach is based on a short-term master key that is present in every node only for a limited amount of time after its deployment. This master key is used to establish the link keys with the neighbors and then it is deleted in order to prevent that the master key is leaked if the node is later compromised. The second approach is based on pre-loading keying material in the nodes before their deployment, but it is done in a clever way so that no assumption on the post-deployment network topology is made and post-deployment addition of new nodes is supported.

5.1.4 Link key establishment using a short-term master key

The establishment of link keys can take advantage of the fact that sensor networks are relatively static networks consisting of stationary nodes. This means that the neighborhood of a node does not change frequently, but it remains more or less the same as it was at the time of its initial deployment. Some nodes may be depleted and die and new nodes may be added to the network occasionally, but this does not result in large and dynamic topology changes. Therefore, it makes sense to discover the neighborhood of the nodes and set up their link keys at the time of their initial deployment.

The link key establishment protocol that we describe in this subsection uses a short-term master key K_{init} that is pre-loaded in every node before its deployment. When the node is deployed, it establishes its link keys with its neighbors, and then it deletes the master key. It is assumed that link keys are established relatively quickly, and the adversary cannot compromise the node before its link keys are established. In other words, by the time the node could be compromised, the master key is already deleted from the node, and the adversary cannot obtain it.

The link key establishment protocol consists of the following four phases: master key pre-loading, neighbor discovery, link key computation, and master key deletion.

The *master key pre-loading* phase is performed before deployment in a secure environment. During this phase, the master key K_{init} is loaded into the nodes, and each node u computes a node master key $K_u = f_{K_{init}}(u)$, where f is some pseudo-random function.

The *neighbor discovery* phase starts right after the deployment of a node. First, the node initializes a timer to fire after some time T_{min}. It then tries to discover its neighbors by broadcasting a HELLO message that contains its identifier and waiting for responses. A neighboring node v that hears the HELLO message of u responds with an ACK message that contains the identifier of v. The ACK message of v is authenticated with the node master key K_v of v. Since node u still possesses the master key K_{init}, it can compute K_v, and it can verify the message authentication code attached to the ACK message.

Once the neighbors are discovered in this way, node u computes its link keys in the *link key computation* phase. The link key K_{uv} between nodes u and v is computed as $K_{uv} = f_{K_v}(u)$. Note that the same key can also be computed by node

v. In addition, no messages need to be exchanged between u and its neighbors in this phase. Note also that node u is not authenticated explicitly to node v. However, each further message that u sends to v will be authenticated with K_{uv}, which proves the identity of u.

Finally, when its timer expires, node u performs the *master key deletion* phase by deleting from its memory K_{init} and each node master key K_v that it computed in the previous phases. It does not delete, however, its own node master key K_u, as this is needed to establish link keys with nodes that may be added later to the network.

Note that this link key establishment protocol can be used when several nodes are deployed at the same time, as well as when a single node is added later to an already deployed network. In the former case, neighboring nodes u and v may send HELLO messages and wait for ACK messages in parallel, which results in the establishment of two link keys K_{uv} and K_{vu} between them. They may decide to keep one of the two keys and delete the other. Alternatively, if node u receives the ACK message of node v before u sends its ACK message to v, then u can avoid sending its ACK message.

5.1.5 Link key establishment with random key pre-distribution

Now, we will describe a set of link key establishment schemes proposed for sensor networks, called random key pre-distribution schemes. As their name suggests, these schemes follow the key pre-distribution approach, but they trade effectiveness and communication overhead for scalability and reduced memory use. In particular, in random key pre-distribution schemes, not every pair of neighboring nodes share a common key initially. This makes it possible to reduce the memory requirement for pre-loaded keys, and thus, the approach becomes scalable and appropriate for sensor networks. At the same time, it is ensured that any two neighboring nodes that initially do not share a key can establish one, with high probability, with some additional communications via intermediate nodes.

The general idea of random key pre-distribution can be traced back to the following variant of the birthday paradox [272]: given a set S of k elements, we randomly choose two subsets S_1 and S_2 of m_1 and m_2 elements, respectively, from S. The probability of $S_1 \cap S_2 \neq \emptyset$ is

$$\Pr\{S_1 \cap S_2 \neq \emptyset\} = 1 - \frac{(k - m_1)!(k - m_2)!}{k!(k - m_1 - m_2)!}. \tag{5.1}$$

For illustration purposes, we plotted the value of expression (5.1) in Figure 5.1, where we set $k = 100$ and $m_1 = m_2 = m$. As we can see, the probability of the two subsets intersecting increases rapidly with m, and it reaches $1/2$ when m is around 8.

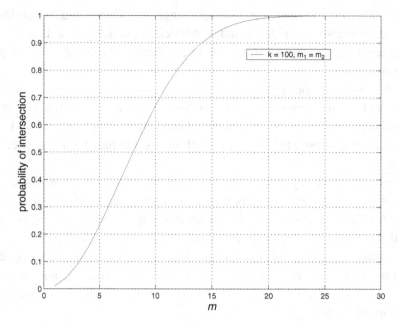

Figure 5.1. The value of expression (5.1) when $k = 100$ and $m_1 = m_2 = m$. As we can see, the probability of the two subsets intersecting increases rapidly with m, and it reaches $1/2$ when m is around 8. In general, the probability will be close to $1/2$ when k is large and m_1 and m_2 are both close to $\sqrt{k}$.

In general, it can be shown that the value of (5.1) will be close to $1/2$ when k is large and m_1 and m_2 are both close to $\sqrt{k}$. The paradox is that we would not expect such a high probability of collision when the size of the selected subsets is only the square root of the original set.

This result can be used in key pre-distribution to considerably decrease the memory requirements imposed on sensor nodes while still maintaining a rather high probability of any two nodes sharing a common key. For this reason, each node is pre-loaded with a random subset of keys selected from a large key pool. Two nodes that have a common key in their subsets are able to communicate securely using the shared key. The probability of this event will be rather high when the number of selected keys is in the order of the square root of the pool size. Thus, we expect that large networks can be supported with a rather limited size memory in sensor nodes.

Below, we elaborate on this idea in more detail. First, we describe a basic scheme and some of its straightforward improvements. Then, we describe an approach to combine random key pre-distribution with threshold cryptography in order to increase the resistance of the scheme to node capture attacks.

The basic random key pre-distribution scheme

The basic scheme works in three phases. In the *initialization phase*, a large pool
S of unique cryptographic keys is randomly generated, and then, for each node, m
keys are selected randomly from S and pre-loaded into the node. This set of m keys
is called the *key ring* of the node. The number k of keys in S is chosen in such a
way that any two nodes will have a common key in their key rings with a certain
probability p (see analysis below).

After the sensors are deployed, the *direct key establishment phase* is performed.
In this phase, the nodes first find out with which of their neighbors they share a
common key. Such key discovery can be implemented by assigning short identifiers
to each key in S before deployment and by having each node broadcast the set of
identifiers that correspond to the keys in the node's key ring. Two neighboring
nodes that discover that they share a common key can then verify that they both
really possess that key by executing a challenge–response protocol. The shared key
is then used to protect the link between the two nodes.

Some pairs of neighboring nodes may not have a common key in their key rings,
and therefore may not be able to set up a secure link in the direct key establishment
phase. In order to remedy this situation, a *path key establishment phase* is performed.
In this phase, neighboring nodes that do not share a key initially establish a shared
key through a path of intermediate nodes where each link of the path is already
secured in the direct key establishment phase. This will work only if the graph,
which consists of the nodes (as vertices) and the secure links created in the direct
key establishment phase (as edges), is connected. As we will see below, this can
be achieved with high probability by appropriately choosing the parameters of the
scheme.

Setting the parameters We use results from random graph theory to set the parameters of the basic scheme. Although sensor networks are not random graphs,
as nodes cannot have communication links with most of the other nodes in the
network, using the random graph metaphor is still useful to give us an idea of the
order of magnitude of the various parameters.

We know from random graph theory [128] that in order for a random graph to be
connected with high probability, the expected degree of the vertices should exceed
a certain threshold. More precisely, in order for a random graph to be connected
with probability c (e.g., c = 0.9999), the expected degree d of the vertices should
be:

$$d = \frac{n-1}{n}(\ln(n) - \ln(-\ln(c))), \qquad (5.2)$$

where n is the number of vertices in the graph.

In our case, the edges of the graph correspond to the secure links created between neighboring nodes in the direct key establishment phase. Recall that p denotes the probability that two nodes have a common key in their key rings. In addition, for a given density of node deployment, let n' be the expected number of neighbors of a node. Then, in our graph of secured links, the expected node degree is $d = p \cdot n'$. Thus, we obtain that, in order for the basic scheme to work, the following should hold:

$$p = \frac{d}{n'}, \tag{5.3}$$

where d is defined in (5.2).

Note that, using (5.1), we can compute p as follows:

$$p = 1 - \frac{((k - m)!)^2}{k!(k - 2m)!}. \tag{5.4}$$

Recall that k is the number of keys in the key pool S, and m is the number of keys in the key rings of the nodes. We can use (5.4) to determine the values of k and m for a given value of p.

Let us consider a numerical example. Let us assume that there are $n = 10\,000$ nodes in the network, and the nodes are deployed in such a way that the expected number of neighbors is $n' = 40$. We want the basic scheme to work with probability $c = 0.9999$. Using (5.2), we can compute that the expected node degree in the graph resulting after the direct key establishment phase should be $d = 18.42$. From this, we obtain $p = 0.46$ using (5.3). Finally, we can use (5.4), to determine the values of k and m. We can check, for instance, that for $k = 100\,000$ and $m = 250$, (5.4) evaluates to approximately 0.5, meaning that a key pool size of $100\,000$ and a key ring size of 250 would be an appropriate choice. Alternatively, we can use (5.4) to determine k if m is given owing to the memory constraints of the sensor nodes. For instance, if the key ring size is limited to $m = 75$ keys owing to memory constraints, then we get from (5.4) that the key pool size should be $k = 10\,000$ to obtain a connected graph after the direct key establishment phase with probability 0.9999.

A brief qualitative analysis We can see that the basic scheme is quite well adapted to the special design constraints for key establishment schemes in sensor networks. First of all, the parameters of the scheme can be adapted to support the memory constraints of the sensor nodes. In addition, setting up pairwise keys does not require intensive computations. Indeed, when the nodes have a common key in their key rings, that common key becomes the shared pairwise key, and no further processing is needed, apart from a simple challenge–response protocol to ensure that the nodes actually possess the key. When two nodes do not have a common key in their key rings, they can establish a shared key through intermediate nodes. This requires some additional processing, because the intermediate nodes must decrypt and re-encrypt the key establishment messages sent between the nodes. However, this must be done only once, at the beginning of the operation of the network. In addition, simulation results in [129] show that the length of the path of the intermediaries is limited to a few hops. This also indicates a moderate communication overhead of the scheme.

The basic scheme does not make any assumptions about the network topology apart from assuming that the expected node degree is known *a priori*. Moreover, the scheme supports the post-deployment introduction of new nodes into the network. For this, the new node must be pre-loaded with its own key ring, and no further action is needed. In particular, the nodes already deployed do not need to be updated, and the new node can use the basic mechanisms (direct and path key establishment) to set up secure links with already deployed nodes.

The disadvantage of the basic scheme is that, by compromising sensor nodes, an adversary obtains keys from the key pool, which may be used to secure links between other, non-compromised nodes. Thus, node capture affects the security of non-captured nodes too. One way to mitigate this problem would be to increase the pool size. In that case, however, the size of the key rings should also be increased in order to ensure the same probability of connectivity of the graph resulting from the direct key establishment phase. The problem is that the size of the key ring is limited by the available memory in sensor nodes, and hence it cannot be arbitrarily increased.

Another related disadvantage is that establishing path keys through captured nodes jeopardizes the secrecy of the recently established key. In order to overcome this problem, compromised nodes must be discovered and excluded from the network rapidly, but discovering that a node is compromised is a very difficult problem in itself.

Finally, yet another disadvantage of the basic scheme is that it does not provide node-to-node authentication. This means that a node can establish shared keys with its neighbors, but it does not know exactly who its neighbors are. Node-to-node authentication would be useful in detecting node replication attacks and in identifying and expelling misbehaving nodes.

q-composite random key pre-distribution

One approach to increase the resilience of the basic scheme against node capture attacks is to use q-composite random key pre-distribution. The q-composite scheme differs from the basic scheme in requiring the nodes to have at least q common keys in their key rings in order to be able to establish a pairwise key. The pairwise key is then computed as the hash of *all* shared keys.

Essentially, the q-composite scheme degenerates into the basic scheme when $q = 1$. Intuitively, when $q > 1$, the probability that two nodes can directly establish a shared key is smaller than the same probability in the basic scheme for the same values of the parameters k and m, because it is less probable to share at least q keys than to share at least one. Thus, in order to maintain the same expected degree of the nodes after the direct key establishment phase (and hence, to ensure secure connectivity), either the size m of the key rings should be increased, or the size k of the key pool should be decreased. However, neither of the above two options is desirable: in the first case, the memory use of the sensors is increased whereas, in the second case, an increased fraction of the keys in the pool is compromised by capturing the same number of nodes. It is true, however, that the latter effect (increased fraction of compromised keys) is counterbalanced by the fact that now, in order for the adversary to compromise a link, it must compromise *all* the keys that have been hashed together to obtain the link key.

The simulation results in [97] show that the q-composite scheme offers greater resilience against node capture than the basic scheme does, only when the number of captured nodes is small, whereas it tends to reveal larger fractions of link keys when a large number of nodes has been captured by the adversary. In effect, by requiring q to be greater than 1, we make it harder for the adversary to obtain sufficient information to compromise links at the beginning, when only a few nodes have been captured. But once a certain amount of information is collected by capturing more nodes, it becomes more and more easy to compromise further links. In other words, the q-composite scheme increases the entry cost of a node capture attack. This makes sense, as it is reasonable to assume that it is more difficult to capture a large number of nodes than to capture only a few of them.

Multipath key reinforcement

Multipath key reinforcement is a technique to strengthen the security of a link key by establishing it through multiple disjoint paths. It can be applied in conjunction with the basic scheme to greatly improve its resilience against node capture. The trade-off is that establishing link keys through multiple paths results in a higher communication overhead.

The operation of multipath key reinforcement is the following. Let us assume that the direct key establishment phase of the basic scheme is performed, and two neighboring nodes u and v have discovered that they have a common key K in their key rings. Instead of simply using this key as the link key between u and v, the nodes will establish their link key in the following way. Node u identifies a set of j disjoint paths to v in the graph resulting from the direct key establishment phase, and sends j key shares $\kappa_1, \kappa_2, \ldots, \kappa_j$ to v such that each key share is sent through a different path. Each key share is protected during transit hop-by-hop, using the keys that are discovered in the direct key establishment phase. Then, both u and v compute the shared link key as $K \oplus \kappa_1 \oplus \cdots \oplus \kappa_j$.

The advantage of multipath key reinforcement is that in order to compromise a link key, the adversary needs to compromise at least one key on every path through which the key shares are transmitted. The simulation results in [97] show that extending the basic scheme with multipath key reinforcement enables it to outperform the q-composite scheme, even when the latter is also extended with multipath key reinforcement. The intuitive reason is that in the q-composite scheme, the trade-off for the increased resilience is the reduced size of the key pool, which undermines the effectiveness of multipath key reinforcement by making it easier for the adversary to build up a critically large collection of compromised keys. As opposed to this, when the basic scheme is extended with multipath key reinforcement, the size of the key pool does not need to be decreased. The cost of the improved resilience in this case is an added overhead in path discovery and key establishment traffic.

Note that multipath key reinforcement can also be used to reinforce path keys that are established between nodes that do not have a common key in their key rings. The operation of the mechanism in this case is similar to the one described above, with the difference that the path key is computed as $\kappa_1 \oplus \kappa_2 \oplus \cdots \oplus \kappa_j$. This will further improve the security of the schemes.

Random key pre-distribution combined with threshold cryptography

As we have seen above, the main problem of the basic random key pre-distribution scheme is that if a node is captured, then all its keys become known to the adversary, and as these keys might have been chosen from the pool by other, non-captured nodes too, their compromise affects the security of the non-captured nodes. We would like to extend the basic scheme in a way that minimizes the effect of capturing a node on other non-captured nodes. In particular, if some key material is leaked, it should not be directly usable by the adversary to learn the key material of other nodes. A possible approach to achieve this is to extend the basic scheme with principles borrowed from threshold cryptography.

The general idea of using threshold cryptography is that capturing less than a certain number of nodes is not sufficient for the adversary to learn anything useful. In order to compromise the links of non-captured nodes, the number of captured nodes must exceed a threshold; hence the name threshold cryptography.

We start with the description of polynomial-based pairwise key pre-distribution, and show how this can be combined with the basic random key pre-distribution scheme later. Let $f(x, y) = \sum_{i,j=0}^{t} a_{ij} x^i y^j$ be a bivariate t-degree polynomial over a finite field $GF(q)$, where q is a large prime number, such that $f(x, y) = f(y, x)$. Each node is pre-loaded with a polynomial share $f(i, y)$, where i is the ID of the node. Any two nodes i and j can compute a shared key. For this, node i evaluates

$f(i, y)$ at point j and obtains $f(i, j)$; similarly, node j evaluates $f(j, y)$ at point i and obtains $f(j, i) = f(i, j)$.

It can be proven that this scheme is unconditionally secure and t-collision resistant. This means that any coalition of at most t compromised nodes knows nothing about the shared keys computed by any pair of non-compromised nodes. In addition, any pair of nodes can establish a shared key, and this incurs no communication overhead (apart from telling the node IDs to each other). The memory requirement of the nodes is $(t + 1) \log(q)$, as each node needs to store a t-degree polynomial over $GF(q)$.

This scheme could be applied in sensor networks, but it has some limitations. In particular, it can only tolerate at most t captured nodes, where the value of t is limited by the memory size of the sensor nodes. This means that t is usually small, and thus the larger the sensor network is, the more likely it is that the adversary can capture more than t nodes.

In order to overcome this problem, we can use the idea of random key pre-distribution; but instead of a pool of keys, now we have a pool of t-degree polynomials. For each sensor node i, we choose a subset of m polynomials from the pool and pre-load into node i the polynomial shares of these m polynomials computed at point i. Two nodes that have polynomial shares of the same polynomial can establish a shared key as described above. It may happen that two nodes that want to establish a shared key have no common polynomials. In this case, they can establish a shared key through a path of intermediate nodes in the same way as path keys are established in the basic random key pre-distribution scheme.

Combining the polynomial-based key pre-distribution scheme with the basic random key pre-distribution scheme combines their advantages and results in a better scheme. In particular, in the combined scheme there is a unique key between each pair of nodes, thus capturing a node does not directly reveal the shared key of any other pair of nodes. In addition, the storage overhead for each node is $m(t + 1) \log(q)$, which differs from the storage overhead of the polynomial based key pre-distribution scheme only in a constant factor m. Although it requires slightly more memory in the sensor nodes, the combined scheme has the advantage that it can tolerate the capture of more than t nodes. The reason is that in order to compromise a polynomial, the adversary needs to obtain $t + 1$ shares of that polynomial. However, owing to the random selection of polynomials, it is very unlikely that $t + 1$ randomly captured nodes have all selected the same polynomial from the pool, and thus collectively have $t + 1$ shares of the same polynomial.

It must be noted, however, that once a polynomial is compromised, every pair of nodes that used the shares of that compromised polynomial to set up a secure link is affected. This means that after the capture of a critically large number of nodes, the security provided by the system starts decreasing abruptly. The advantage of the combined scheme is that it pushes the threshold where the system becomes insecure much higher than in the basic random key pre-distribution scheme and in any of its straightforward extensions (i.e., the q-composite scheme and multipath key reinforcement).

Having described how key establishment can be carried out in sensor networks, we will now move to personal communications.

5.2 Exploiting physical contact

In this and the two subsequent sections, we will explain how authentication and key establishment can be engineered in peer-to-peer communication networks. In this (short) section, we will explain how these operations can be achieved by means of physical contact between the devices. Then we will leverage on a secure side channel such as the one provided by infrared communications. Finally, we will assume that even infrared is not available and show how to achieve peer-to-peer authentication by relying only on radio communications.

Each individual (in this part of the world) possesses a growing number of wireless personal devices: mobile phone, laptop, PDA, remote control devices for appliances, and so on. Some of these devices, such as the remote control of the car locks and the garage door control unit, play an important security role. As a result, one of the problems the user has to solve is to appropriately and securely initialize these devices (and, whenever meaningful, the devices they control). More specifically, the user must be able to tell a given device to "obey" or "become the slave" of another device (already under the user's control).

In a growing number of cases, the problem must be solved in the absence of any trusted server. In addition, the slave devices should not be required to perform complex computations such as modular exponentiations and should not be assumed to have a screen.

To illustrate and solve this problem, Stajano and Anderson rely on the metaphor of the *resurrecting duckling* [357]: a duckling emerging from its egg will recognize as its mother the first moving object it sees that makes a sound. Likewise, a newly purchased device will recognize as its owner the first entity that sends it a secret key. The duckling can "resurrect" in the sense that – in well-defined conditions – it can be *reimprinted* by its mother (its owner), for example if it is transferred to another user.

In order to secure the imprinting or reimprinting operations, the most convenient solution is to use *physical contact* between the master and the slave devices to transfer the secret. Once this operation is completed, the master and the slave can securely communicate over the wireless channel by making use of this secret. Of course, appropriate precautions need to be taken to protect the key in case one of the devices is stolen.

An additional feature of this model is that the ducklings can communicate securely also with each other, independently from the mother. This principle can be used notably to secure communication between sensors.

5.3 Exploiting mobility

As we have mentioned in Part I, traditional mobile networks (such as cellular networks) are secured in a centralized way: each mobile device carries a (symmetric) cryptographic key, provided by the operator at the time of contract signing.

In the previous section, we have explained that the authentication and the establishment of security associations between two devices can be achieved by physical contact. But this solution is not always convenient, because the devices do not necessarily provide the appropriate interfaces, or because users are not necessarily carrying the required cables with them.

In this section, we will thus abandon physical contact and rather rely on a *secure side channel*, as the one provided by infrared communication. We will also take a more global view: instead of focusing on two nodes, we will consider how a *whole* mobile ad hoc network can be secured (hence the authentication and key exchange protocols must be scalable). This is very important, notably in order to secure network-wide mechanisms such as routing, as we will see in Chapter 7.

It is a common belief that peer-to-peer security is more difficult to achieve than traditional security (based on a central trusted authority); moreover, wireless communication and mobility are considered to be at odds with security. Indeed, as we have seen, jamming and eavesdropping are easier on a wireless link than on a wired one, notably because such mischiefs can be perpetrated without physical access or contact. Likewise, a mobile device is more vulnerable to impersonation and to denial-of-service attacks.

Nevertheless, in this section and in the following one, we will show that physical presence is the best way to increase mutual trust and to exchange information in a secure way. Indeed, authentication is straightforward, as users can visually recognize each other (if they meet for the first time, they can be introduced to each other by a common friend whom they trust, or they can check each other's ID).

We will thus show that, far from being a hurdle, mobility can in fact *help* security by enabling basic functions such as authentication and key establishment, even at a full network scale.

5.3.1 *Mechanisms to establish security associations*

In this subsection, we first describe the system model and then we propose the mechanisms for the establishment of security associations.

System model

We assume that each legitimate user has a single device (or "node") and that each node is able to generate cryptographic keys, to check signatures and, more generally, to accomplish any task required to secure its communications (including to agree on cryptographic protocols with other nodes).

We also assume that the adversary can eavesdrop on all radio links and can manipulate messages in all kinds of ways.[3] In contrast with the previous section, we assume that any pair of nodes can communicate over a *secure side channel* (e.g., infrared), provided that they are close enough to each other: the adversary cannot modify messages transmitted over this channel, but we do not require the secure side channel to protect the confidentiality of the exchanged information. Finally, we consider that the adversary can have at her disposal several fake devices.

We will first study the scenario of a self-organized mobile ad hoc network, as defined in Part I. At the end of this subsection, we will consider the presence of a trusted authority.

As mentioned in Part I, if the network is self-organized, it means that there is no infrastructure (hence no PKI), no central authority, no centralized trusted third party, no central server, and no secret share dealer, *even in the initialization phase*; each node is its own authority domain.

In order to establish the security associations, we consider that the nodes can make use of a secure side channel when they get in each other's power range. As we will show in Section 5.3.2, relying exclusively on the mobility of the nodes can lead to a frustratingly low pace of establishment of the security associations. To expedite the process, we introduce the additional, very intuitive notion of *friend*. Two nodes i and j are said to be friends if (i) they *trust* each other to always provide information about themselves and about other nodes they have previously encountered and (ii) they have already established a security association with each other (typically, they know each other's public keys). The security association between friends is assumed to be established (or at least checked) over an out-of-band channel. Note that we do not assume the friend relationship to be transitive, as this would require transitivity of trust. Strictly speaking, this relationship does not even have to be symmetric, yet to simplify the presentation, we will assume this symmetry to hold.

Mechanisms

If a node u possesses a certificate signed by a third party (typically one of her friends), which binds node v with its (v's) public key, then we say that there exists a one-way security association from u to v. Two one-way security associations

---

[3] This is reasonable, as we want the solution to work also for small-sized networks.

between nodes u and v (one in each direction) constitute a two-way security association between the nodes. Likewise, if u and v share a secret key k_{uv}, we say that
there exists a two-way security association between u and v.

If public-key cryptography is used, a (two-way) security association between
two nodes u and v is represented by triplet $(U,\ k_u,\ a_u)$ at the side of v and triplet
$(V,\ k_v,\ a_v)$ at the side of u, where U and V are the names of the users that are
associated with nodes u and v, k_u and k_v are the public keys of u and v, and a_u and a_v
are the node addresses of u and v, respectively. Once nodes u and v have established
a security association between themselves, they can set up secure communication
channels that protect the integrity and confidentiality of the exchanged messages. In
fact, for efficiency reasons, u and v may want to use symmetric-key cryptography for
the protection of their messages. In this case, they establish short-term symmetric
keys (session keys) using the public keys in the security association. In this way,
the nodes establish short-term symmetric-key security associations, which they can
use for example for efficient secure routing.

Similarly, if symmetric-key cryptography is used, a security association between
nodes u and v is represented by triplet $(U,\ k_{uv},\ a_u)$ at the side of v and triplet
$(V,\ k_{uv},\ a_v)$ at the side of u, where k_{uv} is a symmetric key shared by u and v. In
the symmetric-key based approach, we consider security associations to be always
two-way; it is not possible to establish a one-way security association.[4]

When two users meet, they are obviously given the possibility to visually identify
each other. The decision to set up a security association between two nodes is
based on this physical encounter. To support this mechanism, we assume that the
two devices can establish a secure side channel. A secure side channel can only
be point-to-point and works only when the nodes are within a "secure range" of
each other. We consider this assumption to be realistic, as almost all personal
mobile devices are equipped with infrared interfaces (although this tends to be less
true, nowadays). We assume that the activation of the side channel is made by
both users consciously and simultaneously. When activating the side channel, the
users simultaneously associate the name (or the face) of the other person to the
established security association. This operation is very similar to the exchange of
business cards; in fact, it can even be transparently combined with the exchange
of *electronic* business cards (e.g., exchange of vCards between PDAs).[5] These
encounters make it possible for a user to associate a face to a given identity (and

---

[4] In practice, the nodes can derive sub-keys from the shared symmetric key of the security association,
where each sub-key is used in one direction only and perhaps only for a specific security service (e.g.,
either for integrity or for confidentiality, but not for both); this is a policy issue, out of the scope of our
discussion.

[5] http://www.imc.org/pdi/

to a given public key),[6] thus solving many of the classical problems of security in distributed systems (e.g., impersonation attacks and Sybil attacks).

We will now address the public-key approach, and then the symmetric one.

Public-key approach As we assume no authority, each user's device has to generate its public/private key pair(s). Three mechanisms support the establishment of new security associations (Figure 5.2). Mechanism (a) is used when two nodes u and v are in the vicinity of each other, and it consists in u and v exchanging their triplets using the secure side channel. Because the secure side channel ensures the integrity of the exchanged messages, it precludes the possibility of a man-in-the-middle attack. This guarantees a secure binding between the received user name, public key, and node address. In addition, the user can easily verify the validity of the received name because the name should correspond to the person present at the encounter. The node can also verify that the other node indeed possesses the private key that belongs to the received public key by executing a simple challenge–response protocol. Finally, the node address can be verified against the public key. The verification of the node address against the public key is necessary, notably for secure routing. One possible solution is to generate the node address from its public key,[7] by making use of Cryptographically Generated Addresses, described in Chapter 4. In this way, node addresses are bound to public keys in a verifiable way. Note, however, that a malicious node may generate several public keys and the corresponding node addresses and distribute them to other nodes. Whether this is a problem very much depends on how the routing protocol is secured (see Chapter 7).

A possible implementation of the direct establishment of security associations is shown in Figure 5.3.

Users u and v first generate random numbers r_u and r_v, respectively, and exchange, through the secure side channel, their addresses a_u and a_v and the cryptographic hash values $\xi_u = h(r_u \| U \| k_u \| a_u)$ and $\xi_v = h(r_v \| V \| k_v \| a_v)$ of their random numbers and triplets. After this initial exchange, u and v send messages to each other through the radio interface (as they have obtained each other's node address in the first two messages). They exchange their random numbers and triplets, and each of them verifies if the hash value of the received random number and triplet is equal to the received hash value ξ_u (or ξ_v). If this is the case, they can be sure that they have received the random number and the triplet from the party with which

---

[6] If a user wants to establish a security association with a user-independent device (e.g., a printer), she will visually identify the device and bind its identity to the context in which the device operates. Whereas here, we focus on the establishment of security associations between users' personal communication devices.

[7] If the node has several public keys, then the node address is generated from a designated one.

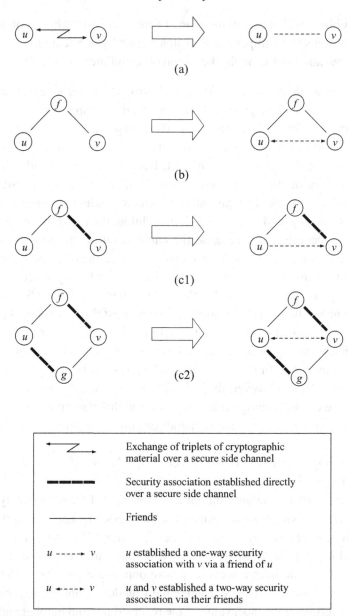

Figure 5.2. Three mechanisms to create new security associations using (a) the secure side channel, (b) a common friend, and (c1, c2) the combination of the first two approaches (mechanism (c1) is used only in the public-key based approach). Used with permission, from [90], © IEEE, 2006.

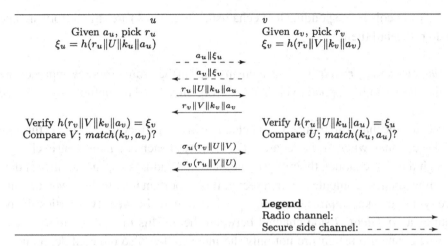

Figure 5.3. Direct establishment of a security association. Used with permission, from [90], © IEEE, 2006.

they exchanged the first messages through the secure side channel. The random numbers serve as nonces and guarantee the freshness of the subsequent messages. Now, both users can verify if the received user name corresponds to the other party and both nodes can verify if the received node address matches the received public key. Finally, the nodes generate and send to each other a signature ($\sigma_u(r_v \| U \| V)$ and $\sigma_v(r_u \| V \| U)$, respectively) on the received random number and on the user names in order to prove that they possess the private keys that belong to the exchanged public keys.

With mechanism (b), two nodes u and v can establish a security association if they have a common friend f. A simple solution is the following: since f knows the triplets of both u and v, it can issue (on request from u and/or v) fresh certificates for both triplets and send them to v and u, respectively, via the network. Both u and v know the public key of f and they also trust f, therefore they can both verify the received certificates and will accept the information therein if the verification is successful.

Mechanism (c1) is a combination of the friendship relationships and the encounters, and they establish only a one-way security association: if nodes u and f are friends and f has obtained the triplet of v in an encounter with v, then f can issue (on request from u) a fresh certificate for the triplet of v and send this certificate to u via the network. As u knows the public key of f, and also trusts f, she can verify the received certificate and accept the received triplet if the verification is successful. A two-way security association between nodes u and v is then established as a combination of two one-way security associations (from u to v and from v to u).

The protocols corresponding to mechanisms (b) and (c1) are straightforward and we do not detail them.

Symmetric-key approach The mechanisms used in the symmetric-key approach are similar: they can be applied to both the self-organized and the authority-controlled networks.

The first mechanism (Figure 5.2, mechanism (a)) is a direct establishment through the side channel: when the nodes are in the vicinity of each other, they can exchange, through the side channel, their user names and node addresses, and additional data that allow them to compute a shared secret. It is important to note, however, that in a pure symmetric-key approach, setting up a shared secret between two parties always requires a *confidential* side channel between them. This means that in this case, the side channel must ensure not only the integrity but also the confidentiality of messages. Like in the public-key implementation, the users can verify the received names through personal encounters. The node addresses, in contrast, can be verified against the received (and verified) names.

Mechanism (b) supports the establishment of security associations between two nodes u and v via a common friend f. By assumption, f already has a security association with both u and v, meaning that it has symmetric keys established with them. In addition, f is trusted by both u and v. Therefore, to establish a session key between u and v, well-known symmetric-key protocols can be used, where f plays the role of the trusted (key) server. The session key can be generated either by f who would send it to both u and v, or by u or v, in which case f would be used as a trusted relay (like in the Wide-Mouth-Frog protocol described in Appendix A).

Finally, mechanism (c2) can be used when two nodes u and v do not have a common friend, or have a common friend f but do not want f to know their shared secret key. Like in the public-key based approach, mechanism (c2) combine the first two mechanisms (encounters and friends). Let us assume that u has a friend f who has already set up a security association with v using the first mechanism. Similarly, let us assume that v has a friend g who has set up a security association with u using the first mechanism. Now u and v can set up a security association using f and g by u generating key contribution k_u and sending it to v via g, v generating key contribution k_v and sending it to u via f, and then both u and v computing a common value k_{uv} from k_u and k_v.

The friend-assisted establishment of a security association, shown in Figure 5.4, illustrates this in more detail. In this protocol, nodes u and v first exchange the names of their friends (to be used in the protocol as trusted relays) and two nonces r_u and r_v (used to guarantee the freshness of subsequent messages). Then, u generates some random key k_u and sends it to v via g (msg3 and msg4), and v generates some random key k_v and sends it to u via f (msg3′ and msg4′). Here, $d_{x \to y}$ is a

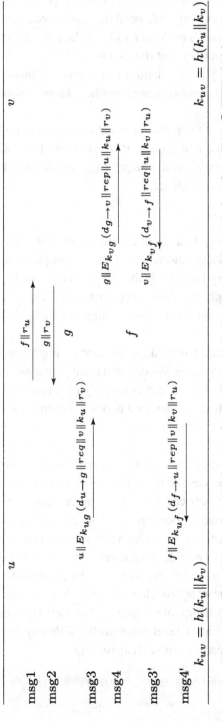

Figure 5.4. Friend-assisted establishment of a security association. Used with permission, from [90], © IEEE, 2006.

direction bit that indicates that the message goes from x to y (and not from y to x).[8] The *req* and *rep* are bits that indicate that the message is a *request* to a friend or a *reply* from a friend, respectively. We need these bits because every node can play either the role of a requesting node (u and v) or the role of a friend (f and g), and thus we must indicate not only that this is a message from x to y but also that x is the requesting node and y is the friend (or vice versa). Finally, u and v compute a common value k_{uv} from k_u and k_v using a publicly known pseudo-random function h (e.g., a hash function).

An interesting feature of the protocol is that it replaces a single trusted party with two parties trusted by one entity each. If f and g are not colluding, then neither of them has enough information to compute k_{uv}. In addition, both u and v trust at least one of them for not colluding.

Presence of an authority

We now assume that there is a central, trusted authority, but that it is not (or at least not always) accessible on-line. A typical example is a vehicular network: neighboring vehicles may need to establish a security association, but do not have the connectivity (or simply the time) to contact a server in order to do so.

The existence of this authority makes things easier, as it can assign a unique identity to each node.

If public-key cryptography is used, the solution is simple: the authority provides a certified public key to each node. We can also assume that each node holds a correct public key of the authority, so that it can verify the correctness of the certificates that other nodes hold. Hence, when two nodes move into the (radio) power range of each other, they will exchange certificates that contain their public keys, and establish a security association.

If the system is restricted to the use of symmetric cryptography, then the protocols we have presented for the case of the self-organized network can be used.

The major difference between the self-organized and the authority-based approach stands in user involvement. In a self-organized approach, users need to establish security associations consciously; on the contrary, in the authority-based approach with public-key cryptography, users do not need to be aware of the establishment of the security associations, as this is done automatically by their nodes. The use of either of these approaches strongly depends on the purpose of the network. Typically, the self-organized approach is useful in securing personal communications on the application level, whereas the authority-based approach is used to secure networking mechanisms such as routing.

---

[8] Note that since messages are always encrypted with a symmetric key k_{xy}, the only possible ambiguity is the direction.

5.3.2 *Performance evaluation*

In this subsection, we provide an estimate of the pace at which security associations are created. We assume that, initially, each node established security associations only with its friends; we further assume that each node has the same number of friends.

In our analysis, we will observe the following values: the convergence $r(t)$, which represents the fraction of the security associations established until time t, and the convergence (meeting) time t_M, which is the time needed to establish all the desired security associations. One additional value of interest is the average meeting frequency $1/t_{IM}$ of nodes. Here, t_{IM} is the node inter-meeting time. This value is important for assessing the frequency of rekeying and the time necessary to perform key revocation.

In the simulations we describe, the Random Waypoint mobility model described in Chapter 2 is used and we extend this model with some new features; we call this new model the *Restricted Random Waypoint* model. As we have seen, in the conventional Random Waypoint model a node chooses its destination and its speed towards this destination randomly. After arriving at the destination, the node pauses for a certain period of time, and then chooses its new destination and its speed. In the Restricted Random Waypoint model, the nodes move in the same way as in the Random Waypoint model, but their choice of destination points is restricted to a number of fixed points on the plane with some probability p. This means that with probability p, a node randomly chooses a point from a finite set of destination points, and with probability $1 - p$, it chooses as its destination a random point on the plane. This model is closer to reality in the sense that users normally do not choose destination randomly, but they rather move to some meeting points (e.g., meeting rooms, lounges, restaurants) where communication between users takes place. If $p = 1$ and if the set of destination points is small, the convergence time will be small as well. On the contrary, if $p = 0$, we have the standard Random Waypoint mobility model and the convergence time will be longer.

In this mobility model, two nodes can establish a security association if they are in the security range of each other (for the self-organized network) or in each other's power range (in the authority-based network). The security range is the maximum range for the secure side channel to be set up; it is significantly smaller than the power range of mobile nodes.

In all simulations, the same simulation area (a 1000 m $\times$ 1000 m square) is used and the number of nodes is set to $n = 100$. When the nodes hit the area border, they bounce off under the same angle under which they hit the border. The node maximum speed is set to 5 m/s (except in one case on Figure 5.5a, where it is 20 m/s), and the minimum speed to 1 m/s. The pause time is set to 100 s.

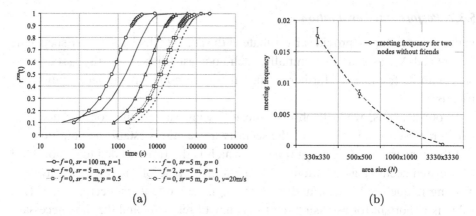

Figure 5.5. Restricted Random Waypoint simulation results; (a) average conver-gence, (b) meeting frequency. Here, f is the number of node's friends, p is the restriction probability (specific parameter of the Restricted Random Waypoint model), v is the node speed, and sr is the range within which the nodes can estab-lish security associations. The results are shown with 95% confidence intervals. Used with permission, from [90], © IEEE, 2006.

In Figure 5.5 we observe the convergence $r^{n \times s}(t)$ and the node meeting frequency. Figure 5.5a shows that the friends mechanism speeds up convergence proportionally to the number of friends. Furthermore, this shows that, as expected, a higher average speed of nodes results in a faster convergence (and therefore a shorter convergence time). The same figure also illustrates another very intuitive result: the convergence is faster if the nodes gather at and around meeting points. It is also interesting to observe that, in the most favorable case (in which the security range is 100 m and the network is controlled by a central authority), 40% of security associations are established in less than 1000 s (17 min). This is an important result, given that, as we will show in Chapter 7, this percentage of security associations is sufficient to support secure routing. Figure 5.5b shows the node meeting frequency (two nodes), in areas of various sizes. We observe that the meeting frequency is inversely proportional to the size of the area.

5.4 Exploiting the properties of vicinity and of the radio link

As we have just seen, authentication and key establishment can be based on the usage of a secure side channel, typically by means of infrared communication. However, as we have mentioned, infrared interfaces are not always available on devices anymore. Consequently, we show in this section how security associations can be established by making use only of the radio link, even in the absence of an authority.

More specifically, we consider the case in which two users, each equipped with a personal device capable of communicating over a radio link, get together and want to establish a shared key. Although they can visually recognize each other, we assume that they do not share any authenticated cryptographic information (e.g., public keys or a shared secret) prior to this meeting. The challenge is the following: *How can the users establish a shared key in a secure way?*[9]

This situation corresponds to the frequent case in which two people get together (e.g., at a meeting, or in the street) and make use of their devices to exchange information, for example their (electronic) business cards. Clearly, the communication between these devices must be properly secured.

Very often, the two users want the security between their devices to be peer-to-peer, thus operating independently from any authority. In practice, this means that the mobile devices must run a protocol to authenticate each other and to protect the data they exchange (to ensure confidentiality and integrity); the latter operation typically requires setting up a symmetric shared key.[10] This key can be used to secure both immediate communications and communications that take place afterwards (e.g., when users exchange e-mail over the Internet).

Assuming that they have visually authenticated each other, we will now show how they can establish a key over the radio link. To this end, we will first precisely define the model we consider, and then describe the protocol that solves the problem. Note that a number of protocols have been proposed to solve this problem, and we will enumerate them in Section 5.7. The solution we will present hereafter requires minimal effort from the users (they do not need to type in anything; all they have to do is to push a button to trigger the unfolding of the protocol and compare a short string of characters).

5.4.1 System model

The Diffie–Hellman (DH) key agreement protocol [366], described in Appendix A, seems to be appropriate for the problem (and the set of assumptions) at hand; the DH key agreement protocol is believed to be secure against a passive adversary (e.g., eavesdropping on a wireless link).[11] Let us briefly review how the DH key agreement protocol works. To agree on a shared key, two users, Alice (A) and Bob (B) proceed as follows. A picks a random secret exponent X_A, and calculates the DH public parameter g^{X_A}, where g is a generator of a group of large order. B does the same, that is, he calculates g^{X_B}. Finally, A and B exchange the public parameters g^{X_A} and g^{X_B} and calculate the shared DH key as $K = g^{X_A X_B} = (g^{X_A})^{X_B} = (g^{X_B})^{X_A}$.

---

[9] This situation is different from the establishment of a shared key between two Bluetooth devices belonging to the same user, which we described in Chapter 1.

[10] In practice, it is recommended to make use of *different* keys for confidentiality, integrity (and, where appropriate, authentication). We will not go to this level of detail here.

[11] This is true if the Computational Diffie-Hellman problem [266] is intractable.

It is well known that the basic version of the protocol is vulnerable to an active adversary who uses a *man-in-the-middle* (MITM) attack. At first glance, it may seem that mounting the MITM attack against wireless devices that communicate over a radio link and are located within the radio communication range of each other can be perpetrated only by a sophisticated attacker. But this is not the case, as we will now explain by a simple example in the framework of Internet protocols.

As we have seen in Subsection 4.2.1, neighbor discovery protocols involve both the MAC and the IP addresses of nodes. In IPv4, the Address Resolution Protocol (ARP) [318] is used by the Internet Protocol to map IP network addresses to the hardware addresses used by a data link protocol (the MAC addresses). An attacker can send spoofed ARP-replies to the victim, who will consequently send all its packets to the attacking machine. In this way, the attacker redirects the traffic between two "legal" machines through an attacking machine, despite the fact that the two legal machines were in radio communication range of each other. In this way, the attacker can perpetrate a MITM attack (by altering the DH parameters). This attack can easily be implemented by making use of publicly available tools for network auditing and penetration testing, such as *dsniff* [352].

Of course, ARP-spoofing is not the only way to mount a MITM attack against wireless devices. Examples of more involved MITM attacks against Bluetooth [353] equipped devices can be found in [204] and [239].

Hence, the goal is to devise mechanisms that prevent the attacker from modifying the DH parameters without being noticed.

Assumptions

We assume each user to be equipped with a computationally constrained personal device (e.g., a PDA). Each device is equipped with a radio transceiver (e.g., IEEE 802.11, which is described in some detail in Chapter 9). We also assume that each device has a human-friendly interface (i.e., a screen and a keyboard).

The solution that we will present makes use of the multiplicative group $\mathbb{G}$ with the generator g. Here, we take $\mathbb{G}$ to be a subgroup of $\mathbb{Z}_p^*$ of prime order q, where $\mathbb{Z}_p^*$ is the multiplicative group of non-zero integers modulo a large prime p. However, the whole treatment here applies to any group in which the Decisional Diffie–Hellman (DDH) problem is hard. These are all groups in which it is infeasible to distinguish between quadruples of the form (g, g^x, g^y, g^{xy}) and quadruples (g, g^x, g^y, g^z) where x, y, z are random exponents. Furthermore, we assume that p and the generator g of $\mathbb{Z}_p^*$, $(2 \leq g \leq p - 2)$ are selected and published. All devices are preloaded with these values.

Concerning the adversarial model, we adopt the Dolev–Yao threat model that we already discussed in Chapter 3. Dolev-Yao is usually unsuitable for wireless networks, yet in this specific case, as we consider only two parties communicating over a single channel and do not address Sybil or replication attacks, the model is appropriate. Thus, we assume that the attacker Mallory (M) controls the radio communication channel: he can obtain any message transmitted over the radio channel. M can initiate a conversation with any other user. However, we assume M to be *computationally bounded*. We further assume that the two parties involved in the communication do trust each other; otherwise, little can be done (a corrupted party can always disclose any secret information received by another party). Whenever we speak of the security of a given protocol, we implicitly assume that the users involved in the protocol (or, more specifically, their devices) are not compromised.

Commitment schemes

Before describing the protocol, we still need to introduce one notion: the *commitment scheme*. The principle of a commitment scheme is the following: (i) a user who commits to a certain value cannot change this value afterwards (we say that the scheme is *binding*), (ii) the commitment is hidden from its receiver until the sender "opens" it (we say that the scheme is *hiding*).

A commitment scheme transforms a value m into a commitment/opening pair (c, d), where c reveals no information about m, but (c, d) together reveal m, and it is infeasible to find $\hat{d}$ such that $(c, \hat{d})$ reveals $\widehat{m} \neq m$. Now, if Alice wants to commit a value m to Bob, she first generates the commitment/opening pair $(c_A, d_A) \leftarrow$ commit(m), and sends c_A to Bob. To open m, Alice simply sends d_A (and m if necessary) to Bob, who runs $\widehat{m} \leftarrow$ open($\hat{c}_A, \hat{d}_A$). We denote with $\hat{x}$ the message at the receiver's side when message x is sent over a public (unauthentic) channel. If the employed commitment scheme is "correct," at the end of the protocol we must have $m = \widehat{m}$. In the security analysis, we assume the use of an ideal commitment scheme.[12] We are now ready to describe the protocol.

5.4.2 Protocol description

Our goal is to ensure the integrity of DH public parameters (g^{X_A}, g^{X_B}) rather than the integrity of the agreed key K. The reason is the following: People build trust in each other when they meet in person; secure communication is usually needed only afterwards (typically when they communicate over the Internet). Clearly, in such a scenario, it is not necessary to compute the shared DH key immediately; this "expensive" computation (typically a modular exponentiation) can be postponed to some later time, when (remote) secure communication is needed. In this way, we reduce the computational burden on the personal devices used during the protocol itself.

---

[12] In particular, we assume that the commitment scheme is *non-malleable*. Informally, an unmalleable commitment scheme means that the attacker is unable to alter a commitment of a targeted party into another (apparently legitimate) commitment. The interested reader may refer to [118] for details on malleability.

The simplest way to check the validity of the exchanged DH public parameters for Alice (A) and Bob (B) is to report the exchanged public parameters g^{X_A} and g^{X_B} to each other and then compare them. The comparison of the exchanged values can be performed by looking at the screen of the communicating party, or by reading aloud the values to be compared. Although this approach provides very strong security, it is clearly impractical because it requires A and B to compare rather large streams of digits. A possible way to make visual (and verbal) verification easier for A and B is to represent the DH public parameters in a more readable form by, for example, significantly reducing the number of digits to be compared by hashing them and potentially encoding the bits in a more readable form; the latter can be achieved by splitting the result of the hash in small groups of bits, and associating a word of the common language (e.g., "house," "dog," etc.) to each possible combination of bits.[13] However, in this way, many different (long) DH public parameters translate to the same (short) bit string (the check value). This may give some advantage to a potential attacker.

Another simple approach consists in first exchanging g^{X_A} and g^{X_B} over the public channel, and in turn, verifying (for example, visually) that $h(g^{X_A} \| g^{\hat{X}_B})$ matches $h(g^{\hat{X}_A} \| g^{X_B})$, where h is a hash function satisfying appropriate security properties, "$\|$" denotes concatenation, and a symbol with a "hat" ("^") designates a value received over the wireless channel. In order for this approach to be usable, the output of the hash function h should be truncated to a relatively short length (e.g., around 50 bits). With this approach, an adversary is successful if she can find values a and b such that $h(g^{X_A} \| a) = h(b \| g^{X_B})$; she is then said to find a *collision* on the truncated output of $h(\cdot)$. Note that it is not sufficient for an adversary to find any collision on $h(\cdot)$. However, the adversary is not constrained to find a *second pre-image* for a single fixed image value g^{X_A} or g^{X_B}; indeed, an adversary controls the inputs to $h(\cdot)$ through the values a and b.[14] Furthermore, the outcome of the used hash function is truncated (e.g., 50 bits long). Therefore, even if $h(\cdot)$ is a second pre-image resistant hash function, this still may not be a sufficient guarantee that the adversary cannot find a collision between truncated $h(g^{X_A} \| a)$ and $h(b \| g^{X_B})$.

In order to make the approach based on string comparison usable, it is essential to make a *proper trade-off between security and usability*. The protocol that we will now describe is called DH-SC (Diffie–Hellman key agreement with String Comparison); it achieves an *optimal* trade-off between security and usability and is provably secure.

The protocol unfolds as shown in Figure 5.6. Most of the operations are carried out by the devices of Alice and Bob. The only operations in which the two (human) characters are consciously involved are (i) the decision to launch the protocol after visual identification and (ii) the verification that $i_A = i_B$.

---

[13] The interested reader can refer for example to RFC 2289 [160] for more information on this solution.
[14] For a given x, x' is said to be a second pre-image if $x \neq x'$ and $h(x) = h(x')$ [266].

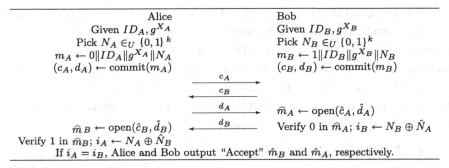

Figure 5.6. Operation of the Diffie-Hellman key agreement protocol with String Comparison (DH-SC). Used with permission, from [81], © IEEE, 2006.

Both Alice (A) and Bob (B) select randomly their secret exponents, respectively, X_A and X_B from the set $\{1, 2, \ldots, q\}$ (q being the order of $\mathbb{G}$) and calculate DH public parameters g^{X_A} and g^{X_B}, respectively. A and B proceed by generating k-bit random strings N_A and N_B, respectively. Finally, A and B calculate commitment/ opening pairs for the concatenations $0\|ID_A\|g^{X_A}\|N_A$ and $1\|ID_B\|g^{X_B}\|N_B$, respectively. Here, 0 and 1 are two public (and fixed) values that are used to prevent a *reflection attack*. ID_A and ID_B are human-readable identifiers belonging to parties A and B (e.g., their e-mail addresses).

The following four messages are exchanged over the radio link. In the first message, A sends to B the commitment c_A. B responds with his own commitment c_B. In turn, A sends out d_A, by which A opens the commitment c_A. B checks the correctness of the commitment/opening pair $(\hat{c}_A, \hat{d}_A)$ and verifies that 0 appears at the beginning of $\hat{m}_A$. If the verification is successful, B sends in the fourth message d_B, by which B opens the commitment c_B. A in turn checks the commitment and verifies that 1 appears at the beginning of $\hat{m}_B$. If this verification is successful, A and B proceed to the final phase.

In the final phase, A and B first generate the verification strings i_A and i_B, respectively, as shown in Figure 5.6 ($\oplus$ is the bitwise "xor" operation). The length of each of these strings is k. Finally, Alice and Bob simply compare i_A and i_B. If they match, Alice and Bob accept each other's DH public parameters g^{X_A} and g^{X_B} and the corresponding identifiers ID_A and ID_B as being authentic. At this stage, Alice and Bob can safely generate the corresponding secret DH key $g^{X_A X_B}$.

Let us now define formally what we mean by a secure protocol.

Definition 5.1 We say that a protocol $\Pi(k, (A, B))$ is a *secure protocol enabling authentication of DH public parameters* between A and B if the (polynomial-time) attacker M cannot succeed in deceiving A and B into accepting DH public parameters different than g^{X_A} and g^{X_B}, except with a satisfactorily small probability $O(2^{-k})$.

To state the result about the security of DH-SC protocol, we need two additional security parameters (k was already introduced before: it is the length of verification strings i_A and i_B). We denote with γ the maximum number of sessions (successful or abortive) of the DH-SC protocol that any party can participate in (during a whole lifetime). We further assume that there are in total n parties in the world that are using the DH-SC protocol. The following result is proven under the assumption that an ideal commitment scheme is used.

Theorem 5.1 *The probability that an attacker succeeds against the DH-SC protocol is bounded by $n\gamma 2^{-k}$. Therefore, for the appropriately chosen parameter k, DH-SC is a secure protocol enabling authentication of DH public parameters.*

Note that the success probability of the attacker, as stated in Theorem 5.1, refers to the success against *any* among all DH-SC protocol runs. In other words, the attacker does not care *which* parties' communication she breaks/influences. On the contrary, the probability that the attacker is successful against a *specific* (targeted) party is only $\gamma 2^{-k}$.

The proof of the theorem and the assessment of the security of the protocol can be found in [81].

Let us give an example of possible values for the above parameters. Assume there are at most $n = 2^{20}$ parties using the protocol and each party can participate in at most $\gamma = 2^{20}$ sessions (successful or abortive) in her lifetime. Then, by choosing $k = 55$ we obtain that the highest probability of success by the attacker (having seen a huge number $n\gamma = 2^{40}$ of protocol runs) is at most $n\gamma 2^{-k} = 2^{-15}$. Note that k also represents the length of the verification strings i_A and i_B to be compared by users. To make this task easier for users, as mentioned before, we can encode the bits in a string of short words from some predefined dictionary. In our specific case, let us call ℓ the number of short words into which we encode the $k = 55$ bits. For example, in order to have $\ell = 5$, where each word is 4 characters long, each user would have to store a dictionary of $2^{\frac{k}{\ell}} = 2^{11} = 2048$ 4-character words. Of course, ℓ can be reduced further by using larger dictionaries.

5.5 Revocation

If public keys and certificates are delivered by an authority, it must be possible, whenever necessary, for the authority to revoke them. This is one of the most difficult problems of public-key cryptography, and it is clear that there is no one-fit-all solution. Indeed, revocation is strongly related to the kind of trust expected between the users and the authorities, and this trust can vary from one application to another.

In order to illustrate this concept, we will rely on one of our running examples, namely the problem of key revocation in a vehicular network, based on the principles introduced in Subsection 2.2.7.

The advantages of using a PKI for vehicular communications are accompanied by some challenging problems, notably certificate revocation. For example, the certificates of a detected attacker or malfunctioning device have to be revoked, i.e., it should not be able to use its keys or if it still does, vehicles verifying them should be made aware of their invalidity.

The most common way to revoke certificates is the distribution of CRLs (Certificate Revocation Lists) that contain the most recently revoked certificates; CRLs are provided to passing by vehicles by roadside units. In addition, using short-lived certificates automatically revokes keys. These are the methods envisioned in the IEEE P1609.2 standard [197]. But there are several drawbacks to this approach. First, CRLs can be very long owing to the enormous number of vehicles and their high mobility (meaning that a vehicle can encounter a high number of vehicles when traveling, especially over long distances). Second, the short lifetime of certificates still creates a vulnerability window. Last but not least, the availability of an infrastructure will not be pervasive, especially in the first years of deployment.

To avoid the above shortcomings, we describe a specific solution, based on a set of revocation protocols called RTPD (Revocation Protocol of the Tamper-Proof Device), RCCRL (Revocation Protocol using Compressed Certificate Revocation Lists), and DRP (Distributed Revocation Protocol). We present the details of RTPD, illustrated in Figure 5.7, and we only outline the main features of RCCRL and DRP; the interested reader may refer to [323, 325] for more details.

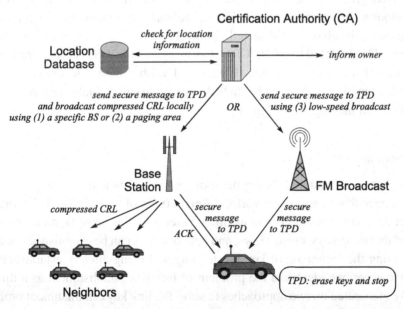

Figure 5.7. Revocation protocol of the Tamper-Proof Device (RTPD). Used with permission, from [325], © IEEE, 2006.

In RTPD, once the CA has decided to revoke all the keys of a given vehicle M, it sends to it a revocation message encrypted with the vehicle's public key. After the message is received and decrypted by the TPD of the vehicle, the TPD erases all the keys and stops signing safety messages. Then, it sends an ACK to the CA. All the communications between the CA and the vehicle take place in this case via a base station. In fact, the CA has to know the vehicle's location in order to select the base station through which it will send the revocation message. If it does not know the exact location, it retrieves the most recent location of the vehicle from a location database and defines a paging area with base stations covering these locations. Then it multicasts the revocation message to all these base stations. In the case when there are no recent location entries or the ACK is not received after a timeout, the CA broadcasts the revocation message, for example, via the low-speed FM radio on a nationwide scale or via satellite.

The RCCRL protocol is used when the CA wants to revoke only a subset of a vehicle's keys or when the TPD of the target vehicle is unreachable (e.g., by jamming or by tampering of the device). Given the expected large size of CRLs in VANETs, the key idea in RCCRL is to use Bloom filters (a probabilistic data structure used to test whether an element is a member of a set) [68]. Thus, the size of a CCRL will be only a few kbytes. RCCRL also relies on the availability of the infrastructure that broadcasts the CCRLs periodically (say, once every 10 min). Compared with RTPD, RCCRL warns the neighbors of a revoked vehicle because they also receive the CCRLs.

The DRP protocol is used in the pure ad hoc mode whereby vehicles accumulate accusations against misbehaving vehicles, evaluate them using a reputation system and, in case misbehavior is detected, report them to the CA once a connection is available. Unlike RTPD and RCCRL, the revocation in DRP is triggered by the neighbors of a vehicle upon the detection of misbehavior. Mechanisms for the detection of malicious data [153] can be leveraged to spot vehicles generating these data (since all messages are signed).

5.6 Summary

In this chapter, we studied the problem of key establishment in sensor networks. We explained that in sensor networks, different types of keys are needed in order to support different types of communication patterns and in-network processing. We showed that node keys, cluster keys, and network keys can be established relatively easily using the technique of key pre-loading and using already established link keys. However, we identified the problem of link key establishment as a difficult one. We elaborated on two approaches to solve the link key establishment problem. The first approach is based on a short-term master key known to every node after its deployment for a limited amount of time. The second approach is based on the idea of random key pre-distribution. We described a basic protocol that uses random key pre-distribution and some of its enhancements.

We then moved to personal communications and explained how physical contact can be used to establish security associations between devices.

Next we considered the case in which users find physical contact (of their devices) impractical, and assumed the presence of a secure side channel, such as the one provided by infrared communications. We explained that mobility can help securing mobile networks. We illustrated this on two scenarios: self-organized networks and networks with an off-line authority. In the first scenario, we showed that the solution is intuitive to the users, as it mimics real-life concepts (physical encounters, friends), and solves some classical problems of security in distributed systems. In the second scenario, a direct establishment of security associations over the (one-hop) radio link solves the security-routing interdependency problem, which we will discuss in Chapter 7.

We then removed the assumption of the presence of a secure side channel and described how two users, moving into the vicinity of each other, can let their devices authenticate each other and set up a security association. We explained that this operation can be securely carried out in spite of the fact that (i) the two devices communicate exclusively over an unsecured radio link and (ii) the two users are assumed not to share any prior information such as mutual certificates.

Finally, we described a way to implement revocation in the case of vehicular networks.

All these mechanisms are of course very useful to protect networking mechanisms (for example, in order to secure routing, as we will see in Chapter 7). It is important to note that they can also be used in order to protect application-level transactions.

5.7 To probe further

A considerable amount of research has been carried out on the topic of establishment of security associations; hence we have organized this section in the same order as the themes addressed in this chapter.

Key establishment in sensor networks In Section 5.1, we explained that an important requirement for key establishment in sensor networks is to support in-network processing. Notably, one of the reasons for setting up link keys and cluster keys is to make in-network data aggregation and passive participation possible. We note that some researchers have explored other ways to support in-network processing based on aggregation of encrypted data without prior decryption (see e.g., [93, 376]).

These concealed data aggregation schemes are promising; however, currently, they support only a limited number of aggregation functions, and they do not allow for passive participation.

The link key establishment approach that is based on a short-term master key was proposed in [404] by Zhu, Setia, and Jajodia as part of LEAP (Localized Encryption and Authentication Protocol), a key management scheme for sensor networks.

The basic random key pre-distribution scheme for link key establishment in sensor networks was proposed by Eschenauer and Gligor in [129]. The q-composite random key pre-distribution scheme and multipath key reinforcement was proposed by Chan, Perrig, and Song as improvements on the basic scheme in [97]. The approach to combine random key pre-distribution with threshold cryptography was proposed by Du, Deng, Han, and Varshney in [122], and by Liu and Ning in [259]. The scheme described in [122] is based on matrices, whereas the scheme proposed in [259] is based on polynomials. In effect, the two proposals are analogous and lead to the same result in terms of improvement with respect to resistance to node capture.

Since the seminal paper [129] of Eschenauer and Gligor on random key pre-distribution in distributed sensor networks, a multitude of papers have been published on pairwise key establishment in sensor networks. A comprehensive survey of these papers (up to 2005) can be found in [82], written by Campete and Yener. They not only describe and classify the various approaches but also identify which contribution is based on which other contributions.

Exploiting physical contact As we have mentioned, Stajano and Anderson have proposed the *resurrecting duckling* security policy model, [356] and [357], in which key establishment is based on the physical contact between communicating parties (e.g., their PDAs).

Exploiting mobility The section of this chapter devoted to the exploitation of mobility is based on [90].

In [403], Zhou and Haas propose a distributed public-key management service for ad hoc networks in which the functionality of the central authority is distributed over a subset of nodes through a threshold cryptography scheme.

Another approach, explored by Čapkun *et al.*, consists in letting each node carry a subset of the trust graph [85, 184]. This approach requires some level of transitivity of trust and was not investigated further.

We should also mention the work of Grossglauser and Tse [156] which shows that mobility can help to increase the per-user throughput in ad hoc networks, and which was a source of inspiration for the solution we have described.

Exploiting vicinity The section of this chapter devoted to the exploitation of mobility is based on a contribution authored by Čagalj, Čapkun, and Hubaux [81].

An approach inspired by the resurrecting duckling security policy model is proposed by Balfanz *et al.* [40]. In that work, the authors go one step further and relax the requirement that the location-limited channel has to be secure against passive eavesdropping; they introduce the notion of a *location-limited channel* (e.g., an infrared link), similar to the secure side channel mentioned in this chapter. A location-limited channel is used to exchange pre-authentication data and should be resistant to active attacks (e.g., man-in-the-middle). Once pre-authentication data are exchanged over a location-limited channel, users switch to a common radio channel and run any standard key exchange protocol over it. Possible options for a location-limited channel include physical contact, infrared, and sound (ultrasound).

Asokan and Ginzboorg propose another solution based on a shared password [30]. They consider the problem of setting up a session key between a group of people (i.e., their computers) who get together in a meeting room and who share no prior context. It is assumed that they do not have access to public key infrastructure or third party key management services. The proposed solution is the following. A fresh password is chosen and shared among those present in the room (e.g., by writing it on a sheet of paper or a blackboard). The shared password is then used to derive a strong shared session key. This approach requires users to type the chosen password into their personal devices.

It is well known that IT security systems are only as secure as their weakest link. In most IT systems the weakest links are the users themselves. People are slow and unreliable when dealing with meaningless strings, and they have difficulties remembering strong passwords. In [315], Perrig and Song suggest using hash visualization to improve the security of such systems. Hash visualization is a technique that replaces meaningless strings with structured images.

In US patent no. 5,450,493 [265], Maher presents several methods to verify DH public parameters exchanged between users. The first method described in [265] is the most relevant for the problem considered in this chapter; other methods are based on certificates and/or shared secrets. A and B first perform the DH key exchange protocol and in turn report to each other values $f(K_A)$ and $f(K_B)$, where K_A and K_B are the shared DH keys as computed by A and B, respectively, and f is a compression function (i.e., f maps a key to 4-digit hex vectors [265]). Unfortunately, this technique has a flaw, that was discovered by Jakobsson [201].

Motivated by this flaw, Jakobsson [201] and Larsson [249] propose two solutions, both based on a temporary secret shared between the two users (one of the solutions is called SHAKE, which stands for *Shared key Authenticated Key Exchange*).

Dohrmann and Ellison [117] propose a method for key verification that is similar to the one described in this chapter. This method is based on converting key hashes to readable words or to an appropriate graphical representation. However, it seems that users are required to compare a substantial number of words (or graphical objects) and this task can take them as much as 24 s according to [117]. This time is significantly reduced when the graphical representation is used.

In [146] and [147], Gehrmann *et al.*, propose a set of techniques to enable wireless devices to authenticate one another via an insecure wireless channel with the aid of the manual transfer of data between the devices. The protocol, which they call MANA II, is similar to the DH-SC protocol described in this chapter, but requires the users to compare a higher number of bits.

Cameras are more and more frequently embedded on mobile devices such as mobile phones. Yet another approach, proposed by McCune, Perrig, and Reiter [270] proposes to make use of the camera to capture the image displayed by another device and perform authentication tests.

We should also mention other key-exchange protocols, proposed primarily for the use in the Internet: IKE [163], JFK [14] and SIGMA [236]. All these protocols involve authentication by means of digital signatures. We also should mention the work of Corner and Noble [106], who consider the problem of transient authentication between a user and her device.

5.8 Questions

(1) What is the purpose of authentication?
(2) In order to support link key establishment between neighboring nodes in a sensor network, we could pre-load in each node $n - 1$ keys, where n is the total number of nodes, such that each pair of nodes share a common key. What are the disadvantages of this approach?
(3) In sensor networks, in order to save energy, sensor nodes spend most of their time in sleeping mode. When new nodes are added to an already deployed network, the new nodes cannot immediately establish link keys with their sleeping neighbors. Why is this a problem if we use the link key establishment scheme based on a short-term master key? How can the problem be solved?
(4) Is it always necessary that both devices have a display when performing key establishment between them? Why?
(5) What is the purpose of a commitment scheme?
(6) Can a key established by DH-SC be reused afterwards for authentication purposes? Why?
(7) Assume we remove the commitment phase from the DH-SC protocol. What kind of vulnerabilities would this modification create?
(8) Is the DH-SC protocol vulnerable to denial-of-service attacks based on jamming? Why? What can the users do about it?
(9) Do you think that the protocol illustrated in Figure 5.3 would be appropriate in vehicular networks? Why? How about the protocol in Figure 5.4?

6

Securing neighbor discovery

Many wireless networking mechanisms, notably routing, require that wireless nodes be aware of their neighborhood. This means that the nodes must know which other nodes they can communicate with directly. The procedure used to acquire this knowledge is called *neighbor discovery*.

In wired networks, neighbor discovery is a simple issue, because neighbor relationships do not change often, and hence routers can be pre-configured with the list of their wired neighbors. In contrast, in mobile wireless networks, the neighbor relationships change dynamically, which makes neighbor discovery an important mechanism. It is particularly important in the upcoming wireless networks that we described in Chapter 2.

Neighbor discovery can be achieved through simple protocols, where a node that wants to determine who its neighbors are broadcasts a neighbor discovery request, and every node that receives this request responds with a neighbor discovery reply. Receiving a reply means that the requesting node and the responding node can hear each other's transmission. In other words, they can communicate with each other directly, and hence they should consider each other as neighbors. The neighbor discovery protocol is sometimes called "hello protocol," and the request and the reply are called "hello messages."

An adversary can try to thwart the successful execution of the neighbor discovery protocol, for instance, by jamming the communication between two nodes. In this way, the adversary achieves that two nodes, which otherwise could communicate directly, cannot establish a neighbor relationship. Blocking the links between many pairs of nodes in this manner can have serious consequences to the connectivity of the network, and hence indirectly to upper layer protocols, such as routing.

Unfortunately, it is quite difficult to eliminate this attack. The usual way to prevent jamming is to use spread spectrum communications. But, when the nodes execute the neighbor discovery protocol, they usually have no common context that could be used to determine the frequency hopping sequence to be used. It is possible, for instance, that the nodes meet each other for the first time, but even if they do not, identifying the other already needs the exchange of messages. This means that the neighbor discovery request should be broadcast using a pre-determined, public hopping sequence that is vulnerable to jamming. Note that the more links the adversary wants to remain undiscovered, the more effort she needs to invest in the attack, in the sense that she must be physically present at many points in the network. In addition, the adversary usually tries to avoid being detected, and jamming large parts of the network almost certainly results in adversary detection. Therefore, such a jamming attack is likely to affect only a limited number of nodes.

Besides preventing two nodes from establishing a neighbor relationship, the adversary can also try to arrange that two far away nodes, which otherwise could not communicate directly with each other, believe that they are neighbors. One way to achieve this is identity spoofing: a node controlled by the adversary can use the identity of a legitimate node and establish neighbor relationships with other nodes in the name of that legitimate node. Identity spoofing can be prevented by building cryptographic entity authentication mechanisms into the neighbor discovery protocol. Entity authentication is a widely studied problem in traditional networks, and therefore we will not address it here; some of the basics can be found in Appendix A.

Another way to create false neighbor relationships that cannot be prevented by cryptographic mechanisms solely, is to install *wormholes* in the network. In this chapter, we study this problem in detail. We begin by explaining what a wormhole is and how it can be used to mount severe denial-of-service attacks. Then, we give an overview of some of the approaches that are proposed in the literature to detect wormholes.

6.1 The wormhole attack

A wormhole is an out-of-band connection, controlled by the adversary, between two physical locations in the network. The two physical locations representing the two ends of the wormhole can be at any distance from each other; however, the typical case is that this distance is large. The out-of-band connection between the two ends can be a wired connection or it can be based on a long-range, directional wireless link. The adversary installs radio transceivers at both ends of the wormhole. Then, she transfers packets (possibly selectively) received from the network at one end of the wormhole to the other end via the out-of-band connection and there re-injects the packets into the network.

The effect of a wormhole on neighbor discovery is that some nodes that would not be neighbors otherwise can establish a neighbor relationship due to the presence of the wormhole. This is illustrated in Figure 6.1. More importantly, the wormhole can have devastating effects on upper layer protocols, especially on routing, as we will see below.

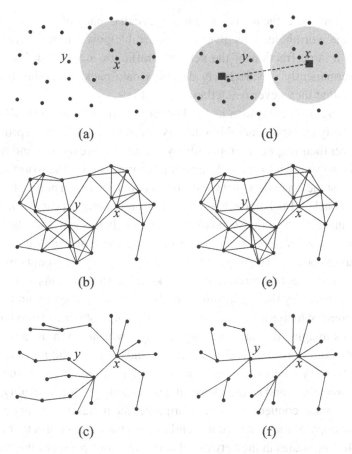

Figure 6.1. Illustration of the effect of a wormhole on neighbor discovery and routing. Part (a) shows a set of wireless nodes that are placed randomly on the plane. The gray disk around node x represents its communication range. For simplicity, we assume that the size of the communication range of every node is the same. Part (b) shows the neighbor relationships between the nodes, and part (c) shows the minimum length routes to node x from all other nodes in the network assuming the neighbor relationships in part (b). The route length is measured in the number of the hops. Part (d) illustrates a wormhole, where the transceivers of the adversary are denoted by black rectangles, and the out-of-band connection is represented by the dashed line. In part (e), nodes x and y become neighbors owing to the presence of the wormhole and the fact that the adversary relays the packets of the neighbor discovery protocol between them. Part (f) shows the minimum length routes to node x from all other nodes in the network assuming the neighbor relationships in part (e). One can observe that in part (f), many of the nodes reach node x via the wormhole.

Clearly, the wormhole affects route discovery mechanisms that operate on the connectivity graph. For instance, link state routing protocols, such as the Optimized Link State Routing (OLSR) protocol [103], search for the shortest paths in the connectivity graph that is constructed locally by each node using the information obtained from periodic link state update messages of the other nodes. With a well-placed wormhole, the adversary can achieve that many of these shortest paths go through the wormhole. This gives a considerable power to the adversary, who can monitor a large fraction of the network traffic, or mount a denial-of-service attack by permanently or selectively dropping data packets passing through the wormhole so that they never reach their destinations.

Some routing protocols, such as the Dynamic Source Routing (DSR) protocol [210], do not rely on explicit neighbor discovery mechanisms. In these protocols, the nodes discover their neighbors implicitly by means of route request and route reply messages. However, these protocols are equally vulnerable to the wormhole attack. In DSR, for instance, a node x that wants to discover a route to a node y broadcasts a route request packet. Each node that receives this route request for the first time appends its identifier to the request and re-broadcasts it. In this way, the whole network is flooded with copies of the route request packet, some of which will eventually reach the target node y. Each request packet that reaches node y contains the list of identifiers of the nodes that processed the packet; this list represents a route between x and y, discovered by the algorithm. A route reply is returned by node y to each request received, which follows the reverse of the route obtained from the request.

Now, let us imagine that an adversary sets up a wormhole in the network, such that one end of the wormhole is close to node x and the other end is close to node y. When the adversary receives a route request originating from node x, she tunnels it through the wormhole and re-broadcasts it near to node y. Most probably, owing to the fast out-of-band connection of the wormhole, the nodes near to node y receive this tunneled copy of the route request earlier than the other copies that follow the normal multi-hop routes in the network. Therefore, those copies of the request will be discarded as duplicates later when they arrive near to node y. As a result, the route discovery protocol will be unable to discover routes between x and y other than those going through the wormhole. To some extent, this is even worse than in the case of link state routing protocols, where the nodes, being aware of the topology of the network, can at least try alternative (potentially suboptimal) routes when they realize that the route through the wormhole provides an unacceptable level of throughput.

The wormhole attack is also dangerous in other types of wireless applications where direct, one-hop communication and physical proximity play an important role. An example is a wireless access control system for buildings, where each door is equipped with a contactless smart card reader, and they are opened only if a valid contactless smart card is presented to the reader. The security of such a system depends on the assumption that the personnel carefully guard their cards. Thus, if a valid card is present, then the system can safely infer that a legitimate person is present as well, and the door can be opened. Such a system can be defeated if an adversary can set up a wormhole between a card reader and a valid card that could be far away, in the pocket of a legitimate user: the adversary can relay the authentication exchange through the wormhole and gain unauthorized access. The feasibility of this kind of attack has been demonstrated in [221].

Before we leave this section and begin the presentation of some countermeasures for wormholes, we must emphasize that, although the wormhole attack can have strong effects on routing, it is essentially an attack against neighbor discovery. In particular, in order to mount a wormhole attack, the adversary does not need to control nodes in the network, hence she can stay "invisible" at the routing layer. It is sufficient for the adversary to install simple radio transceivers at the two ends of the wormhole that operate in the physical layer and function as repeaters. Because of the broadcast nature of wireless communications, the adversary can overhear packets that are transmitted in the proximity of her radio transceivers; she can capture these packets, and transfer them through the wormhole.

We must also note that the adversary does not need to understand what she transfers through the wormhole. Indeed, the adversary does not even need to wait to receive the entire packet before she starts to transfer it to the other end of the wormhole; she can operate on a bit-by-bit basis. This means that a wormhole attack can be effective even if packets are encrypted; in this case, the adversary transfers the encrypted bits through the wormhole, without breaking any cryptographic keys.

There exists an attack that has similar effects on routing to the wormhole attack, but in contrast to the wormhole attack, it is carried out in the routing layer. This attack is called a *tunneling attack*. In the tunneling attack, the adversary controls some corrupted nodes in the network. When one of her controlled nodes receives a route request packet, the adversary puts the entire request packet in the payload part of a normal data packet and sends this data packet to another adversarial node using the normal multi-hop forwarding mechanism of the network. The receiving adversarial node takes the route request out of the data packet and processes it as if it had received it via its radio interface. All this is similar to the way in which IP packets originating from one part of a virtual private network (VPN) are tunneled through gateways to another part of the same VPN; hence the name tunneling.

Although the effects of the tunneling attack and the wormhole attack on routing protocols are similar (essentially, routes are shortened and made more attractive in both cases), there are some important differences between the two attacks. In order to carry out the tunneling attack, the adversary needs to have corrupted nodes in the network. Thus, the adversary is visible at the network layer, unlike in the wormhole attack. In contrast, in the tunneling attack, there is no need for an out-of-band connection between the devices of the adversary, but they can communicate by using the network itself.

As we mentioned before, from an architectural point of view, the tunneling attack can be considered as an attack at the routing layer, whereas the wormhole attack is carried out at the physical layer. For this reason, we discuss the two attacks in different chapters of the book. Specifically, the tunneling attack is addressed in Chapter 7, where security of routing is discussed, and we address the wormhole attack in the context of neighbor discovery in this chapter.

6.2 Wormhole detection mechanisms

In the rest of this chapter, we study how wormholes can be detected. Broadly, the different detection mechanisms fall into two classes: the centralized mechanisms and the decentralized ones. In the centralized approach, data collected from the local neighborhood of every node are sent to a central entity. The central entity uses the received data to construct a model of the entire network, and tries to detect inconsistencies in this model that are potential indicators of wormholes. In the decentralized approach, each node constructs a model of its own neighborhood using locally collected data; hence no central entity is needed, which is a big advantage of this approach. We note, however, that in some applications, central entities are inherently present in the network. One example is a sensor network, where the base stations are in a position to collect data from the nodes, and thus they can play the role of the central entity. In this kind of network, the centralized approach can be acceptable too. In the following, we first present some techniques that use the centralized approach, and then we give an overview on some of the mechanisms that use the decentralized approach.

6.2.1 Centralized approaches

The central entity tries to detect the wormholes by identifying inconsistencies in the constructed model. The kinds of inconsistencies that might appear in the model, due to the presence of wormholes, depend on the nature of the local information provided by the nodes. We illustrate this with two examples.

In the first example, the nodes report only the list of their believed neighbors to the central entity. In this case, the model constructed by the central entity consists of the connectivity graph of the network. A crucial observation is that a wormhole *always* increases the number of edges in the connectivity graph, as it introduces new neighbor relationships. This increase in the number of edges changes the properties of the connectivity graph in a detectable way with respect to some expectations that are based on basic assumptions about the system (e.g., the distribution of node positions, the communication range of the nodes, etc). The main idea of the first mechanism is to detect the changes in the connectivity graph using statistical methods.

In the second example, the nodes also estimate the distances from their believed neighbors and send their neighbor list with the estimated distances to the central entity. In this case, the model constructed by the central entity is a virtual layout of the network. The crucial observation here is that a wormhole contracts the virtual layout in certain regions, because it makes some nodes appear to be neighbors where in reality these nodes are far away from each other. The main idea of the second mechanism is to detect these contractions by visualizing the virtual layout.

Statistical wormhole detection

Let us assume that the network consists of n nodes placed in a flat area of size S uniformly at random.[1] Let us further assume that the nodes are static (e.g., the network is a static sensor network), and their communication range r is fixed and it is the same for every node. Then, we can compute the probability that a node has exactly k neighbors ($0 \leq k < n$) as

$$p(k) = \binom{n-1}{k} \cdot q^k \cdot (1-q)^{n-1-k}, \tag{6.1}$$

where

$$q = \frac{r^2 \pi}{S}. \tag{6.2}$$

Hence, although the random variables representing the node degree of the different nodes in the network are not independent, in a dense network we expect that the distribution of the node degrees is close to the binomial distribution with parameters $n-1$ and q.

Now, let us assume that an unsophisticated adversary establishes a wormhole in the network that functions as a perfect repeater: every bit overheard at one end of the wormhole is transferred to the other end and re-transmitted there. Such a wormhole allows every pair (x, y) of nodes such that x resides in the communication range of one of the wormhole's transceivers and y resides in the communication range of the other transceiver to set up a neighbor relationship. Thus, assuming that the communication range of the wormhole's transceivers is the same as that of the nodes, the number of believed neighbors of the nodes within the range of the wormhole will double on average. Therefore, the degree distribution that the central entity can observe in the connectivity graph that is constructed from the nodes' neighborhood information will be distorted with respect to the binomial distribution derived above.

In order to illustrate this phenomenon, we performed a simple experiment. We placed $n = 300$ nodes uniformly at random in a rectangular area of size 500×500 square units, and we set $r = 54$ units. In part (a) of Figure 6.2, we plotted in gray the expected histogram of the node degree (induced by the binomial distribution), and in black the observed histogram when a randomly placed wormhole was present in

---

[1] In fact, the nodes are not necessarily placed randomly in the field. However, any known structure would make the detection of the wormhole easier. Therefore, we discuss here the case where the locations of the nodes are random.

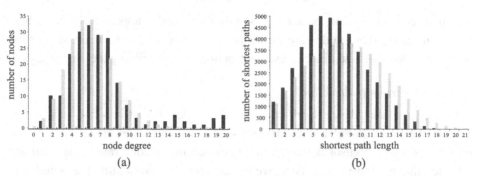

Figure 6.2. Results of two experiments, where $n = 300$ nodes were placed uniformly at random in a rectangular area of size 500×500 square units, r is set to 54 units. In part (a), the gray bars show the expected histogram of the node degree, and the black bars show the observed histogram when a randomly placed, perfectly repeating wormhole is present in the network. There is a clear difference between the two histograms. In particular, the black histogram shows that there are several nodes with unexpectedly high node degrees, a sign that indicates the presence of a wormhole. In part (b), the gray bars show the expected histogram of the length of the shortest paths in the network, which we obtained by measuring the lengths of the shortest paths in randomly generated networks with the same parameters as above. The black bars show the observed histogram of the length of the shortest paths when there is a randomly placed wormhole in the network, which created a single new link in the connectivity graph. The difference between the histograms is clearly observable: the black histogram shows that when the wormhole is present, shorter paths are more likely. Used with permission, from [72], with kind permission of Springer Science and Business Media.

the network. One can clearly observe the difference between the two histograms. As we expected, the black histogram shows that if there is a (perfectly repeating) wormhole in the network, then there are a few nodes with unexpectedly high node degrees in the connectivity graph.

In order to defeat the detection mechanisms based on the verification of the node degree distribution in the connectivity graph, a more sophisticated adversary would not connect each pair of nodes within the range of the wormhole, but it would rather allow for the creation of only a small number of false neighbor relationships. But even if it creates just a few new links, the length of the shortest paths between many pairs of nodes can decrease significantly, especially if the wormhole's out-of-band connection spans over a long distance. This is justified by the result of another experiment, which is shown in part (b) of Figure 6.2. In this experiment, the parameters were the same as in the previous case, but the wormhole created only a single link between two randomly selected nodes within its range. The black bars show the observed histogram of the length of the shortest paths between all pairs of nodes when the wormhole is present. The gray bars show the expected histogram of the length of the shortest paths in wormhole-free networks, which we obtained by measuring the lengths of the shortest paths in randomly generated networks with the same parameters as above. Again, the difference between the histograms is clearly observable. The black histogram shows that shorter paths are more likely when the wormhole is present.

The notion of "clearly observable difference" between histograms is not precise enough to build a wormhole detection algorithm on it. We need a mathematically rigorous way of deciding if two data samples originate from the same distribution or from different ones. Fortunately, there exist standard statistical tests, such as the χ^2-test, for this purpose.

The main disadvantage of the statistical method that we described above is that, although it detects the presence of wormholes with high confidence, it does not locate them. In other words, it tells us that there are probably wormholes in the system, but it does not tell us where they are set up and exactly which nodes are affected. Therefore, to some extent, it only does half of the job. The method described in the following subsection overcomes this problem.

Wormhole detection with multi-dimensional scaling

Another centralized wormhole detection approach is based on augmenting the connectivity information with (possibly inaccurate) distance estimations between neighboring nodes. However, in return for the increased complexity introduced by the distance estimation requirement, this technique allows for the localization of wormholes.

The main idea here is to reconstruct a virtual layout of the network and identify inconsistencies in it. For this reason the connectivity information and the inaccurately estimated distances between the neighbors are fed into a multi-dimensional scaling (MDS) algorithm, that tries to determine a virtual position for every node in such a way that the constraints induced by the connectivity and the distance estimation data are respected. As the distances estimations can be inaccurate, the algorithm has a certain level of freedom in "stretching" the nodes within some error bounds.

Now, let us suppose that an adversary has installed a wormhole in the network and has created fake links in the connectivity graph between far away nodes. If the estimated distances between the affected nodes are much larger than the nodes' communication range, then the wormhole is detected immediately. Hence, the adversary must also falsify the distance estimation and arrange that the estimated distances between the nodes affected by the wormhole become credible (i.e., smaller than the communication range plus the maximum distance estimation error). This will result, however, in a distortion in the virtual layout constructed by the MDS algorithm; in particular, the layout will be contracted between the affected nodes. By visualizing the virtual layout or by computing appropriate indicator values, the distortion can be detected and the wormhole can be located by identifying the affected nodes.

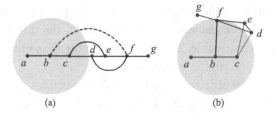

(a) (b)

Figure 6.3. Wormhole detection by constructing a virtual layout of the network based on the connectivity information and the inaccurate distance measurements between the nodes that believe they are neighbors. For easier understanding, the main idea is illustrated on a one-dimensional network. Part (a) of the figure shows the real placement of the nodes. The gray disk represents the communication range of node b. The lines represent the established neighbor relationships. The dashed line between nodes b and f represents a fake neighbor relationship created by an adversary with the help of a wormhole. Part (b) shows the virtual layout of the network reconstructed from the inaccurate distance measurements of the neighboring nodes. As nodes b and f are neighbors, their distance must be smaller than the communication range. However, this constraint makes it impossible to fit the nodes on a straight line. In other words, the virtual layout is contracted between nodes b and f with respect to the real layout of the nodes. By visualizing the virtual layout, the contraction can be identified and the wormhole can be detected.

Figure 6.3 illustrates how the wormhole contracts the virtual layout constructed by the MDS algorithm. For simplicity, we illustrate the main idea in the case of a one-dimensional network, but the same approach works in two or in three dimensions. Part (a) of Figure 6.3 shows the real placement of the nodes. The gray disk represents the communication range of node b. The lines represent the established neighbor relationships. The dashed line between nodes b and f represents a fake neighbor relationship created by an adversary with the help of a wormhole. Part (b) of the figure shows the virtual layout of the network reconstructed from the inaccurate distance measurements of the neighboring nodes. Since nodes b and f are neighbors, their distance must be smaller than the communication range. However, this constraint makes it impossible to fit the nodes on a straight line. In other words, the virtual layout is contracted between nodes b and f with respect to the real layout of the nodes. The distortion in the virtual layout can be detected by a human operator if the layout is visualized. The detection can also be automated; the interested reader is referred to [369] for the details.

6.2.2 Decentralized approaches

The advantage of decentralized wormhole detection mechanisms is that they do not require a central entity in the system, and therefore, they can be used in a wider range of applications. In this subsection, we give a brief overview of the main approaches proposed in the literature.

Wormhole detection based on distance estimation

A straightforward idea for wormhole detection is to estimate the real physical distance between the nodes that are believed to be neighbors. If the estimated distance is larger than the nodes' communication range, then the nodes are likely connected through a wormhole, and they should not consider each other as neighbors. Distance estimation can be done by the nodes themselves locally, so it can form the basis of decentralized wormhole detection approaches.

One wormhole detection approach based on the idea of distance estimation is called *packet leashes*, and it consists of two mechanisms: geographical and temporal leashes. The main idea of both mechanisms is to add some information to the packets that restricts their maximum allowed transmission distance. Allegorically, the added information keeps the packet on a leash, hence the name of the mechanisms. A geographical leash is based on location information, and it allows the receiver of the packet to determine an upper bound on its distance to the sender. A temporal leash is based on timing information, and it ensures that the packet has an upper bound on its lifetime. Indirectly, however, this also ensures an upper bound on the distance between the sender and the receiver, because the packet cannot travel faster than the speed of light.

Both types of leashes can be used for wormhole detection, because they allow the receiver of the packet to detect whether the sender is further away than the nodes' communication range. More precisely, the receiver can determine only an upper bound on its distance to the sender. However, if this upper bound is greater than the nodes' communication range, then the receiver should not accept the packet. In this way, packets that arrive through a wormhole are always rejected.

Packet leashes can be added to the packets of the neighbor discovery protocol when the nodes use such mechanisms explicitly for setting up their neighbor relationships. In this case, the application of packet leashes prevents the establishment of fake neighbor relationships. When no explicit neighbor discovery mechanism is used in the system, packet leashes can still be added to the packets of the routing protocol, in order to prevent the undesirable effects of wormholes on routing, which we described in Section 6.1.

Now, we describe the operating principles of geographical and temporal packet leashes in more detail.

As we mentioned above, geographical leashes are based on location information. It is assumed that each node is aware of its own location, which can be determined using GPS (Global Positioning System) or some other positioning mechanism (e.g., [86, 250, 88, 89]). It is further assumed that the nodes maintain loosely synchronized

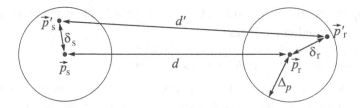

Figure 6.4. Computation of the upper bound on the real distance d' between the sender and the receiver when using geographical leashes and assuming that the nodes are static. Because of the inaccuracy of positioning, the real locations $\vec{p}_s{'}$ and $\vec{p}_r{'}$ of the sender and the receiver, respectively, can be anywhere within a range of Δ_p around the locations $\vec{p}_s$ and $\vec{p}_r$ that are determined by the positioning system. Let $d = ||\vec{p}_r - \vec{p}_s||$, $\delta_s = ||\vec{p}_s{'} - \vec{p}_s||$, and $\delta_r = ||\vec{p}_r{'} - \vec{p}_r||$. Then, we have that $d' \leq d + \delta_s + \delta_r \leq d + 2\Delta_p$.

clocks. When sending a packet, the sender includes its location $\vec{p}_s$ in the packet and the time t_s of sending. When receiving a packet, the receiver compares these values to its own location $\vec{p}_r$ and the reception time t_r. Given a maximum positioning error Δ_p, a maximum clock synchronization error Δ_t, and an upper bound v_{max} on the speed of the nodes, the receiver can compute an upper bound on the real distance d' between the sender and itself at the time of receiving the packet as follows:

$$d' \leq d + 2\Delta_p + 2v_{max}(t_r - t_s + \Delta_t), \qquad (6.3)$$

where $d = ||\vec{p}_r - \vec{p}_s||$ is the distance between the locations $\vec{p}_r$ and $\vec{p}_s$. Because both nodes could have some positioning error, d' could be larger than d. However, if the nodes are static, then d' must be smaller than $d + 2\Delta_p$, as illustrated in Figure 6.4. This explains the second term in (6.3). In addition, if the nodes are not static, then they could diverge during the packet transmission time. Given that the maximum clock synchronization error is Δ_t, when the sender's clock shows t_s, the time at the receiver can only be $t_s - \Delta_t$. Therefore, the time that elapses between sending and receiving the packet is upper bounded by $t_r - (t_s - \Delta_t)$. During this time, the nodes can diverge at most $2v_{max}(t_r - t_s + \Delta_t)$. This explains the third term in (6.3).

Temporal leashes require that the nodes have tightly synchronized clocks, such that the maximum difference Δ_t between any two nodes' clocks is in the order of a few hundred nanoseconds. This precision can be achieved with some of today's technologies such as LORAN-C [275], WWVB [276], or GPS. When sending a packet, the sender includes in the packet the time t_s of sending the first bit of the packet. When receiving a packet, t_s is compared with the time t_r of receiving the first bit of the packet at the receiver. More precisely, the receiver computes an upper bound on its distance d' to the sender as

$$d' \leq v_{light}(t_r - t_s + \Delta_t), \qquad (6.4)$$

where v_{light} is the speed of light. In order for this upper bound to be useful, $v_{light} \Delta_t$ must be much smaller than the communication range of the nodes. That is the reason the nodes' clocks must be tightly synchronized (i.e., Δ_t must be very small).

A potential problem with the temporal leash mechanism described above is that when using a contention based medium access control protocol, the sender cannot know exactly when the first bit of the packet will be sent. For instance, if the IEEE 802.11 protocol is used, then the sender cannot know the starting time of the transmission until approximately one slot time (20 μs) before the transmission really begins. This might be too short for timestamping the packet, especially if the timestamps are authenticated with digital signatures (see next paragraph). We can try to solve this problem by using more efficient authentication mechanisms (e.g., Schnorr signatures, or symmetric key MACs) or by increasing the minimum packet length such that the computation of the signature can be completed during the transmission of the packet payload.

Both geographical and temporal leashes require that the packets carrying the leashes are authenticated and their integrity is protected, because otherwise an adversary can modify or forge a leash and jeopardize the distance estimation. Origin authentication and integrity protection can be based on digital signatures or on symmetric key MACs. The advantage of digital signatures is that they provide broadcast authentication, and therefore they can be used efficiently for protecting neighbor discovery beacons, route discovery messages, or link state updates; all of which are usually broadcast messages. The disadvantage of digital signatures is that they are several orders of magnitude slower than symmetric key MAC computations, and speed is critical, especially in the case of temporal leashes. Although MACs can be computed faster, they cannot be used efficiently to protect broadcast messages (see Appendix A for more details).

One way to solve this problem is to use TESLA [314] with Instant Key-disclosure (TIK) to authenticate temporal leashes in packets. TESLA combines the advantages of digital signatures and MACs. Its description can be found in Appendix A, and therefore, we will not detail it here. Instead, we briefly present the main idea of the TIK protocol.

The TIK protocol is based on the observation that the authentication delay of TESLA can be removed in an environment where the nodes' clocks are tightly synchronized. TESLA requires that the MAC value of the packet is received earlier by the receiver than the time at which the TESLA key used for computing the MAC is disclosed by the sender. This can be achieved by sending the MAC value at the beginning of the transmission and disclosing the TESLA key at the end of the same transmission, as shown in Figure 6.5. The receiver's clock shows $t_r + \tau_{mac}$ when it received the entire MAC. The sender's clock shows $t_s + \tau_{mac} + \tau_{pkt}$ when it starts disclosing the key; at the same moment, the time at the receiver can

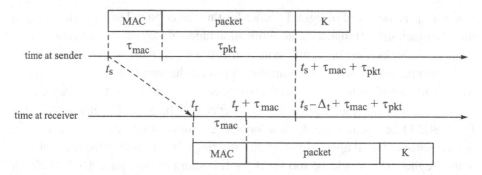

Figure 6.5. Illustration of the main idea of the TIK protocol (i.e., TESLA with Instant Key-disclosure). The sender sends the MAC of the packet at the beginning of the transmission and discloses the TESLA key with which the MAC was computed at the end of the same transmission. The TESLA condition is satisfied if the receiver receives the MAC value earlier than the time at which the sender starts disclosing the TESLA key. The receiver can verify this by checking if $t_r < t_s - \Delta_t + \tau_{pkt}$ holds. Note that the receiver knows t_s from the timestamp placed in the packet by the sender. Used with permission, from [180], © IEEE, 2003.

be $t_s - \Delta_t + \tau_{mac} + \tau_{pkt}$ at the least. Hence, if the receiver finds that $t_r + \tau_{mac} < t_s - \Delta_t + \tau_{mac} + \tau_{pkt}$, where t_s is known to the receiver from the temporal leash in the packet, then the TESLA condition is satisfied (i.e., the full MAC is received before any bit of the key with which it was computed is released), and the receiver can start the verification of the MAC essentially without any delay. Clearly, in order for this to work, very precise timings are needed and, in particular, Δ_t must be very small (or otherwise packets need to be extremely long).

Although packet leashes provide an effective solution to the wormhole detection problem, they have some disadvantages that prevent their usage in certain environments. The main disadvantage of the geographical leash mechanism is that it requires the nodes to be equipped with GPS receivers or to be able to determine their location in some other way. Integrating a GPS receiver in every node can be prohibitively expensive in some applications, for instance, in sensor networks. In addition, GPS has known problems in an indoor environment. Other positioning mechanisms could be used, but their security must also be ensured. The main disadvantage of the temporal leash mechanism is that it requires very tight clock synchronization, which might not be possible to achieve in some environments.

Another approach that is also based on distance estimation between the nodes, but does not require any clock synchronization or localization mechanisms, is based on the concept of *distance-bounding*. The main idea of distance-bounding is simple but very powerful. It is based on the facts that electro-magnetic waves propagate nearly with the speed of light and with current technology it is easy to measure local timings with nanosecond precision. The distance-bounding technique essentially consists of a series of rapid bit exchanges between the two nodes. Each bit sent by the first node is considered to be a challenge for which the other node is required to send a one bit response immediately. By locally measuring the time between sending out the challenges and receiving the responses, the first node can estimate its distance to the other node, assuming that the messages travel with the speed of light and the processing delay at the other node is negligible.

Note that the estimated distance is only an upper bound on the real distance between the nodes, because the second node could be closer, but it can delay the responses in order to appear to be further. Even if the nodes are trusted for not delaying their responses, an active adversary can delay the messages between the parties, and hence, the estimated distance will still be just an upper bound on the real distance. However, in the case of a wormhole attack, the adversary's goal is not to make the two nodes believe that they are far away from each other. On the contrary, the adversary wants the two nodes to believe that they are within each other's range, when in reality they are not. In order to achieve that the estimated distance is smaller than the nodes' real distance, the adversary should arrange that the messages travel faster than the speed of light, which is impossible. Thus, distance-bounding can be used for wormhole detection.

We slightly modify the above described distance-bounding technique such that it allows both nodes to measure the distance between them simultaneously and it uses symmetric key cryptographic primitives for authentication purposes. In order for this to work, it is assumed that each pair of nodes share a symmetric key. We call the resulting protocol Mutual Authenticated Distance-bounding, or shortly MAD.

Let x and y denote the two nodes in the protocol, and let their shared key be k_{xy}. We will denote the message authentication function controlled by the key k_{xy} by $mac_{k_{xy}}$. The operation of the protocol is summarized in Figure 6.6, and it is explained as follows.

- **Initialization phase**
  Both x and y generate uniformly at random two numbers. The numbers of x are denoted by r and r', and the numbers of y are denoted by s and s'. Numbers r and s are ℓ bits long, and r' and s' are ℓ' bits long (i.e., $r, s \in \{0, 1\}^{\ell}$ and $r', s' \in \{0, 1\}^{\ell'}$) Both x and y compute a commitment to the generated numbers by using a collision resistant one-way hash function H: $c_x = H(r||r')$ and $c_y = H(s||s')$. Finally, x sends c_x to y and y sends c_y to x. Note that the random numbers can be generated and the commitments can be computed well before running the protocol.
- **Distance-bounding phase**
  Let the bits of r and s be denoted by r_i and s_i ($i = 1, 2, \ldots, \ell$), respectively. The following two steps are repeated ℓ times, for $i = 1, 2, \ldots, \ell$:

- x sends bit α_i to y immediately after it received β_{i-1} from y (except for α_1 which is sent without receiving any bit from y), where $\alpha_1 = r_1$ and $\alpha_i = r_i \oplus \beta_{i-1}$ for $i > 1$;
- y sends bit $\beta_i = s_i \oplus \alpha_i$ to x immediately after it received α_i from x.

Node x measures the times between sending α_i and receiving β_i, and y measures the times between sending β_i and receiving α_{i+1}. From the measured times, they both estimate their distance.

- **Authentication phase**

  Node x computes the bits $s_i = \alpha_i \oplus \beta_i$, and the MAC

$$\mu_x = mac_{k_{xy}}(x||y||r_1||s_1|| \cdots ||r_\ell||s_\ell).$$

Similarly, y computes the bits $r_1 = \alpha_1$ and $r_i = \alpha_i \oplus \beta_{i-1}$ for $i > 1$, and the MAC

$$\mu_y = mac_{k_{xy}}(y||x||s_1||r_1|| \cdots ||s_\ell||r_\ell).$$

Finally, x sends $r'||\mu_x$ to y and y sends $s'||\mu_y$ to x. Node x verifies that the commitment c_y and the MAC μ_y of y are correct, and y verifies that the commitment c_x and the MAC μ_x of x are correct.

In the above protocol, the MAC ensures the authenticity of the exchange: both x and y can believe that they ran the distance-bounding phase with the other, and thus, the distance that they estimate is really the distance between x and y. Committing to r and s in the initialization phase ensures that the protocol is successful only if exactly the bits of r and s are exchanged. As r and s are random, an adversary cannot try to cheat x by predicting the bits of s and responding earlier than y, and similarly it cannot cheat y either. More precisely, the probability that such an attack succeeds is $2^{-\ell}$ and hence decreases exponentially in ℓ.

The advantage of MAD is that it does not require the localization of the nodes or the synchronization of their clocks. MAD still requires, however, special hardware in the nodes in order to quickly switch the radio from receive mode into send mode. In addition, it needs a special medium access control protocol that allows for the transmission of bits without any delay.

Wormhole detection using position information of anchors

In the previous subsection, we saw a straightforward way of using the location information of the nodes for wormhole detection. To be more precise, we described the concept of geographical packet leashes and an implementation that uses the nodes' location data for estimating the real distances between them. We argue, however, that obtaining the location data of every node is not feasible in many applications, because it requires either a GPS receiver in every node or complex

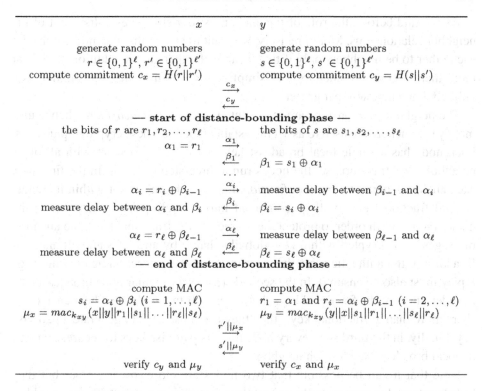

Figure 6.6. The Mutual Authenticated Distance-bounding (MAD) protocol. Used with permission, from [84], © ACM, 2003.

positioning mechanisms that need to be secured. In this subsection, we describe a wormhole detection mechanism that requires only a few specialized nodes to be aware of their locations. These specialized nodes can be viewed as *anchors* that help other nodes to set up their neighbor relationships in a secure way.

We assume that the nodes are randomly deployed in some area and they are static. The random deployment with node density λ can be modeled as a spatial homogeneous Poisson point process [108] with rate λ. In this model, the probability that there are exactly k nodes within a region of size S is

$$\frac{(\lambda S)^k}{k!} e^{-\lambda S}. \tag{6.5}$$

We assume that a small fraction of the nodes are specialized and they are aware of their own locations. These nodes are the anchors. We further assume that the transmission range R of the anchors is larger than the transmission range of the regular nodes. The anchors are also deployed randomly, but their density λ^* is much smaller than the density of the regular nodes.

As we said before, the role of the anchors is to assist the establishment of the neighbor relationships. More precisely, we want to ensure that two nodes consider each other to be neighbors only if they hear each other, and in addition, they hear more than T common anchors. For simplicity, we assume that threshold T is a publicly known system parameter.

The neighbor relationships are represented by *local broadcast keys*, hence ultimately the anchors are used to assist the establishment of these cryptographic keys. Each node has a single local broadcast key that it wants to share with all of its neighbors. For this purpose, the nodes run a three-step protocol. In the first step, each anchor generates a random *fractional key* and broadcasts it within its range. Several fractional keys will be combined into a single pairwise key, hence the name fractional. In order to protect the secrecy of the fractional keys, the anchors' messages are encrypted with a key globally shared by all nodes and all anchors. In addition, the authenticity of the messages and the protection against message replay must also be ensured. In the second step, every regular node broadcasts the key identifiers of the fractional keys that it hears. If two nodes that hear each other share more than T fractional keys, then they use those keys to generate a pairwise key. Finally, in the third step, every node uses its pairwise keys to securely unicast its local broadcast key to each neighbor.

Note that it can be possible that two nodes are close to each other but they hear less than T common anchors, and therefore they cannot establish a neighbor relationship. The probability P_{fail} of such an event can be computed as follows. Let us consider Figure 6.7. The two nodes are x and y, and their distance is d. The anchors heard by node x are located in a disk of radius R around x, and similarly for y. The common anchors are those in the intersection of the two disks, which is represented by the shaded area A_{cmn} in the figure. The probability that we are interested in is the probability that there are no more than T anchors in A_{cmn}:

$$P_{\text{fail}} = \sum_{k=0}^{T} \frac{(\lambda^* S_{\text{cmn}})^k}{k!} e^{-\lambda^* S_{\text{cmn}}}, \tag{6.6}$$

where S_{cmn} is the size of A_{cmn}.

Figure 6.8 shows P_{fail} as a function of the relative distance d/R between x and y. The different curves belong to different values of λ^*. T is set to $\lceil (1/3)\lambda^* R^2 \pi \rceil$, which means that the threshold number of common anchors required to establish a neighbor relationship is one third of the number of the anchors heard by the nodes on average. $d/R = 1$ means that the distance of the nodes equals the range R of the anchors. Recall, however, that the communication range of the regular nodes is smaller than R. Therefore, if the nodes hear each other directly, then their distance is smaller than R, and the probability of not being able to set up a neighbor relationship is small, as illustrated by Figure 6.8.

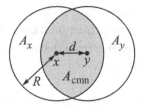

Figure 6.7. Two nodes x and y can become neighbors only if they hear each other and they hear more than a threshold number of common anchors. The anchors heard by node x are located on a disk A_x of radius R around x, and similarly for y. The common anchors are those in the intersection of the two disks, which is represented by the shaded area A_{cmn} in the figure. Used with permission, from [251], © IEEE, 2005.

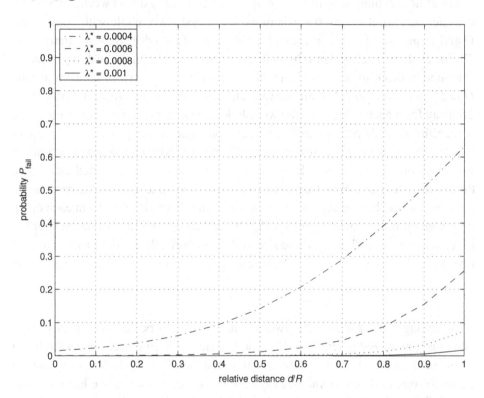

Figure 6.8. The probability P_{fail} of two nodes not being able to set up a neighbor relationship as a function of their relative distance d/R. P_{fail} is defined in (6.6). The different curves belong to different values of λ^*. T is set to $\lceil (1/3)\lambda^* R^2 \pi \rceil$.

The careful reader might have noticed that the local broadcast key establishment protocol presented above does not prevent wormhole attacks yet. Indeed, an attacker can tunnel the messages of some node and some anchors through a wormhole, and in this way, it can achieve that far away nodes hear each other and more than T

common anchors. In order to prevent this, each anchor also puts its location data in the message in which it broadcasts its fractional key. The nodes can use the location data in the messages that they receive from the anchors to detect wormhole attacks based on the following two principles.

(1) A node cannot hear two anchors that are $2R$ apart from each other because any anchor heard by a node must lie within a range of radius R around the node.
(2) A node cannot receive the same message twice from the same anchor, because the messages sent by the anchors are encrypted, and each anchor includes a one-time password in every message that it sends.

Let us now explain why these principles can be used to detect wormholes. First, consider part (a) of Figure 6.9, where we illustrated a wormhole with transceivers O and D, and a node x that is located in such a way that it directly hears transceiver D. The anchors directly heard by node x are those that lie in the disk A_x of radius R around x. In addition, x also hears the anchors in the disk A_O of radius R around O due to the wormhole. If there are two anchors in A_x and A_O that are further than $2R$ away from each other, then the wormhole is detected based on Principle 1. The probability P_1 of detection is not easy to compute but we can give a lower bound of it as follows. Consider the shaded areas A'_x and A'_O, which have a distance of $2R$ from each other. If there is at least one anchor in each of these shaded areas, then the attack is detected. Note that this event does not include all possible cases when there are two anchors in A_x and A_O that are further than $2R$ away from each other, thus, it yields only a lower bound on P_1. The probability that at least one anchor lies in A'_x is $1 - e^{-\lambda^* S'_x}$, where S'_x is the size of A'_x. Similarly, the probability that at least one anchor lies in A'_O is $1 - e^{-\lambda^* S'_O}$, where S'_O is the size of A'_O. Hence, we get that

$$P_1 \geq (1 - e^{-\lambda^* S'_x})(1 - e^{-\lambda^* S'_O}). \tag{6.7}$$

Assuming a fixed distance between x and O, it can be shown that this lower bound is maximized when $S'_x = S'_O$. The left side of Figure 6.10 shows the lower bound on P_1 when A'_x and A'_O are selected such that $S'_x = S'_O$ holds. The different curves belong to different values of λ^*. We can observe, on the one hand, that the probability of detection is very close to 1 when the distance between x and O is larger then $1.5\,R$. On the other hand, below this distance the detection probability drops abruptly.

However, when the distance between x and O is smaller than $2R$, we can use Principle 2. Consider part (b) of Figure 6.9. When x and O are closer than $2R$, the disks A_x and A_O overlap. If there is an anchor in the intersection A_{xO}, and the adversary transfers every message blindly from one end of the wormhole to the other end, then the message carrying the fractional key of that anchor is heard twice

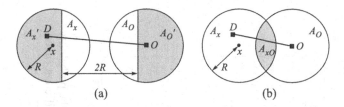

Figure 6.9. Part (a) of the figure illustrates a wormhole with transceivers O and D, and a node x that is located in such a way that it directly hears transceiver D. The anchors directly heard by node x are those that lie in the disk A_x of radius R around x. In addition, x also hears the anchors in the disk A_O of radius R around O due to the wormhole. The shaded areas A'_x and A'_O have a distance of $2R$ from each other. If there is at least one anchor in each of these shaded areas, then the attack is detected based on Principle 1. Part (b) illustrates the case when x and O are closer than $2R$, and the disks A_x and A_O are overlapping. If there is an anchor in the intersection A_{xO}, then the wormhole can be detected based on Principle 2. Used with permission, from [251], © IEEE, 2005.

by x: first directly and then from transceiver D who receives it from O through the wormhole. Thus, the wormhole can be detected based on Principle 2.

The probability P_2 of detection is equal to the probability that there is at least one anchor in A_{xO} that can be computed as follows

$$P_2 = 1 - e^{-\lambda^* S_{xO}}, \tag{6.8}$$

where S_{xO} is the size of A_{xO}. The right side of Figure 6.10 shows P_2 as a function of the relative distance d/R between x and O. The different curves belong to different values of λ^*. We can observe that the detection probability is close to 1 when the distance between x and O is not larger than 1.5 R.

As the wormhole can be detected based on any of the two principles, the overall detection probability is very close to 1 irrespective of the distance between x and O.

Wormhole detection with directional antennas

Let us assume that each node in the network is equipped with a directional antenna. Every antenna has n, non-overlapping zones, and each zone has a spanning angle of $2\pi/n$; hence the zones collectively cover the entire area around a node. When a node is idle, it listens to the carrier in omni-directional mode. When it receives a message, it determines the zone in which the received signal strength is maximal and uses that zone to communicate with the sender. An important assumption is that the orientation of the zones is always established with respect to the Earth's median, and therefore all nodes use the same orientation irrespective of their physical locations and their own orientations. This can be achieved in modern antennas with

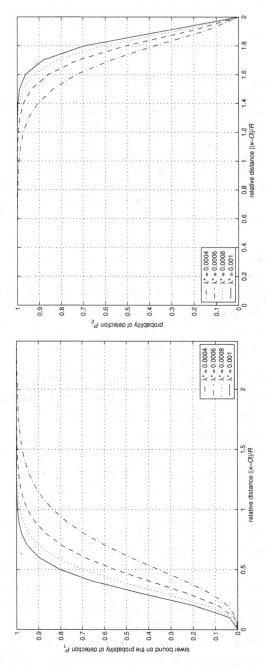

Figure 6.10. The lower bound on the probability P_1 of detecting a wormhole based on Principle 1 (left side) and the probability P_2 of detection based on Principle 2 (right side) as a function of the relative distance d/R between x and O. The different curves belong to different values of λ^*.

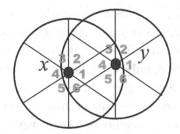

Figure 6.11. When two nodes are within each other's communication range, they must hear each other's transmission from opposite directions. For instance, if $n = 6$ and a node x hears a node y in zone 1, then y hears x in zone 4.

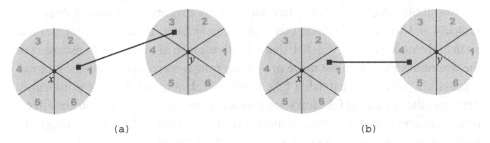

(a)   (b)

Figure 6.12. If two nodes communicate (unknowingly) through a wormhole, then they may or may not hear each other in opposite zones. Part (a) of the figure illustrates the case when they do not hear each other in opposite zones. Part (b) illustrates the case when they hear each other in opposite zones despite the presence of the wormhole. Used with permission, from [176], © 2004.

the help of a magnetic needle that always remains collinear to the Earth's magnetic field.

The main idea is that when two nodes are within each other's communication range, they must hear each other's transmission from opposite directions. For instance, if $n = 6$ and a node x hears a node y in zone 1, then y hears x in zone 4, as illustrated in Figure 6.11. In the following, we will denote the zone in which x hears y by Z_{xy}, and the opposite zone by $\bar{Z}_{xy}$. Hence, if x and y are within each other's range, then $Z_{yx} = \bar{Z}_{xy}$ holds. However, if nodes x and y communicate through a wormhole, then this condition is not always satisfied. This is illustrated in Figure 6.12(a), where node x hears node y (through the wormhole) in zone 1, but y hears x (through the wormhole) in zone 3.

Based on this observation, one could use the following neighbor discovery protocol for establishing neighbor relationships between the nodes. We assume that x and y already share a key. First, the initiator x of the neighbor discovery broadcasts a hello message that contains its identifier. This is done by sweeping through the zones and transmitting the hello message in every direction. When a node y receives a hello message in zone Z_{yx}, it sends a response to x in the same zone, where the response contains the identifiers of y and x, the zone identifier Z_{yx}, and a random number R. Apart from the identifier of y, the response is encrypted with a key shared by x and y, and hence, only x can decrypt it. When x receives a response in zone Z_{xy}, it decrypts it and verifies whether $Z_{xy} = \bar{Z}_{yx}$. If this equality holds, then it sends R back to y as a confirmation that the verification was successful. Two nodes consider each other neighbors only if they have successfully run this protocol.

Note that the above protocol does not always detect that x and y are communicating through a wormhole. In order to see this, consider Figure 6.12b, where node x hears node y (through the wormhole) in zone 1 and y hears x (through the wormhole) in zone 4. Thus, they successfully execute the neighbor discovery protocol and wrongly conclude that they are neighbors. Indeed (on average) one sixth of the node pairs (x, y), such that x is in the range of one of the transceivers of the wormhole and y is in the range of its other transceiver, can execute the protocol successfully and will establish a fake neighbor relationship. Therefore, though it decreases the number of fake links, the protocol does not really eliminate the effects of the wormhole, because even a single fake link can make the routes through the wormhole appear shorter than other routes in the network.

In order to overcome this problem, the nodes can cooperate and help each other to detect the wormhole. The idea is based on the observation that if two nodes x and y are real neighbors, then every node that both x and y can communicate with must be able to run the protocol successfully with both x and y. On the other hand, if x and y are not real neighbors, then there could be a node v that they both can communicate with (possibly via a wormhole), but v cannot run the neighbor discovery protocol successfully with either x or y. Thus, v can play a *verifier* role and help establish the legitimacy of the neighbor relationship between x and y.

There are certain conditions that must be met to be a valid verifier. First, we observe that if y hears v in the same zone in which it hears x (i.e., $Z_{yv} = Z_{yx}$), then y could hear both x and v through the wormhole (see Figure 6.13a). This means that x and v could be real neighbors, and therefore the wormhole cannot be detected using v as a verifier. Hence, we require that for a valid verifier $Z_{yv} \neq Z_{yx}$ holds. Moreover, we can also observe that even if $Z_{yv} \neq Z_{yx}$, if v hears x in the same zone in which y hears x (i.e., $Z_{vx} = Z_{yx}$), then they could both hear x through the wormhole's transceiver (see Figure 6.13b). If, in addition, x happens to hear the other transceiver of the wormhole in zone $\bar{Z}_{yx}$, then x can establish neighbor relationships with both y and v. Thus, we require that for a valid verifier $Z_{vx} \neq Z_{yx}$ holds too.

We extend the neighbor discovery protocol with the use of verifier nodes as follows. The first three steps of the verified neighbor discovery protocol are the

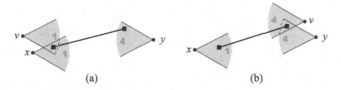

(a)   (b)

Figure 6.13. The conditions for being a valid verifier. Part (a) illustrates that if
node y hears v in the same zone in which it hears x, then y could hear both x and
v through the wormhole. Hence, we require that for a valid verifier $Z_{yv} \neq Z_{yx}$
holds. Part (b) illustrates that if v hears x in the same zone in which y hears x (i.e.,
$Z_{vx} = Z_{yx}$), then they could both hear x through the wormhole's transceiver. If,
in addition, x happens to hear the other transceiver of the wormhole in zone $\bar{Z}_{yx}$,
then x can establish neighbor relationships with both y and v. Thus, we require
that for a valid verifier $Z_{vx} \neq Z_{yx}$ holds too.

same as the steps of the simple protocol we described above. So let us assume
that x and y successfully ran the first three steps. Then, y broadcasts a verification
request in all of the zones except for Z_{yx} (as nodes in that zone cannot be verifiers in
any case due to the first condition of being a valid verifier). The verification request
includes the identifier of Z_{yx}, so prospective verifiers can check whether they satisfy
the second condition of being a valid verifier. Nodes that receive the verification
request and satisfy the conditions of valid verifiers respond with an encrypted
message. This message confirms that the verifier heard x in a zone different from
Z_{yx} and successfully ran the first three steps of the protocol with x (which means
that $Z_{xv} = \bar{Z}_{vx}$). Finally, if at least one verifier responds to y, then y accepts x
as a neighbor and sends a confirmation message to x, who can then accept y as a
neighbor.

Let us assume that v is a valid verifier. The first condition of being a valid verifier
(i.e., $Z_{yv} \neq Z_{yx}$) ensures that if y hears x through the wormhole, then it hears v
directly (as it hears it from another zone). In addition, the second condition (i.e.,
$Z_{vx} \neq Z_{yx}$) ensures that if y hears x through the wormhole, then x cannot run the
first three steps of the protocol successfully with both y and v. This would require
that $Z_{xy} = \bar{Z}_{yx}$ and $Z_{xv} = \bar{Z}_{vx}$ hold, which cannot be the case for the following
reasons: Because x is at the other end of the wormhole, it hears both y and v in the
same zone, so $Z_{xy} = Z_{xv}$. This means that if both $Z_{xy} = \bar{Z}_{yx}$ and $Z_{xv} = \bar{Z}_{vx}$ holds,
then $\bar{Z}_{yx} = \bar{Z}_{vx}$ should be true. But this is impossible owing to the second condition
that says that $Z_{vx} \neq Z_{yx}$. All this means that if y hears x through the wormhole,
then no valid verifier will respond to the verification query, and therefore x and y
will not become neighbors.

However, there is still a problem with this neighbor discovery protocol, which
manifests itself if x and y are just beyond the communication range (so they should
not be neighbors) but there is a valid verifier that they can both hear directly. Such

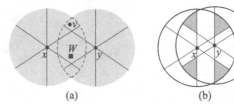

a situation is illustrated in Figure 6.14a. In this case, an adversary can place a repeater W between x and y and relay the messages of the neighbor discovery protocol between them. Node y can use v as a verifier, because v satisfies the conditions. In addition, v responds to y's verification request, because by being in the range of x, it could have run the first three steps of the protocol with x successfully. Thus, x and y will believe each other neighbors.

In order to prevent also this kind of attack, we must further strengthen the conditions for valid verifiers, and we must require that if Z_{yv} is adjacent to Z_{yx}, then Z_{xv} is not adjacent to Z_{xy}, and vice versa, if Z_{xv} is adjacent to Z_{xy}, then Z_{yv} is not adjacent to Z_{yx}. Figure 6.14b illustrates the region where valid verifiers are located when x and y are close to each other.

We must note that it can happen that two nodes are within each other's communication range, but there are no potential verifier nodes that they can use (i.e., the shaded areas in Figure 6.14b are empty). In this case, the nodes cannot set up a neighbor relationship and we lose a potential link. Clearly, the probability of losing a link between nodes x and y depends on the density of the network. Simulation results in [176] show that when the nodes have around 10 other nodes within the range of their directional antennas, around 58% of all potential links are lost, and 5.3% of the nodes become completely disconnected owing to the strict constraints on verifier nodes. Even for a more dense network, when nodes have on average 32 other nodes within the range of their directional antennas, around 40% of the potential links are still lost and 0.03% of the nodes become disconnected.

Losing links is not desirable, because it reduces the robustness of the network in case of link failures, and increases the average length of the routes in the network. Another disadvantage of this approach is that it requires the nodes to be equipped with directional antennas. This is an assumption that cannot be satisfied in many applications. Moreover, the protocol proposed above detects only a single wormhole; when there are several of them, then y and v can hear x through different wormholes, and the protocol can be executed successfully. We leave the construction of an example, as an exercise, for the reader.

6.3 Summary

Neighbor discovery is a basic mechanism that is essential to the operation of many wireless networks. In this chapter, we identified two attacks on neighbor discovery: (i) preventing the creation of a neighbor relationship between two nodes that are within each other's power range, and (ii) making two nodes that are not within each other's power range believe that they are neighbors. The first attack can be implemented by jamming. However, a large scale jamming adversary can be easily detected. The second attack can be implemented by setting up wormholes in the system. Wormhole attacks do not seem to be easily detectable at first sight, but in this chapter we described some techniques that can be used to detect them.

We classified wormhole detection techniques into two groups: centralized and decentralized approaches. In the centralized approach, data collected from the local neighborhood of every node are sent to a central entity. The central entity uses the received data to construct a model of the entire network and tries to detect inconsistencies in this model that are potential indicators of wormholes. In the decentralized approach, each node constructs a model of its own neighborhood using locally collected data; hence no central entity is needed.

We discussed two centralized wormhole detection mechanisms: one based on statistical hypothesis testing and another based on multidimensional scaling and visualization. Moreover, we discussed several decentralized mechanisms for wormhole detection, including techniques based on distance estimation, the availability of position information, and the use of directional antennas. We also discussed the advantages and the disadvantages of all these techniques.

6.4 To probe further

The centralized wormhole detection mechanism based on statistical hypothesis testing was proposed by Buttyán, Dóra, and Vajda in [72]. In their paper, more details about this method can be found, including simulation results that demonstrate its effectiveness.

The application of multidimensional scaling for wormhole detection was proposed by Wang and Bhargava in [369]. They described how this mechanism works in two dimensions. Multidimensional scaling itself is a technique that was originally developed in social sciences, but later it was also adopted to solve positioning problems in wireless networks (see e.g., [349]).

Packet leashes are proposed by Hu, Perrig, and Johnson in [180]. As we mentioned, geographical packet leashes rely on position information. Secure positioning is discussed in several papers, including [250, 88, 89].

Wormhole detection based on distance bounding and the MAD protocol are proposed by Čapkun, Buttyán, and Hubaux in [84]. The idea of distance bounding itself originates from Brands and Chaum [66], who developed it to prevent the *mafia fraud*, a sophisticated attack aimed at stealing money from innocent people using a fake ATM (Automated Teller Machine).

The wormhole detection technique based on position information of anchor nodes is proposed by Lazos *et al.* in [251]. In their paper, the anchors are called *guards*.

Directional antennas have been recognized as a powerful way of increasing the capacity and the connectivity of wireless networks. Their application for wormhole detection is proposed by Hu and Evens in [176].

6.5 Questions

(1) The main disadvantage of the statistical wormhole detection method described in Subsection 6.2.1 is that, although it detects the presence of a wormhole with high confidence, it does not locate it. How could the statistical approach be used to identify the nodes that are affected by the wormhole?

(2) Let us consider the TIK protocol described in Subsection 6.2.2. Let us assume that the maximum clock synchronization error is 180 ns and the maximum communication range of the nodes is 250 m. Compute the minimum packet length required by the TIK protocol in order for the TESLA condition to be satisfied.

(3) A disadvantage of the MAD protocol described in Subsection 6.2.2 is that it needs several rounds of rapid bit exchanges. Can you think of a way to perform distance bounding by a single exchange of multiple bit messages?

(4) What is the purpose of combining the next bit to be sent to the other party with the last received bit in the MAD protocol? Can you construct an attack against a modified version of the protocol where the bits are sent independently from the received bits?

(5) Consider the wormhole detection method based on directional antennas that we presented in Subsection 6.2.2. Try to construct an example with two wormholes that are not detected by the described method.

7

Secure routing in multi-hop wireless networks

As we have described in Chapter 2, some of the upcoming wireless networks use multi-hop wireless communications. In those networks, the nodes have two roles: they act as end-systems and they also perform routing functions. This means that routing control messages are sent over wireless channels. Moreover, because of the lack of their physical protection, some of the routers could be corrupted and not follow the routing protocol faithfully. This can have undesirable effects on the operation of the network. In extreme cases, the operation of the entire network can be disabled by attacking the routers and manipulating the messages of the routing protocol. This chapter is devoted to this problem. More precisely, we study the problem of securing the routing protocol in two kinds of multi-hop wireless networks: mobile ad hoc networks and wireless sensor networks.

7.1 Routing protocols for mobile ad hoc networks

A large amount of work on routing in mobile ad hoc networks has been carried out in the research community, which has resulted in a multitude of routing protocols. One way to classify ad hoc network routing protocols is illustrated in Figure 7.1. As we can see, there exist *topology-based* routing protocols and *position-based* routing protocols. Topology-based protocols are based on traditional routing concepts, such as maintaining routing tables or distributing link-state information, but they are adapted to the special requirements of mobile ad hoc networks. Position-based protocols use information about the physical locations of the nodes to route data packets to their destinations.

Topology based protocols can be further classified into two groups: *proactive* and *reactive* protocols. Proactive routing protocols try to maintain consistent, up-to-date routing information within the system so that at any time, every node knows how to route packets to all other nodes in the network. In contrast to this, in the case of reactive routing protocols, a route is established between a source and a

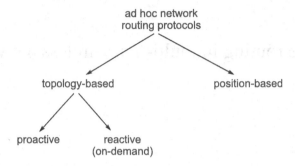

Figure 7.1. Classification of ad hoc network routing protocols. Topology-based protocols are based on traditional routing concepts, such as maintaining routing tables or distributing link-state information, but they are adapted to the special requirements of mobile ad hoc networks. Position-based protocols use information about the physical locations of the nodes to route data packets to their destinations. Topology-based protocols can be proactive or reactive. Proactive protocols try to maintain consistent, up-to-date routing information within the system. In contrast to this, reactive protocols establish a route between a source and a destination only when it is needed. For this reason, reactive protocols are also called on-demand protocols.

destination only when it is needed (i.e., when the source wants to send something to the destination). For this reason, reactive protocols are also called *on-demand* protocols.

Proactive protocols usually require periodic exchanges of routing information among the nodes. If only a few pairs of nodes communicate with each other, then most of the periodically exchanged information is useless (in the sense that it is never used), and hence proactive protocols can waste a lot of resources unnecessarily. But as routing information is always (nearly) up-to-date and available, packets can be sent to any destination virtually with no delay. In contrast to this, in the case of reactive or on-demand protocols, the nodes use their resources for setting up routes only when they are really needed. At the same time, it may well happen that when a node wants to communicate with another node, no working route to that other node is available, and the communication must be delayed until such a route is discovered. There exist some *hybrid* protocols that try to combine the advantages of the proactive and the reactive approaches. Typically, hybrid protocols use the proactive approach to maintain up-to-date routing information at each node regarding the node's local neighborhood (e.g., up to a certain number of hops), and they use the reactive approach when routes to far away destinations are needed.

As we mentioned above, position-based ad hoc network routing protocols use information about the physical locations of the nodes to route data packets to their destinations. In general, each node is aware of its own location (by means of GPS or some other positioning service) and obtains the location information of other nodes via a *location service* that is provided by the nodes themselves in a distributed manner. When sending a data packet, the source obtains the location of the destination from the location service, and it includes this information in the header of the packet. Then, each intermediate node makes routing decisions based on its own location and the location of the destination obtained from the packet header.

The advantage of position-based routing is that the nodes do not need to maintain routing information or to discover routes explicitly, and therefore the control overhead of these protocols tends to be smaller. However, there is still some overhead associated with the operation of the location service and with the retrieval of the location information of the destinations. The disadvantage of position-based routing is that they rely on additional hardware in each node or some other mechanisms by which the nodes can determine their own location. Another disadvantage is that position-based routing protocols must cope with voids (i.e., geographic areas where no node can be found), which somewhat complicates their operation.

Providing a comprehensive description of all existing ad hoc network routing protocols is obviously out of the scope of this book. Our goal is rather to highlight the basic operating principles of some mainstream routing protocols in order to allow the understanding of the security implications and the secure routing protocols presented later.

7.1.1 On-demand source routing

As an illustrative example of on-demand source routing protocols, we briefly describe the operation of the Dynamic Source Routing (DSR) protocol. DSR was among the very first routing protocols proposed for mobile ad hoc networks, and its design has been highly influential to other similar protocols proposed later. A detailed description of DSR can be found in [210].

DSR is a source routing protocol which means that every data packet carries the list of those nodes in its header that the packet should traverse in order to reach its destination. When a node receives such a data packet, it first verifies if it is the destination of the packet. If not, then the node verifies if its identifier is in the list carried by the packet, and if so, it forwards the packet to the next node in the list, which must be its direct neighbor (strict source routing). Otherwise the packet is dropped.

One main advantage of source routing is that it is trivial to detect routing loops just by identifying repeating values in the list of node identifiers in the packet header. Another advantage is that forwarding nodes do not need to maintain up-to-date routing information in order to be able to forward the packet towards the destination, because that information is available directly from the packet header. Finally, a third advantage of source routing is that every node that receives or overhears a packet can learn routing information from the packet header and cache it locally for future use. The disadvantage of source routing is the communication overhead resulting from carrying the whole route in the packet header, which limits the applicability of this approach in highly resource constrained environments, such as sensor networks, and in large networks where routes can be very long.

When using source routing, the source of a packet must know a full route to the destination before actually sending the packet. In DSR, such a route can be available to the source from its local route cache. If an appropriate route is not found in the cache, then the source uses the *route discovery* mechanism of DSR to dynamically discover a route to the destination. In addition to route discovery, DSR has a *route maintenance* mechanism that allows the source to detect if a route that it is trying to use is broken.

Basic DSR route discovery

DSR route discovery is based on flooding the entire network with a route request and returning some route replies. A route request message contains the identifiers of the source and the destination, and a record listing the identifiers of every intermediate node that forwarded this particular request message. Each request also has a request identifier, which, together with the identifier of the source, uniquely identifies the request and allows the intermediate nodes to detect and discard duplicates.

The source generates a route request message with a new request identifier and an empty list of forwarding nodes and broadcasts it to its neighbors. Each intermediate node that receives a copy of the request verifies that it has not received that request before. If the request has already been received, then it is dropped. Otherwise, the intermediate node appends its identifier to the list of identifiers in the request and re-broadcasts the request to its neighbors. This procedure is repeated until the request reaches the destination.

The destination generates a route reply by copying the recorded list of identifiers from the route request into the route reply. The route reply is then unicast back to the source. For this, the destination needs a route to the source. It could have such a route already in its route cache. Otherwise, if bi-directional links can be assumed, then the destination can obtain such a route by reversing the list of identifiers received in the route request. If links cannot be assumed to be bi-directional, then the destination must use DSR route discovery to obtain a route to the source; but in this case, it piggybacks the route reply on the route request in order to avoid an infinite recursion. When the source receives the route reply, it extracts the route from it and caches it locally in its route cache. Then it uses the route to send the data packets that were buffered during the execution of the route discovery procedure.

Basic DSR route maintenance

DSR requires each intermediate node to make sure that the data packet that it is forwarding reaches the next hop. This requirement can be satisfied in several ways. Firstly, the data link layer protocol can provide an acknowledgement for each delivered data packet. Secondly, the intermediate node can overhear the transmission of the packet by the next intermediate node, which serves as a passive acknowledgement. Finally, if none of these options is available, then the intermediate node can request the next intermediate node to send an acknowledgement packet (which can take a multi-hop route if the link between the two nodes is not bi-directional).

If no acknowledgement arrives for a given packet, then the intermediate node tries to re-transmit it for some time. If all attempts are unsuccessful, then the intermediate node generates a route error message, indicating that the link to the next intermediate node is not functioning, and sends this error back towards the source of the packet. The source and each of the intermediate nodes that forward the error invalidate the routes that contain this broken link in their route caches. Then, the source can try to send the data packet via an alternative route if it has some in its cache, otherwise it initiates a new route discovery.

DSR optimizations

The basic operation of DSR, as described above, can be extended with many optimizations to further improve the performance of the protocol. Such optimizations include: the caching of overheard routing information; replying to route requests by intermediate nodes using their cached routes; effectively expanding the local cache with the caches of neighboring nodes by sending a non-propagating route request (with hop limit equal to 0) and allowing neighboring nodes to reply from their caches; packet salvaging; automatic route shortening; increased spreading of route error messages to reduce the number of invalid route reply packets generated by intermediate nodes that are not aware of a broken link; and caching negative information (e.g., a broken link) in the route caches. We must note, however, that such optimizations often greatly increase the complexity of the protocol, and thus make it more vulnerable to different attacks. For this reason, secure routing protocols that are based on the same principles as DSR usually do not apply them.

7.1.2 On-demand distance vector routing

Another group of on-demand protocols do not use source routes in packets, but make routing decisions based on traditional routing tables. However, those tables are updated only in an on-demand manner. In particular, routing information for destinations that are not in active communications is not maintained. A well-known example of such an on-demand, routing table based protocol is the Ad hoc On-demand Distance Vector (AODV) protocol. A detailed description of AODV can be found in [312]; here we give only a brief overview.

In AODV, each node maintains a routing table where each entry of the table contains information related to a particular destination, including the following: the identifier of the destination, the number of hops needed to reach that destination, the identifier of the next hop on the route towards the destination, the list of precursor nodes that can forward packets to the destination via the node maintaining this routing table, and a destination sequence number (which helps to identify and discard out-of-date routing information and ensures the loop-freedom of the protocol). When an intermediate node receives a packet to be forwarded to a given destination, it looks into its routing table to see who is the next hop towards that destination and then forwards the packet to that next hop node. This procedure is repeated until the packet reaches its destination.

Obviously, this works only if the routing tables of the source and the intermediate nodes contain a valid entry for the destination of the packet. In AODV, this is ensured through a route discovery procedure similar to that of the DSR protocol. In other words, a route request is flooded in the network, and a route reply is sent back to the source; the difference is that instead of updating route caches, here the nodes update routing tables upon processing route request and route reply messages.

AODV route discovery

When a source wants to send a data packet to a destination, and it does not have a valid entry for that destination in its routing table, then it generates and broadcasts a route request message. This route request contains the identifiers of the source and the destination, a hop count, and two sequence numbers, the first of which is the current sequence number of the source, and the second is the last known sequence number of the destination. Each node has a single sequence number, which is incremented after each detected change in the node's neighbor set. The route request also contains a broadcast identifier, which plays a role similar to the role of the request identifier in DSR (i.e., it helps intermediate nodes to detect and discard duplicates of the same request).

When an intermediate node receives a route request, it first determines if it is a duplicate or not. Duplicates are silently discarded. If the request is not a duplicate, then the node checks if it has a valid entry in its routing table for the destination indicated in the request. If it does not have a valid entry, or it has a valid entry with a sequence number smaller than the destination sequence number in the request, then the node re-broadcasts the request after incrementing the hop count in it. On the other hand, if the intermediate node does have a valid entry for the destination with a sequence number at least as large as the destination sequence number in the request, then it generates a route reply. Obviously, if the request reaches the destination, then it will also generate a route reply.

Besides the processing described in the previous paragraph, upon receipt of a route request message, an intermediate node creates or updates the entry in its routing table that corresponds to the *source* of the request. In fact, if such an entry already exists, it is updated only if its sequence number is smaller than the sequence number of the source received in the request, or if the two sequence numbers are equal, but the length of the new route indicated by the hop count in the request is smaller. When an entry is created or updated, the destination identifier of the entry is set to the identifier of the source of the request, the length of the route in the entry is set to the hop count in the request, the next hop is set to the identifier of the node from which the request was received, and the sequence number of the entry is set to the sequence number of the source in the request. This entry will be needed, if eventually the intermediate node receives a route reply that should be forwarded back to the source.

As we mentioned above, a route reply can be generated by either the destination or an intermediate node that has a valid entry in its routing table for the destination. The route reply contains a destination sequence number and a hop count. If the reply is generated by the destination, then the sequence number in the reply is set to the current sequence number of the destination and the hop count is set to zero. If the reply is generated by an intermediate node, then the sequence number and the hop count in the reply are set to the sequence number and the hop count in the entry that corresponds to the destination in the routing table of the intermediate node. In addition, the intermediate node generates a so-called gratuitous route reply, which it sends to the destination. This route reply message will set up the necessary state in the routing tables of the intermediate nodes between the intermediate node that generated the reply and the destination, so that these nodes will be able to forward data packets from the destination back to the source.

The route reply message intended for the source is then forwarded back on the reverse path taken by the route request. The processing of the route reply by the intermediate nodes is very similar to the processing of the route request. In particular, the hop count in the reply is incremented before the reply is passed on. Moreover, each intermediate node creates or updates the routing table entry corresponding to the destination by setting the hop count in the entry to the hop count in the reply, the next hop field in the entry to the node from which it received the reply, and the sequence number in the entry to the destination sequence number in the reply. In addition, the precursor list of the entry corresponding to the destination is extended with the node to which the route reply is forwarded, and the precursor list of the entry corresponding to the source is extended with the node from which the route reply was received. These updates are made only if the destination sequence number in the reply is greater than the sequence number in the entry corresponding to the destination, or if the two sequence numbers are the same, but the length of the route indicated by the hop count in the reply is smaller than that currently stored in the entry.

AODV route maintenance

AODV also has a route maintenance mechanism that uses route error messages such as the ones used in DSR. When a node detects a broken link to the next hop while attempting to forward a data packet, it invalidates the routing table entries corresponding to those destinations that were reachable through this failed next hop. Then it generates a route error message that contains the list of those destinations that became unreachable and sends it to the nodes in the precursor lists of the invalidated entries. A node receiving a route error verifies if it uses the sender of the message as the next hop towards the destinations listed as unreachable in the error message. If this is the case for some destinations, then the node invalidates the corresponding routing table entries and sends a similar error message to the nodes in the precursor lists of the invalidated entries.

7.1.3 Proactive routing

Proactive routing protocols maintain up-to-date routing information for all possible destinations in the network. These protocols are usually based on a periodic exchange of routing information, and they have two types: link-state protocols and distance vector based protocols.

In *link-state* protocols, each node periodically floods the network with a message that contains the state of the links of that node. As these messages are propagated in the entire network, each node learns the link-state information of every other node, and thus each node has a full view of the network topology. Then, centralized shortest path algorithms can be used locally by each node to determine the best route to all other nodes in the network.

In contrast to this, in *distance vector* based protocols, the nodes execute a distributed shortest path algorithm to determine the best route to every other node in the network. For this purpose, each node periodically sends its current routing table to the neighboring nodes. Thus, each node obtains the routing information known by its neighbors. By inspecting the routing tables of its neighbors, a node can discover that there is a better route to some destination than the route that has been known so far by the node. In this case, the node updates its routing table to incorporate the new information. By repeating the routing table exchange and routing table update steps, the system converges to a stable state, where each routing table contains correct routing information.

At first sight, it seems that because of their periodic nature, proactive routing protocols have too much overhead to be applicable in mobile ad hoc networks. It turns out, however, that some optimized versions of them can work pretty well under certain circumstances. In particular, if a large and frequently changing set of random pairs are communicating, then maintaining routes to all possible destinations is not so much of an overhead anymore, and the proactive approach can even outperform the reactive one.

OLSR

An example of a proactive routing protocol proposed for mobile ad hoc networks is the Optimized Link State Routing (OLSR) protocol. Its full description can be found in [103]; here we summarize only its main characteristics.

OLSR is a link-state protocol and, as such, it periodically floods the entire network with control messages containing link-state information. However, OLSR minimizes the control overhead induced by flooding by using only selected nodes, called *multipoint relays* (MPRs), to retransmit control messages. The set of MPRs of a given node is a subset of its neighbors that are selected in such a way that they cover (in terms of radio range) all strict two-hop neighbors (i.e., those nodes that can be reached in two hops and that are not neighbors themselves) of the node. The nice thing about MPRs is that requiring only them to participate in the flooding significantly reduces the number of retransmissions of a given control message, while the way they are selected ensures that all nodes in the network will receive the message.

A second optimization used by OLSR is that it floods only partial link-state information. Indeed, it can be shown that in order to compute the shortest paths between any pair of nodes, it is sufficient that each MPR declares only the links to its MPR selectors (i.e., those neighbors that selected it as MPR).

There are two basic types of messages in OLSR: HELLO messages and TC (topology control) messages. HELLO messages are local broadcast messages that are received by the neighbors of the sender, but they are not retransmitted. TC messages are global broadcast messages that are flooded in the network by MPR nodes.

The HELLO message sent by a given node A contains the list of its believed neighbors. For each neighbor in the list, the state of the link to that neighbor is indicated. In addition, the neighbors that are selected as MPRs by A are marked as such.

When a node B receives such a HELLO message, it learns a whole lot of information from it. First of all, it learns that A is its neighbor (if it has not known that yet). If *B* is listed as a neighbor in the HELLO message, then *B* learns that *A* considers it as a neighbor, and thus there must be a symmetric link between them; otherwise the link is asymmetric because *B* hears *A*, but not vice versa. Assuming that *B* is listed in the HELLO message, *B* learns the state of the link between *A* and *B*. If, in addition, *B* is marked as an MPR, then now it knows that *A* selected it as MPR, and hence, *A* is in *B*'s MPR selector set. Finally, by looking at the list of neighbors of *A*, *B* learns about its two-hop neighborhood. To summarize, HELLO messages in OLSR are used for link-state sensing, neighbor detection, two-hop neighbor detection, and MPR signaling. Indirectly, HELLO messages are also used in MPR selection, because the nodes can determine their MPR set based on the neighborhood and two-hop neighborhood information that they obtained from the HELLO messages.

TC messages are sent in the network to advertise links. They contain a list of advertised neighbors. Only MPR nodes send and retransmit TC messages, and a TC message must contain at least those neighbors that have selected the sender node as an MPR. Based on the advertised links in TC messages, each node in the network reconstructs a (partial) topology of the network and builds a routing table that contains forwarding information for all possible destinations in the network. This routing table is then used for routing data packets towards their destinations.

DSDV

Another proactive ad hoc network routing protocol is the Destination-Sequenced Distance Vector (DSDV) protocol [313]. The main novelty of DSDV with respect to other distance vector based protocols is the application of sequence numbers that prevents routing loops.

DSDV is a predecessor of AODV, and thus, it has a similar sequence number mechanism. In DSDV too, each entry of a routing table is tagged with the most recent sequence number known for the destination to which the entry belongs. Similarly, periodic routing updates also contain sequence numbers for each destination in the update. When a node receives a routing update, for each destination in the update, the node prefers the newly advertised route if the sequence number in the update is greater than the sequence number known by the node for that destination, or if the two sequence numbers are equal, but the routing metric in the update indicates that the newly advertised route is shorter than the one known by the node. If none of these conditions is satisfied, then the update for the given destination is ignored. Like in AODV, sequence numbers are increased, when a change in the state of a link of the node is detected.

Another optimization in DSDV is that besides full updates listing all destinations, a node can also send *incremental updates* that list only destinations for which the route has changed since the last full update sent by the given node.

7.1.4 Position-based routing

In position-based routing protocols, the source of a data packet includes in the packet header the location of the destination. This information is used by inter-mediate nodes to route the packet towards the destination. Based on the packet forwarding strategy used by the intermediate nodes, three types of approaches can be distinguished: greedy forwarding, restricted directional flooding, and hierarchical protocols.

Greedy forwarding

In greedy forwarding, it is assumed that each node is aware of its own location and the locations of its neighboring nodes. The former is obtained by means of GPS or some GPS-free localization service. The location information of the neighbors can be learnt by using periodic, local, one-hop broadcast messages, called beacons, in which each node announces its own location. In addition, recall that each data packet carries the approximate position of its destination.

Greedy forwarding means that, upon receipt of a data packet by an intermediate node, the packet is forwarded to a neighbor that is closer to the destination than the forwarding node itself. However, there can be several such neighbors, and there are different strategies to choose the next hop from them. These strategies are illustrated in Figure 7.2 and explained below.

One intuitively appealing strategy is to forward the packet to the neighbor that makes the largest progress towards the destination (i.e., the node which is the closest to the destination). This strategy is called MFR (Most Forward within Radius) [362], and it tries to minimize the number of hops taken by the packet on its way to the destination.

Another, less intuitive strategy is to forward the packet to the nearest neighbor that is still closer to the destination than the forwarding node. This strategy is called NFP (Nearest with Forward Progress) [171]. NFP only makes sense when the nodes can control their transmission power; in this case, NFP minimizes the probability of packet collisions, and thus, the average progress of packets can be higher for NFP than for MFR.

Yet another strategy is to forward the packet to the neighbor that is the closest to the straight line between the forwarding node and the destination. This strategy is called compass routing [235], and it tries to minimize the spatial distance that the packet travels.

Finally, the forwarding node could select the next hop randomly from the set of neighbors that are closer to the destination than the forwarding node itself [287]. This strategy minimizes the number of operations required to forward the packet,

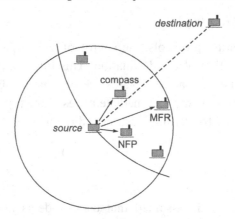

Figure 7.2. Illustration of the operation of greedy forwarding strategies. MFR (Most Forward within Radius) selects the neighbor that is the closest to the destination and, in this way, minimizes the number of hops taken by the packet. NFP (Nearest with Forward Progress) selects the nearest neighbor that is still closer to the destination; when the nodes can control their transmission power, this strategy can minimize the probability of packet collisions. Compass routing selects the neighbor that is the closest to the straight line between the forwarding node and the destination, and it minimizes the spatial distance that the packet travels. Finally, the next hop can be selected randomly; this can be a good strategy if the location information of the neighbors is inaccurate. Used with permission, from [269], © IEEE, 2001.

and it can be advantageous when the location information of the neighbors is inaccurate.

No matter what forwarding strategy is used, routing protocols based on greedy forwarding must cope with the problem of dead-ends: it can happen that an intermediate node receiving the data packet has no neighbor that is closer to the destination than the node itself, therefore it cannot pass on the packet. To recover from this situation, the protocol can try to construct a planar sub-graph of the graph that represents the ad hoc network and then use a planar-graph traversal algorithm to find a path to the destination. An example of this approach is the *face routing* algorithm [62] which has many variants including GFG (Greedy-Face-Greedy) [63], GPSR (Greedy Perimeter Stateless Routing) [218], and the GOAFR+ (Greedy Other Adaptive Face Routing) family of algorithms [240, 241].

Restricted directional flooding

The idea of restricted directional flooding algorithms is that an intermediate node re-broadcasts a data packet only if it lies "in the direction of the destination." In order for the intermediate node to decide if it lies in the good direction, it is sufficient to know (besides its own location) the location of the destination and the

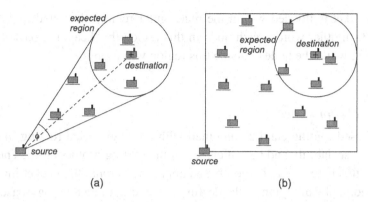

(a) (b)

Figure 7.3. Illustration of the operation of the restricted directional flooding algorithms DREAM (a) and LAR (b). In DREAM, first an expected region of the destination is calculated. Then, the direction to the destination is defined by the line between the forwarding node and the center of the destination's expected region, and the angle ϕ. Each neighbor of the forwarding node that lies within this angle must re-broadcast the packet. These calculations are repeated by each intermediate node that receives the packet until it reaches the destination. In LAR, the source of the data packet calculates an expected region of the destination, and then the packet is flooded within the rectangular region illustrated in the figure. Used with permission, from [269], © IEEE, 2001.

location of the previous intermediate node. This information can be included in the packet header by the source and the previous intermediate node, respectively, thus, there is no need for periodic beacons to learn the location information of all neighbors.

One approach to determine if a node lies in the direction of the destination is called DREAM (Distance Routing Effect Algorithm for Mobility) [42], and it is illustrated in part (a) of Figure 7.3. DREAM first calculates an expected region of the destination of the packet. This is a disk, the center of which is the approximate location of the destination obtained from the packet header, and the radius is a function of the maximum speed of the nodes and the time elapsed since the generation of the packet by the source (known from a timestamp in the packet header). Then, the direction to the destination is defined by the line between the forwarding node and the center of the destination's expected region, and the angle ϕ. Each neighbor of the forwarding node that lies within this angle is in the good direction, and must re-broadcast the packet. Note that ϕ increases as the packet gets closer to the destination, which helps to cope with the problem that the location information of the destination inserted in the packet header by the source becomes less accurate as time elapses (assuming that the nodes move).

Another approach is called LAR (Location Aided Routing) [223]. In LAR the source of the data packet calculates an expected region of the destination, and then the packet is flooded within the rectangular region illustrated in part (b) of Figure 7.3. In other words, each node in this region that hears the packet will re-broadcast it, while the nodes outside this region will drop it.

Hierarchical approaches

Position-based routing can be combined with topology-based routing in order to increase the scalability and the efficiency of the routing protocol [56]. The typical way to do this is to use a position-based approach to route the packet into the es-timated geographical region of the destination, and to use a topology-based proto-col locally within that region to actually find the destination. This hybrid approach counters the disadvantages of the pure position-based and the pure topology-based approaches: the scalability problems of topology-based protocols are mitigated by using them only locally, within a limited geographical area, whereas the prob-lem of inaccurate location information in position-based routing is mitigated by using a topology-based protocol to finally route the packet to the destination. An-other advantage of hierarchical approaches is that they can be used in applica-tions where not every node can determine its own location: those location unaware nodes can route packets to location aware nodes, so-called location proxies, by using topology-based routing.

Location services

As we have seen above, in position-based routing protocols, the source of a data packet must be able to obtain the location information of the destination. This in-formation is usually provided by a location service. Although some position-based routing protocols specify it explicitly, the location service does not actually need to be part of the routing protocol, but it can be provided as a separate mechanism that is used by the routing protocol as an external service. Because of its rather loose connection to routing itself, we do not describe the operation of location services here.

7.2 Attacks on ad hoc network routing protocols

Routing is a fundamental service in any kind of network; hence an ideal target for attacks. In this section, we describe why and how ad hoc network routing proto-cols can be attacked. The general discussion will be followed by some illustrative examples.

7.2.1 Setting the scene

Before indulging into the attacks themselves, it is useful to elaborate on the general objectives and capabilities of an adversary in the context of ad hoc network routing protocols. That is what we do in this subsection.

General objectives of attacking routing

Attacks against routing protocols can have the following three objectives:

- increase adversarial control over the communications between some nodes;
- degrade the quality of the service provided by the network;
- increase the resource consumption of some nodes (e.g., CPU, memory, or energy).

We must note that these objectives are not fully independent from each other. Firstly, the ultimate goal of increasing the resource consumption of some nodes is typically to degrade the quality of the service provided by the network. For instance, the result of overloading some nodes with excessive traffic could be that they start dropping legitimate traffic, or they deplete their batteries and the network becomes disconnected and dysfunctional. Nevertheless, it makes sense to distinguish these two objectives, because some attacks, as we will see, aim at increasing resource consumption in a brute force manner without any particular projection on how this would degrade the quality of service; it is intuitively simpler to think of these attacks as pure resource-consumption attacks. In addition, there are attacks that degrade the quality of service without actually increasing the resource consumption of the nodes. Secondly, the ultimate goal of increasing adversarial control over the communications between some nodes can also be the degradation of the quality of service, although this is not always the case.

We must also note that achieving any of the three objectives listed above can result in achieving another objective as a side effect. For instance, one way to achieve increased adversarial control over the communications between some nodes is to divert some traffic through particular routes, but in this case, the resource consumption of the nodes on those routes is also increased, because those nodes must now support some extra traffic.

Outsider vs. insider adversary

The adversary can interfere with the routing protocol in two ways: from outside and from inside. An *outsider* adversary can attack the communications of some nodes, which is made easy by the usage of wireless channels. This usually means eavesdropping, jamming, and injecting fabricated or replayed messages into the network. In addition to all these, an *insider* adversary controls some nodes in the network. We refer to these nodes as adversarial nodes. Adversarial nodes can exhibit any Byzantine behavior.

It is quite reasonable to assume that ad hoc networks are subject to *both* outsider and insider attacks. First of all, the nodes are usually not physically protected, so they can be captured and compromised by the adversary. In addition, in civilian scenarios, the adversary can acquire some nodes in a legitimate way (e.g., she can buy them). Indeed, in civilian scenarios, the term adversary refers to a set of misbehaving users that are otherwise legitimate. Hence, in the following section, we assume that the adversary is part of the network and that she launches attacks from adversarial nodes.

Attack mechanisms

An attack against routing is a specific combination of some attacking mechanisms aiming at achieving one or more of the objectives introduced above. Those attacking mechanisms include classical mechanisms such as eavesdropping, replaying, modifying and deleting control packets (i.e., packets containing routing information). In addition, an adversary can try to fabricate control packets containing fake routing information, or it can create control packets under a fake identity: the former is called packet forgery, and the latter is usually referred to as spoofing.

Data packets can also be eavesdropped, replayed, or modified, but these misdeeds are typically not considered to be routing security issues; they must be prevented or detected at higher (end-to-end) or lower (hop-by-hop) layers. Dropping data packets maliciously, however, is considered to be an attacking tool, which targets the packet forwarding function of routing.

Besides the classical attack mechanisms, there are mechanisms that can even be useful in some applications, but they can be misused in the context of routing to implement various attacks. Such a mechanism is *tunneling*. Tunneling means that two potentially remote adversarial nodes pass control packets back and forth between each other by encapsulating them into normal data packets and using the multi-hop routing service offered by the network to transfer those data packets. Recall that in our terminology, a tunnel is not a wormhole (see also Chapter 6 on wormholes), although both have similar effects on routing. However, a wormhole operates in the physical layer, and it does not require the adversary to control nodes in the network; a wormhole can be implemented with two simple radio transceivers connected through an out-of-band channel. By definition, tunneling means that an adversary can send data packets between its nodes, therefore, these nodes must be addressable, and hence, present at the routing layer.

7.2.2 Types of attacks

Using the attack mechanisms mentioned in the previous subsection, the adversary can mount the following types of attacks against routing protocols:

- route disruption;
- route diversion;
- creation of incorrect routing state;
- generation of extra control traffic; and
- creation of a gray hole.

Below, we describe each of these attacks in more details.

Route disruption

In a route disruption attack, the adversary prevents a route from being discovered between two nodes that are otherwise connected. In other words, there exists a route between the two victim nodes, but due to the adversary, the routing protocol is unable to discover it. The primary objective of this attack is to degrade the quality of service provided by the network. In particular, the two victims cannot communicate, and other nodes can also suffer and be coerced to use suboptimal routes.

There are various ways to implement a route disruption attack. In the case of topology-based routing protocols, if the adversary controls a set of nodes that form a vertex cut in the network, then it is fairly easy to prevent the discovery of any routes between the two parts of the network by dropping all control packets sent from one part into the other. Another way to mount a route disruption attack is to forge error messages that would invalidate the correct routing state in some victim nodes, thereby effectively preventing them from being able to communicate with some other nodes.

A subtle way of implementing a route disruption attack against some on-demand routing protocols is to combine tunneling and the deletion of control packets. Let us assume, for instance, that the protocol works in the following way. When a source wants to discover a route to the destination, it floods the network with a route request. When the destination receives the first copy of the request, it does not respond immediately, but it waits until it receives a copy of the request from each of its neighbors. Then, it selects the neighbor with the best routing metric (e.g., the neighbor from which it received the request with the smallest hop count or shortest list of identifiers), and sends a single route reply back to the source through that neighbor. In this case, the adversary can set up a tunnel somewhere between the source and the destination and make the route through this tunnel appear to the destination as the route with the best routing metric.[1] Subsequently, when the single route reply passes through the tunnel on its way back to the source, the adversary drops it. In this way, the source will never be able to discover a route to the destination.

In the case of on-demand protocols that drop duplicates of a given route request and use predictable request identifiers (e.g., a sequence number), route disruption can be achieved by predicting the next request identifier of a victim node and flooding the network with a spoofed route request containing that identifier. As a result, when the victim wants to discover a new route, its route request will be perceived as a duplicate and dropped even by legitimate nodes. This can prevent the discovery of any routes by the victim.

Similarly, link-state protocols also use duplicate detection to control flooding, which can be exploited in a route disruption attack. For instance, an adversarial node that re-broadcasts the link-state update of a victim node can change the state of all links of the victim to asymmetric, pointing from the victim to its neighbors. As a result, some nodes that receive this modified link-state update message will not be able to find a route to the victim. In addition, even if these nodes also receive the correct link-state update on an alternative path later, they will drop it as a duplicate.

In position-based routing protocols, there is no explicit route discovery, but the adversary can still prevent a source from being able to communicate with a destination by falsifying the location information of the destination. This can be achieved by spoofing location update messages and coercing the location service to respond to location queries with false information, or it can be achieved by forging or modifying a response from the location service.

Route diversion

In a route diversion attack, the adversary does not prevent the establishment of routes, but it achieves that some established routes are diverted. This means that, owing to the presence of the adversary, the protocol establishes routes that are different from those that it would establish, if the adversary did not interfere with the execution of the protocol.

The objective of route diversion can be to increase adversarial control over the communications between some victim nodes. In this case, the adversary tries to

---

[1] Note that a tunnel appears to be a single link in the network topology for the routing protocol. In particular, the intermediate nodes that implement the tunnel do not increase the hop count (in distance vector based protocols) or extend the identifier list (in source routing protocols) in the route request, as it is encapsulated into a regular data packet, and handled as such by the intermediate nodes. Therefore, route request messages that passed through the tunnel carry better metrics than those that traveled in the normal way.

achieve that the diverted routes contain one of the nodes that it controls or a link that it can observe. Then, the adversary can eavesdrop or modify data sent between the victim nodes easier. A particularly efficient way to divert routes through nodes under adversarial control is to set up a tunnel. As routes through the tunnel appear to be shorter, many pairs of communicating nodes can choose the tunneled routes, thereby allowing the adversary to access their communications easier.

Another objective of route diversion can be to increase the resource consumption of some nodes. For instance, if routes are diverted through a tunnel, as described above, then the nodes close to the two ends of the tunnel will receive a higher amount of transit traffic, and so they must use more resources to forward that traffic. Alternatively, by modifying or forging routing messages, the adversary can divert many routes directly through a victim node.

Finally, route diversion can aim at increasing the length of the discovered routes, and thereby, increasing the end-to-end delay between some nodes, which can be viewed as a degradation of quality of service.

In topology-based routing protocols, route diversion can be implemented by forging or manipulating control packets. For instance, in source routing protocols, the adversary can change the list of identifiers in route reply messages. In distance vector based protocols, the adversary can decrease hop count values in control messages, whereas in link-state protocols, she can change the state of a link in a link-state update message from symmetric to asymmetric. All these modifications will result in the establishment of diverted routes by some nodes.

Even the simple dropping of control packets can cause the diversion of routes. For instance, assume that there are only two routes between a source and a destination, a long one and a short one, and the latter goes through an adversarial node. Furthermore, assume that an on-demand routing protocol is used. Then, by simply not re-broadcasting the route request, the adversary can prevent the discovery of the shorter route and divert the traffic between the source and the destination through the longer one.

In position-based routing protocols, an adversary can lie about its position and appear to be closer to a given destination than other nearby nodes. In this way, some routes can be diverted, and the adversary can increase its control over the communications of some nodes. Larger detours can be created by pretending that the data packet encountered a dead-end and forcing the protocol to enter into recovery mode. Besides diverting the packet, this will also lead to increased resource consumption associated with the execution of the recovery algorithm.

Creation of incorrect routing state

Another type of attack aims at jeopardizing the routing state in some nodes so that the state appears to be correct but, in fact, it is not, and thus, data packets routed using that state will never reach their destinations. One example of this attack is when the route discovery procedure of a source routing protocol returns a non-existent route to the source; as a consequence, data packets using this non-existent route will be dropped when they reach the first non-existent link in the route. Another example is the creation of a routing loop. In this case, some packets will be forwarded in a cycle until their hop count reaches the maximum allowed value, and at the end, they are discarded. Distance vector based protocols are particularly vulnerable to this kind of attack, because the nodes do not have a full view of the whole network (unlike in case of link-state protocols) or of the entire route (unlike in the case of source routing protocols). Yet another example is when the network is disconnected, but the routing state in some nodes falsely indicates that each destination is reachable. Again, data packets are starting to be forwarded, but they are eventually dropped, because some destinations are indeed unreachable.

We must note that routing information can become incorrect due to mobility even if no adversary tries to interfere with the protocol. For instance, links can be broken, which results in source routes becoming non-functional. All protocols proposed for mobile ad hoc networks have a mechanism to cope with this situation (usually based on sending route error messages), meaning that some inherent protection against the attack consisting of creating incorrect routing state already exists in those protocols. The problem is that these mechanisms are not designed with malicious attacks in mind, but they assume random errors due to mobility and node failures. In particular, the reaction time of these mechanisms is quite poor, and a considerable amount of resources could be wasted before the attack is detected.

Hence, one objective of creating incorrect routing state is exactly to increase the resource consumption of some nodes: the victims will use their incorrect state to forward data packets, until they learn that something goes wrong. Obviously, another objective is to degrade the quality of service.

Incorrect routing state can be created by spoofing, forging, modifying, or dropping control packets. For instance, in source routing protocols, an adversarial node can simply overwrite the list of identifiers accumulated in route request messages, returned in route reply messages, or included in the header of data packets. In distance vector based protocols, routing loops can be created by manipulating or forging routing messages in such a way that the resulting routing tables of some nodes contain a loop. In link-state protocols, a link-state update message can be modified so that a non-functional link appears to be functional for some nodes, who can then select routes that contain that link. Position-based routing protocols seem to be more resistant to this kind of attack, because intermediate nodes do not store routing states.

Generation of extra control traffic

As we have seen, many routing protocols flood the entire network with control packets. An attack aiming at increasing resource consumption can exploit this fact by injecting spoofed control packets into the network. In on-demand protocols, a spoofed route request can be flooded in this way; similarly, in link-state protocols, a spoofed link-state update can be flooded in the network. In distance vector based protocols, a spoofed routing update message can cause a sequence of "triggered" updates propagating in the entire network.

Position-based routing protocols again seem to be more resistant to this attack, because they do not use control packets. However, the attacker can send forged or spoofed location update messages to the location service. The new location information will be distributed among some nodes in the network, and this generates some extra control traffic. Similarly, the adversary can initiate excessive location information retrieval, which will also generate some extra control traffic.

Setting up a gray hole

All the attacks that we have discussed so far target the route establishment function of routing. Whereas the gray hole attack is concerned with the packet forwarding function. In a gray hole attack, an adversarial node selectively drops data packets that it should forward. If all data packets are dropped, then the gray hole degenerates into a black hole.

The primary objective of the gray hole attack is to degrade the quality of service. In particular, the packet delivery ratio between some nodes can decrease considerably. The gray hole can be detected by protocols above routing (e.g., at the transport layer), and the source can try to use an alternative route, however, this can incur some delay if no alternative route is currently available and a new route must be established. In addition to degrading the quality of service, a gray hole will also waste the resources of those nodes that forward the data packets that are finally dropped by the adversarial node.

A gray hole attack is trivial to implement: The adversarial node just needs to participate in the route establishment and then, when it receives data packets for forwarding, it drops them. However, with a bit of more sophistication, much more harm can be caused. In particular, the adversary can first set up a tunnel and divert traffic towards itself; then it drops packets.

All routing protocols that use a single path to route a packet to its destination are vulnerable to this attack. Yet, the restricted directional flooding protocols that we discussed in the context of position-based routing are inherently resistant to gray hole attacks.

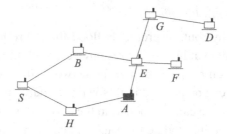

Figure 7.4. The network topology used in our examples that illustrate attacks against DSR and AODV. The network contains a single adversarial node A which is represented by the black node in the figure. The interconnection of the nodes represents the neighbor relationships; two nodes are considered to be neighbors if they can hear each other's transmission.

7.2.3 Some examples

In the previous subsection, we described the major types of attacks against ad hoc network routing protocols. In this subsection, we present specific examples that illustrate how these attacks can be mounted against the DSR and the AODV protocols. All examples assume the network topology depicted in Figure 7.4.

Route disruption

Let us consider the DSR protocol first. The source S initiates a route discovery towards destination D by flooding the network with a route request. Let us assume that the route caches of all nodes are empty. The adversarial node A can always prevent route S, H, A, E, G, D from being discovered by dropping the route request received from S or by dropping the route reply that contains this route. However, the adversary can also prevent the discovery of route S, B, E, G, D. For this, A must arrange that E receives the route request from A earlier than from B, as in that case, E drops the request received from B as a duplicate. Node A can, for instance, keep the channel constantly busy to prevent E from receiving anything from B. In this way, the adversary can achieve that none of the existing two routes between S and D are discovered by DSR.

In the case of AODV, the adversary can prevent the discovery of the routes between S and D by manipulating the hop count value in the route request message. In particular, the adversarial node A can set the hop count value in the route request message received from H to 0. In this way, E will believe that the next node on the shortest route to S is A, and therefore, E will forward route reply messages destined to S to A. By dropping these messages, A can prevent the discovery of any existing routes between S and D.

Route diversion

In the DSR protocol, route diversion can be performed as follows. Let us assume again that S initiates a route discovery towards D. When A receives the route request from H, it responds with a fake route reply that contains the route S, H, A, D. This fake reply is sent to S. Since the fake route reply contains a shorter route than the one discovered by the protocol (i.e., S, B, E, G, D), S may decide to use the route S, H, A, D. Thus, the adversary can successfully divert the communication between A and D. In order to stay invisible, A must modify the source route S, H, A, D to S, A, E, G, D in each subsequent data packet that is sent by S to D.

Route diversion in AODV can be achieved by manipulating the hop count value in routing control messages as described above in the route disruption attack against AODV. In particular, if the adversarial node A does not drop the route reply message originating from D in the above route disruption attack, then route S, H, A, E, G, D will be discovered by the protocol, which can be viewed as a diverted route, as the route S, B, E, G, D would be shorter.

Creation of incorrect routing states

In the case of DSR, creating incorrect routing states means that the adversary coerces the source of a route discovery to accept and cache a non-existent route to the destination. For instance, when node S initiates a route discovery in order to communicate with D, A does not re-broadcast the route request received from H, but it waits until it hears another copy of the same request from E. This copy of the request contains the route S, B, E. Afterwards, A generates a route reply containing the route S, B, E, A, D, and sends it to E. The reply is routed back to S via node B, and S caches the non-existent route S, B, E, A, D.

In the AODV protocol, the adversary can create an incorrect routing state in some nodes by manipulating the destination sequence number or the hop count in the routing control messages. S initiates a route discovery by flooding the network with a route request. A receives the route request from H, and increments the destination sequence number before re-broadcasting it in the name of F. As a result, E sets F as its next-hop towards S, since F is a neighbor of E, and the falsified route request appears to be fresher than the correct one coming from B. Thus, E gets into an incorrect state, as there is no route to S via F.

7.3 Securing ad hoc network routing protocols

In the previous section, we saw what kind of attacks are possible against ad hoc network routing protocols. In this section, we elaborate on techniques that can be used to defend against some of those attacks. In terms of presentation, we follow the strategy of the previous section: we first introduce applicable countermeasures in general, explain why they are needed, and then we illustrate how these counter-measures can be used by presenting some specific ad hoc network routing proto-cols that are designed with security in mind. A benefit for the reader is that this latter part on specific protocols can also be used as an independent quick reference on secure routing protocols proposed for mobile ad hoc networks.

7.3.1 Countermeasures

Origin authentication of control packets

As we have seen in the previous section, many of the attacks against routing proto-cols are based on spoofing or modifying control packets. The usual way to thwart these types of misdeeds is to authenticate packets. But there are two questions that arise here. First, who should authenticate the control packets? Second, who should be able to verify the authenticity of control packets?

Regarding the first question, it seems to be natural to require that control packets are authenticated by their originators. This would make it possible to detect spoof-ing. In addition, as message authentication mechanisms usually provide integrity protection functions, it would also make it possible to detect the modification of control packets.

However, the effectiveness of control packet origin authentication can also de-pend on the way we handle the second question. In order to see this, let us consider an on-demand routing protocol and assume that each control packet is authenti-cated by its originator (i.e., by the source, in case of route requests, or the des-tination, in case of route replies). Let us further assume that only the target (i.e., the destination or the source) of the control packet can verify its authenticity.[2] What we achieve with this is that spoofed or modified route requests will not be responded to by the destination, and spoofed or modified route replies will not be accepted by the source. In other words, we prevent the creation of an incorrect rout-ing state (e.g., inserting a non-existent route in the route cache) in the source and the destination. However, the adversary can still increase the resource consump-tion of the nodes, because spoofed or modified route requests and route replies are still processed and forwarded by intermediate nodes (because they cannot detect that they are spoofed or modified); in addition, this work is superfluous, because those spoofed or modified control packets will be discarded by their targets any-way. Spoofed route requests are especially harmful, because the entire network

---

[2] This is the case when authentication is based on a shared symmetric key between the source and the destination.

will be flooded with them. In addition, if the intermediate nodes update their routing state based on the control packets that they forward (e.g., they cache routes observed in forwarded route replies), then the adversary can successfully create incorrect routing state in intermediate nodes despite the fact that control packets are authenticated.

Thus, a useful design principle can be the following:

- each control packet should be authenticated by its originator;
- in order to prevent the creation of incorrect routing state in the nodes, each node that updates its routing state as a result of processing a control packet should be able to verify the authenticity of the packet;
- in order to prevent the generation of extra (and superfluous) control traffic, each node that processes, and re-broadcasts or forwards a control packet should be able to verify its authenticity.

Since, typically, a routing control packet is processed by several nodes in the network, the authentication mechanism should enable *broadcast authentication*. In other words, the originator should authenticate the control packet in such a way that all the nodes that will process and act upon it will be able to verify its authenticity. However, the originator usually does not know in advance which other nodes will process the packet. Hence, in practice, authenticity should be verifiable by every node. An authentication mechanism that makes this possible is the digital signature; the symmetric key equivalent of this is TESLA [314].

We must note here that verifying digital signatures (or MACs, in the case of TESLA) also consumes resources. So an adversary can attempt to take advantage of this by injecting fake control packets in the network; these packets will be dropped, but trying to verify the signatures on them will increase the resource consumption of the nodes. The good thing is that these forged packets are not propagated in the network, but they are caught immediately by the first legitimate node that verifies their signature. Hence, the effect of the attack is localized: although some limited number of nodes might suffer, the rest of the network is saved from processing useless control traffic. We must also mention that computation consumes less energy than communication, therefore it makes sense to trade-off increased computational overhead for decreased communication load.

Control packet origin authentication using digital signatures or TESLA is an effective way to secure link-state protocols, because in those protocols, control packets do not change while being broadcast in the network. However, proactive distance vector protocols and on-demand routing protocols need additional protection mechanisms, as we will see below.

Protection of mutable information in control packets

In many routing protocols, notably in on-demand protocols, the intermediate nodes add information to the control packets before forwarding or re-broadcasting them. For instance, in on-demand source routing protocols, the intermediate nodes extend the list of identifiers in route request packets with their own identifiers. Likewise, in on-demand distance vector protocols, the hop count field in the control packets is updated by each intermediate node. Since other nodes will act upon this added information, it must also be protected somehow from being forged and modified. However, control packet origin authentication will not solve this problem, because the information that we are talking about is added after the originator sends the control packet.

So what can we do? Well, we can apply the same principles as before, and require that each node that adds information to a control packet should authenticate the added information in such a way that each other node that acts upon this information is able to verify its authenticity. This sounds simple, but it becomes complicated when we take a closer look.

First of all, there are *traceable* additions and *untraceable* additions. Extending the list of identifiers accumulated in the route request is a traceable addition, because each modification preserves the previous state of the packet, therefore, anyone can see who added information to it. In contrast to this, incrementing the hop count in control packets is an untraceable addition of information, because it is impossible to tell just from the hop count value who contributed to it. Needless to say, traceable and untraceable additions require different protection mechanisms; it should also be clear, that protecting untraceable additions is much harder.

A seemingly simple solution for authenticating traceable additions to a control packet is that each intermediate node that adds information re-signs the entire updated packet. However, there are (at least) three problems with this approach. Firstly, the signature and the added information can be removed. For instance, imagine that when a node adds its identifier to the list of identifiers in a route request, it signs the entire request to authenticate its added information and to link it to the rest of the packet. An adversarial node can still remove identifiers and corresponding signatures from the end of the list and, in this way, can manipulate the routing information. In addition, such removal will be undetected, because all the remaining signatures verify correctly. Indeed, recall that being traceable means that the packet preserves its previous states, and the adversary can exploit this by enforcing a previous correct state.

There are no easy ways to work around this problem. Some proposals (e.g., Ariadne, described later in this chapter) use a hash value in the packet, which is re-hashed by each intermediate node, thereby introducing an untraceable element in the packet, which prevents the adversary to revert a previous correct state. However, this per-hop hash approach does not provide a perfect solution, as we will see later. Another interesting countermeasure against removing signatures from the end of a signature list is to replace the signature list with an *aggregate signature*. Aggregate signatures make it possible to compact multiple signatures from different parties into a single signature in such a way that it is very hard to remove any of the signatures from the aggregate signature, but at the same time, anybody can still verify who signed the message.

A second problem of authenticating traceable additions by re-signing the entire control packet is that it increases the resource consumption of the nodes considerably. Let us consider, for instance, a route request of an on-demand source routing protocol. As it is flooded in the network, not only each node has to sign it, but potentially (if intermediate nodes cache the routes learned from route requests, or simply to prevent the propagation of a modified request), each node has to verify *every* signature in the request, and the number of signatures grows by each hop taken by the request.

In order to overcome this problem some protocols (e.g., SDSR [216] and endairA [12]) avoid signing the route request. In some other protocols (e.g., in SRP [299] and in Ariadne [182]), the intermediate nodes are actually not required to verify the authenticity of the information added by other intermediate nodes to control packets. This has some disadvantages. For instance, in this case, an adversary can increase the resource consumption of the intermediate nodes by modifying a control packet, as this modification will not be detected by the intermediate nodes. In addition, in this case, intermediate nodes should not be allowed to update their routing state as a result of processing control packets, because updating the routing state based on unauthenticated information can lead to an incorrect state. However, by not allowing state update in intermediate nodes, the effectiveness of the routing protocol is somewhat decreased. The advantage, however, is that intermediate nodes do not perform any verifications, apart from performing control packet origin authentication and, instead of digital signatures, they can authenticate the information that they add to control packets by using more efficient MACs that are verifiable by the target of the control packets (assuming that the target shares a key with every intermediate node that processed the packet). Note that the target typically acts upon the control packet (e.g., updates its route cache), therefore it needs to authenticate all information in it. This approach is followed by Ariadne with MACs.

We must mention here that using symmetric key MACs requires the establishment of shared keys between the nodes. We can use traditional session key establishment protocols for this purpose (e.g., protocols based on a key distribution center), but there is an interesting problem that arises here: key distribution messages must also be routed somehow, but at the same time, they are needed for setting up the routing infrastructure securely. Thus, there seems to be a vicious circle of requirements here. One way to solve this problem is to use another approach to route key distribution messages, such as broadcasting them blindly in the network. Another approach is to distribute some of the keys out-of-band (e.g., manually, or using some of the techniques described in Chapter 5), and then use these available keys for authenticating routing messages and setting up the remaining keys at the same time. This can be achieved by piggybacking key establishment information on routing control packets that are secured with the already established keys; an example of this approach is BISS (Building secure routing out of an Incomplete Set of Security associations) [87].

The third problem with authenticating traceable additions by re-signing the control packet is that, in fact, authentication will not solve every problem. In particular, adversarial nodes can add incorrect information to control packets, and then sign them. Recall that the adversary is an insider, who can possess some signing keys. As a result, the packet will be verified as authentic, but the information inside can still be incorrect, leading to the creation of incorrect routing state in some nodes. This is a tough problem. An approach to mitigate it would be to verify the consistency of the information received in a control packet by cross-checking it with other control packets. Unfortunately, this solution might not be applicable in general.

Let us now investigate how to deal with untraceable additions to control packets. Typically, untraceable additions are used by distance vector protocols (both proactive and reactive). In reactive distance vector protocols, intermediate nodes increase the hop count in control packets (i.e., in route requests and route replies). In proactive distance vector protocols, control packets are not forwarded or rebroadcast explicitly, but still nodes broadcast their routing tables, which can cause some changes in the routing tables of their neighbors who will then broadcast their routing tables, and so on. Thus, the principle is the same and, in both cases, the same problem arises: the nodes update their routing state based on untraceable information (received in control packets or accumulated in the routing tables of their neighbors).

In essence, this is a problem related to trust. Let us explain why through an example. Assume that a node A receives a control packet from one of its neighbors B indicating that the hop count to a given destination is x. In order to accept the packet, A must believe that B behaved correctly and it received the control packet from one of its neighbors C with a hop count $x - 1$. However, this is not enough; A must also believe that C behaved correctly and it received the control packet from some other node D with a hop count $x - 2$, and so on. Thus, A must trust the whole chain of nodes that processed the control packet; however, owing to the untraceability condition, it does not even know who those nodes are. In other words, the whole thing works only if A can trust every node in the network, but it cannot, because it knows that the adversary is an insider who controls some of the nodes. Note that exactly the same problem arises in proactive distance vector protocols.

There is no bullet-proof solution to this problem. Some protocols (e.g., SAODV [391] and SEAD [177]) use per-hop hashing. The idea is the following: control packets (in the case of SAODV) or routing table entries (in the case of SEAD) contain not only a hop count, but also a hash value that is initialized by the originator of the given control packet or the destination corresponding to the given routing table entry. In the case of SAODV, each intermediate node that forwards or re-broadcasts a control packet, increments the hop count and computes the one-way hash of the hash value that it received in the control packet. Likewise in SEAD, when a node updates an entry in its routing table, it increments the hop count and hashes the hash value that it received in the corresponding entry of the routing table of its neighbor. As a result, adversarial nodes cannot decrease the hop count in control packets (in SAODV) and in routing table entries (in SEAD) that they process, because they cannot invert the hash function. However, they can always increase the hop count, and this can also lead to creating incorrect routing state in other nodes.

Another approach is to eliminate hop counts as a routing metric. One interesting solution is to replace the hop count with the packet propagation delay. For instance, the originator of a control packet can put a timestamp in the packet when sending it. Each intermediate node that receives the packet can compute the time that the packet traveled until it arrived at the given node (assuming synchronized clocks). Then, a routing table entry can be updated if a faster route is discovered, and the routing metric in the entry would be set to the computed delay. An example of a protocol that follows this approach is ARAN (described later in this chapter). Actually, using the delay as a routing metric is a good idea, because the adversary cannot transfer data faster than the speed of light, therefore attacks (at least those attempting to shorten routes) are inherently limited. The disadvantage is that this approach needs synchronized clocks. In addition, the packet propagation delay between two nodes can vary in time, and it might not be the same in both directions.

To summarize, there is no perfect solution to protect mutable information in routing control packets, and hence, different protocols choose different trade-offs. Link-state protocols are very advantageous from this point of view, because they have no mutable information in their control packets. Similarly, position-based protocols are also advantageous, because they have no control packets at all.

Detecting tunnels

Recall that tunneling means that two adversarial nodes pass routing control packets back and forth between each other by encapsulating them into normal data packets. The result of this is that the two adversarial nodes appear to be neighbors in the routing topology, while in reality, they may be far away from each other. The effect of the tunneling attack on routing is therefore similar to the effect of wormholes. In particular, routes through the tunnel appear to be shorter and may be preferred by some nodes, which increases the adversarial control over the communications of those nodes. Note, however, that tunneling is a network layer attack, because it requires at least two adversarial nodes that are addressable and that can communicate with each other using the network itself. In contrast to this, wormholes do not require presence at the network layer, and they use out-of-band channels to connect the transceivers of the adversary.

A first interesting observation with respect to tunnels is that they rely on the very same mechanism that they try to subvert (i.e., routing). The problem is that, because of the tunnel, some nodes store incorrect routing state. This may also be true for the nodes that participate in the forwarding of the tunneled traffic. At the end, the incorrect routing state in these nodes may result in the disruption of the communication through the tunnel. In other words, the tunnel may collapse upon itself.

Let us assume, for instance, that node F is the first node that should forward the tunneled traffic from adversarial node A to another adversarial node A'. Since A and A' report themselves neighbors, F may conclude that the shortest route from itself to A' is the route through A. Thus, when A sends tunneled messages towards A' via F, F will send those messages back to A. However, the adversary may solve this problem by introducing a third adversarial node midway between A and A', and sending all tunneled traffic between A and A' through this third node.

Let us now consider how tunneling could be detected. First of all, since wormholes and tunnels are similar, some wormhole detection approaches may be successfully adopted to detect tunnels. In particular, some centralized wormhole detection approaches that we described in Subsection 6.2.1 can be easily adopted to detect tunneling attacks against link-state routing protocols. In this case, the tunneling adversarial nodes report each other as neighbors in their link-state update messages. A central entity that can collect all link-state update messages is able to re-construct the believed topology of the network, and it can detect inconsistencies in this topology. For instance, if the adversarial nodes are far away from each other, then they will not have common neighbors in the re-constructed network topology. This is suspicious, as in a typical network topology, neighboring nodes likely have common neighbors.

Some decentralized wormhole detection approaches can also be adopted to detect tunnels. Let us consider, for instance, on-demand source routing protocols where the routing control packets carry the list of nodes that constitute a route or a route segment. If each node is aware of its own geographic position, then, besides its identifier, each intermediate node can also place its position in the control packets. Thus, by inspecting the position information in a control packet, non-corrupted nodes can detect shortcuts in the route or route segment carried by the packet. Clearly, this approach needs other countermeasures as well to prevent the adversary from modifying the position information in the routing control packets.

Another approach is based on the observation that a virtual link in the topology created by a tunnel is actually a multi-hop route, and therefore, the delay on this virtual link must be much higher than the delay on a real link. Now, there are two strategies. First, each node could measure the round trip time to its three-hop neighbors explicitly.[3] From the measured timings, the node can deduce whether its three hop neighbors are reachable through a tunnel or not, and if so, which link may be the tunneled one. Alternatively, the routing protocol could use the end-to-end delay between a source and a destination as a routing metric instead of the hop count. In this way, the tunnel may not be detected explicitly, but its effect is greatly reduced, because sending routing control packets through the multi-hop tunnel cannot be significantly faster than sending them through similar alternative routes. An example of a distance vector routing protocol that uses this principle is ARAN [337], which we discuss later in Subsection 7.3.2.

Finally, when a control packet is tunneled from one adversarial node to another, it is placed in the payload part of a data packet. Hence, the control packet "disappears" at one end of the tunnel and "re-appears" at the other end. Such a disappearing and re-appearing packet may be detected by the neighbors of the two ends of the tunnel. In particular, if the nodes monitor the incoming and outgoing traffic of their neighbors, then an overheard control packet which is not re-transmitted by a neighbor may be an indication for a tunneling attack. Similarly, if a node overhears the transmission of a control packet, but no one in the neighborhood heard this control packet before, then the packet may have arrived through a tunnel. Unfortunately, this kind of monitoring is not very reliable as we will see soon in the next subsection where we discuss countermeasures against gray holes. Another disadvantage is that continuous monitoring requires the nodes to run in promiscuous mode and listen to the channel all the time, which consumes a lot of energy.

---

[3] Note that measuring round trip times to direct neighbors or to two-hop neighbors would not help to detect the virtual link.

Combating gray holes

Gray holes are attacks against the packet forwarding function of routing. Consequently, countermeasures must aim at the protection of data packets rather than
control packets. There are two main approaches to combat against gray holes. The
first one consists in using multiple, preferably disjoint routes between the source
and the destination to send data packets. The idea is that even if forwarding is
disrupted by the adversary on some of these routes, the packet can still be delivered through the remaining routes. This is a robust approach, but it induces
increased resource consumption in intermediate nodes, as they have to forward
multiple copies of the same data packet. This overhead can be reduced by using special coding schemes by which the packet can be split into smaller pieces
in such a way that a threshold number of pieces (but not all) is sufficient to re-
construct the entire packet. In this case, not the packet itself, but its smaller pieces
must be sent through multiple routes to the destination. The adversary can drop
some of the pieces, but if enough of them are received by the destination, then
the packet is successfully delivered. This approach still has some overhead, as the
pieces must be redundant to be able to tolerate the loss of some of them. However, what we lose on redundancy, we gain in terms of robustness and, at the end,
the nodes waste more resources (by forwarding packets that are never delivered)
in the single path approach than in the case of using multiple routes. As an example of multi-path forwarding, we present the SMT protocol [302] later in this
section.

One can also use a "detect and react" approach to mitigate the effect of gray
hole attacks. In this case, the idea is first to detect if a node does not forward data
packets, and then, to select routes that avoid the misbehaving node. Gray holes
can be detected by requiring each node to monitor the activities of its neighbors.
By doing so, a correctly behaving node can detect that one of its neighbors has
received a packet that it should forward, but it does not. This kind of monitoring
can be implemented by putting the network interface of the nodes in promiscuous
mode (most interface cards allow this) and by listening to everything in the wireless
channel. If a node does not overhear the retransmission of a packet by its neighbor,
then that neighbor can be suspected to misbehave. This sounds simple, but the
devil is always in the details. First of all, this approach requires the nodes to run
in the promiscuous mode, which consumes much more energy than allowing the
nodes to go into the sleep mode when they have nothing else to do. Secondly, it
turns out that this kind of monitoring is not very reliable. Below, we list some
cases where either a correctly behaving node is falsely identified as misbehaving
or a misbehavior is not detected.

For illustration purposes, let us assume that a packet should be forwarded by
nodes A, B, and C – in this order. For the first example, let us further assume that

A forwarded the packet. When B forwards the packet, A can receive something at the same time from another node X that B does not hear. Thus, the transmissions of X and B will collide at A, and A will not hear that B forwarded the packet. Hence, A will falsely believe that B is misbehaving. For this reason, the nodes should not decide definitely about their neighbors' behavior after just one experience, but they should monitor their neighbors for an extended period of time. Typically, a node would be identified as a misbehaving node (i.e., a gray hole) only if it is perceived as dropping packets with a rate higher than a predefined threshold. This, however, allows the adversary to drop packets with a rate below that threshold.

For the second example, let us assume again that A forwarded the packet. When B also forwards it, the transmission can collide with some other transmission at C. Hence, A will believe that B forwarded the packet, but in fact, C did not receive it. Then, B can skip the retransmission, because A will not accuse it for misbehavior.

For the third example, suppose that the nodes can control their transmission power, and that A is closer to B, than B to C. Then, B can forward the packet with a power that allows A to overhear B's transmission, but does not allow C to receive the packet. Again, A will falsely believe that B behaves correctly.

Finally, let us assume that B and C are colluding, and B does not report when C drops a packet. Again, neither B nor C will be identified as a misbehaving node.

Instead of local neighbor monitoring, misbehaving nodes can also be detected in an end-to-end manner by requiring the destination and the intermediate nodes on the route of a packet to send acknowledgements to the source. This approach requires that the nodes use source routing, and therefore, the source knows the entire route to the destination. The idea is the following: the destination is required to return an acknowledgement for every packet that it receives successfully. Based on these acknowledgements, the source keeps track of the loss rate in a time window of a given size. If the loss rate exceeds a threshold, the source starts a binary search on the route to identify the misbehaving node, or more precisely, the link that causes the delivery failure of the packets. For this, the source adaptively specifies a list of intermediate nodes in the subsequent packets that should also return an acknowledgement for the packets that they successfully processed. These nodes are called *probe nodes*. First, one probe node is selected in the middle of the path between the source and the destination. If the acknowledgements arrive from this node but not from the destination, then the bad link must be between the probe node and the destination. Otherwise, if the acknowledgements do not arrive from the probe node either, then the bad link must be between the source and the probe node. Once the sub-path that contains the bad link is identified, a new probe node is specified in the middle of that sub-path. This procedure is continued until the sub-path that contains the bad link is narrowed down to a single link, which must be the bad link. The misbehaving node can be either end of the identified bad link.

We must note that the approach described above requires the lifetime of the routes to be sufficiently long so that there is enough time to identify the bad link. In addition, the source must be able to authenticate the acknowledgements sent by the probe nodes.

Now, let us assume that the gray hole is detected with a reasonable level of accuracy. The next step is to invoke some reaction mechanism. This typically involves the distribution of information about misbehaving nodes to other nodes in the network, so that the nodes can avoid the gray holes when setting up routes. Warnings about a gray hole can be distributed in the entire network. In this case, it is assumed that the nodes maintain reputation values for all other nodes in the network, and when they receive notifications about gray holes they update these values by lowering the reputation of the misbehaving nodes. The reputation values are then used in the route establishment process (i.e., routes traversing nodes with good reputation are preferred).

Building up reputation can be a lengthy process, but it is possible to expedite it by letting the nodes exchange their reputation values. In this case, if A trusts B, and B says that C is not very trustworthy in terms of packet forwarding, then A can try to avoid routes containing C, even if it has no previous experience with C at all. This principle of "recommendations" is very intuitive and well-known from real life, but it is quite difficult to implement it in ad hoc networks correctly. The main problem is that the adversarial nodes can try to frame some other nodes by disseminating bad reputation reports about them. Hence some kind of a trust model must be used to govern the maintenance of reputation values, which carefully combines each node's own experience with reports from other nodes.

Later in this section, we describe two examples of the "detect and react" approach: Watchdog and Pathrater [267], and the ODSBR protocol [35]. The former uses neighbor monitoring, whereas the latter is based on the adaptive acknowledgement scheme that we described above.

7.3.2 Specific examples of secure routing protocols

SRP

SRP (Secure Routing Protocol) was proposed in [299] as a secure on-demand source routing protocol for mobile ad hoc networks based on symmetric key cryptography. The design of SRP was driven by the observation that due to the mobility of the nodes, and hence the volatility of the routes, it would be impractical to require that the source or the destination shares keys with all intermediate nodes on a route. Therefore, in SRP, only the source and the destination share a key, which simplifies the key management considerably. This results in a strict end-to-end

$S \rightarrow *$	:	$(rreq,\ S,\ D,\ id,\ sn,\ mac_S,\ [\ ])$
$F_1 \rightarrow *$	:	$(rreq,\ S,\ D,\ id,\ sn,\ mac_S,\ [F_1])$
$F_2 \rightarrow *$	:	$(rreq,\ S,\ D,\ id,\ sn,\ mac_S,\ [F_1,\ F_2])$
$D \rightarrow F_2$	:	$(rrep,\ S,\ D,\ id,\ sn,\ [F_1,\ F_2],\ mac_D)$
$F_2 \rightarrow F_1$	:	$(rrep,\ S,\ D,\ id,\ sn,\ [F_1,\ F_2],\ mac_D)$
$F_1 \rightarrow S$	:	$(rrep,\ S,\ D,\ id,\ sn,\ [F_1,\ F_2],\ mac_D)$

Figure 7.5. Operation example of SRP and format of the SRP messages. The identifier of the source is S, the identifier of the destination is D, and the identifiers of the intermediate nodes are F_1 and F_2. id is a randomly generated query identifier, sn is a query sequence number maintained by S and D, mac_S is the MAC generated by S that covers the fields $rreq$, S, D, id, and sn, and mac_D is the MAC generated by D that covers the fields $rrep$, S, D, id, sn, and (F_1, F_2).

exchange of routing control information between the source and the destination, and end-to-end authentication of routing control packets. SRP introduces another useful design principle as well: the avoidance of optimizations. This means that in SRP, the intermediate nodes do not send replies to route discovery messages (only on behalf of the destination) and they do not cache information from overheard routing control packets.

The operation of SRP and the format of SRP messages are illustrated in Figure 7.5. The source generates a route request message and broadcasts it to its neighbors. The integrity of this route request is protected by a MAC that is computed with a key shared between the source and the destination. Each intermediate node that receives the route request for the first time appends its identifier to the request and re-broadcasts it. The MAC in the request is not checked by the intermediate nodes (as they do not know the key with which it was computed), and the nodes do not append their own MACs either (i.e., they do not authenticate the route request). When the route request reaches the destination, it contains the list of identifiers of the intermediate nodes that passed the request on. This list is considered as a route found between the source and the destination.

The destination verifies the MAC of the source in the request. If the verification is successful, then it generates a route reply and sends it back to the source via the reverse of the route obtained from the route request.[4] The route reply contains the route obtained from the route request, and its integrity is protected by another MAC generated by the destination with a key shared between the destination and the source. Each intermediate node passes the route reply to the next node in the route (towards the source) without modifying it. When the source receives the reply it verifies the MAC of the destination, and if this verification is successful, then it accepts the route returned in the reply.

---

[4] SRP assumes that links are bi-directional.

S	:	$h_S = MAC_{SD}(rreq, S, D, id)$
$S \rightarrow *$	:	$(rreq, S, D, id, h_S, [\,], [\,])$
F_1	:	$h_{F_1} = H(F_1, h_S)$
$F_1 \rightarrow *$	:	$(rreq, S, D, id, h_{F_1}, [F_1], [sig_{F_1}])$
F_2	:	$h_{F_2} = H(F_2, h_{F_1})$
$F_2 \rightarrow *$	:	$(rreq, S, D, id, h_{F_2}, [F_1, F_2], [sig_{F_1}, sig_{F_2}])$
$D \rightarrow F_2$	:	$(rrep, D, S, [F_1, F_2], [sig_{F_1}, sig_{F_2}], sig_D)$
$F_2 \rightarrow F_1$	:	$(rrep, D, S, [F_1, F_2], [sig_{F_1}, sig_{F_2}], sig_D)$
$F_1 \rightarrow S$	:	$(rrep, D, S, [F_1, F_2], [sig_{F_1}, sig_{F_2}], sig_D)$

Figure 7.6. Operation example of Ariadne with signatures and format of the Ariadne messages. The source is S, the destination is D, and the intermediate nodes are F_1 and F_2. id is a randomly generated query identifier, H is a publicly known one-way hash function, and MAC_{SD} is a MAC function used with the key shared between S and D. The sig_{F_1}, sig_{F_2}, and sig_D are digital signatures of F_1, F_2, and D, respectively. Each signature is computed over the message fields that precede the signature.

The destination can receive several route requests that belong to the same route discovery process,[5] and it sends a reply to each of these requests. It is assumed that the source waits for some time (defined by a timeout parameter), and then it outputs the set of routes collected from all the replies it received.

SRP is a very efficient protocol, as route request and route reply messages contain only a single MAC value; moreover, these MAC values are not processed by the intermediate nodes. At the same time, if it is combined with a secure neighbor discovery protocol, then SRP provides protection against attacks aiming at route disruption, route diversion, and the creation of incorrect routing state. To be more precise, SRP resists against attacks mounted by an adversary from a single adversarial node, or from multiple non-colluding nodes, but not against attacks mounted by colluding adversarial nodes (see Questions at the end of this chapter).

Ariadne

Ariadne is proposed in [182] as a secure on-demand source routing protocol for ad hoc networks. Ariadne comes in three different flavors corresponding to three different techniques used for data authentication. More specifically, authentication of routing messages in Ariadne can be based on TESLA [314], on digital signatures, or on standard MACs. Ariadne with digital signatures is conceptually the simplest among these three versions, therefore, we begin the presentation with that variant.

The operation of Ariadne with digital signatures is illustrated in Figure 7.6. As we can see, there are two main differences between Ariadne and SRP. First, in

---

[5] As the neighbors of the destination re-broadcast the request at most once, the destination can receive at most as many requests as the number of its neighbors.

Ariadne, not only do the source and the destination authenticate the messages, but the intermediate nodes also insert their own digital signatures in route requests. Second, Ariadne uses per-hop hashing to prevent removal of identifiers from the accumulated route in the route request.

The source generates a route request message and broadcasts it to its neighbors. The route request message contains the identifiers of the source and the destination, a randomly generated request identifier, and a MAC computed over these elements with a key shared by the source and the destination. This MAC is hashed iteratively by each intermediate node together with its own identifier using a publicly known one-way hash function. The hash values computed in this way are called per-hop hash values. Each intermediate node that receives the request for the first time re-computes the per-hop hash value, appends its identifier to the list of identifiers accumulated in the request, and generates a digital signature on the updated request. Finally, the signature is appended to a signature list in the request, and the request is re-broadcast.

When the destination receives the request, it verifies the per-hop hash by re-computing the MAC of the source and the per-hop hash value of each intermediate node. Then, it verifies all the digital signatures in the request. If all these verifications are successful, then the destination generates a route reply and sends it back to the source via the reverse of the route obtained from the route request.[6] The route reply contains the identifiers of the destination and the source, the route and the list of digital signatures obtained from the request, and the digital signature of the destination on all these elements. Each intermediate node passes the reply to the next node on the route (towards the source) without any modifications. When the source receives the reply, it verifies the digital signature of the destination and the digital signatures of the intermediate nodes (for this, it needs to re-construct the requests that the intermediate nodes signed). If the verifications are successful, then it accepts the route returned in the reply.

Ariadne with TESLA is similar to Ariadne with digital signatures, but instead of signatures, the intermediate nodes compute MACs on the route request with their current TESLA key (see Appendix A for details on the operation of TESLA). The advantage of this is that MACs can be computed more efficiently than digital signatures. However, the disadvantage is that the application of TESLA introduces some delay in the route discovery process, which may not be desirable in dynamic networks. When the destination receives the route request, it verifies that the TESLA keys that were used in the request have not been disclosed yet (i.e., their indicated disclosure time is still in the future). The destination also verifies the per-hop hash value received in the request by iteratively computing all the per-hop hash values

---

[6] Here, it is again assumed that links are bi-directional.

S	:	$h_S = MAC_{SD}(rreq,\ S,\ D,\ id)$
$S \to *$	:	$(rreq,\ S,\ D,\ id,\ h_S,\ [\ ],\ [\ ])$
F_1	:	$h_{F_1} = H(F_1, h_S)$
$F_1 \to *$	:	$(rreq,\ S,\ D,\ id,\ h_{F_1},\ [F_1],\ [mac_{F_1}])$
F_2	:	$h_{F_2} = H(F_2, h_{F_1})$
$F_2 \to *$	:	$(rreq,\ S,\ D,\ id,\ h_{F_2},\ [F_1,\ F_2],\ [mac_{F_1},\ mac_{F_2}])$
$D \to F_2$	:	$(rrep,\ D,\ S,\ [F_1,\ F_2],\ mac_D)$
$F_2 \to F_1$	:	$(rrep,\ D,\ S,\ [F_1,\ F_2],\ mac_D)$
$F_1 \to S$	:	$(rrep,\ D,\ S,\ [F_1,\ F_2],\ mac_D)$

Figure 7.7. Operation example of Ariadne with standard MACs. It is assumed that each intermediate node shares a key with the destination D. mac_{F_1} and mac_{F_2} are MACs computed by F_1 and F_2, respectively, with the keys that they share with D. Each MAC is computed over the message fields that precede the MAC.

of the intermediate nodes. If these verifications are successful, then the destination returns a route reply, which includes the MACs of the intermediate nodes obtained from the request.

Each intermediate node that receives the route reply waits until it can disclose the TESLA key that it used to compute the MAC for the corresponding route request. Then, it appends this TESLA key to the reply before passing it on to the next intermediate node. In this way, when the route reply arrives at the source, it contains all the TESLA keys needed to verify the MACs of the intermediate nodes (obtained from the request and inserted also in the reply). The source can thus authenticate all intermediate nodes.

Finally, when Ariadne is used with standard MACs, then it is assumed that each intermediate node shares a key with the destination. Figure 7.7 illustrates the operation and the messages of Ariadne with standard MACs. The source generates a route request and broadcasts it in the network. Each intermediate node computes a MAC on the request with the key that it shares with the destination. Hence, all these MACs can be verified by the destination when it receives the request. In addition, the per-hop hash mechanism is used too, in order to prevent the removal of MACs from the end of the MAC list in the request. If the verifications of the MACs and the per-hop hash value by the destination are successful, then the destination generates a reply, which contains the discovered route, and it is protected by a MAC computed by the destination with a key that it shares with the source. Each intermediate node passes the reply to the next node on the route without any modification. When the reply arrives at the source, it verifies the MAC of the destination. Note that in this case, the source cannot authenticate the intermediate nodes, but it must trust the destination to perform this authentication correctly. In addition, the intermediate nodes can authenticate neither the route request nor the route reply.

In [182], an optimized version of Ariadne is proposed, which does not use a per-hop hash value and a MAC list in the route request, but it uses instead a single MAC that is updated by the intermediate nodes iteratively. In this optimized version of Ariadne, the route request re-broadcast by the ith intermediate node F_i has the following format:

$$(rreq, \ S, \ D, \ id, \ [F_1, \ldots, F_{i-1}, F_i], \ mac_{F_i}),$$

where mac_{F_i} is a MAC computed by F_i with the key that it shares with D on the route request that it received from F_{i-1}:

$$(rreq, \ S, \ D, \ id, \ [F_1, \ldots, F_{i-1}], \ mac_{F_{i-1}}).$$

This optimized version is more efficient than the basic protocol in terms of computational and communication overhead. First, there is no need for the per-hop hash mechanism anymore, because the MACs computed by the intermediate nodes can play the same role as the per-hop hash values in the original protocol. Second, route requests are shorter, because they do not contain a per-hop hash value and they contain only a single MAC instead of a MAC list. Finally, this optimized version seems to be more secure than the original version that uses a MAC list, because the adversary cannot access the individual MACs of the intermediate nodes in the same way as it can in case of a MAC list, and therefore, MACs cannot be removed from the route request at the adversary's will.

In the basic Ariadne protocol, a route request is not authenticated until it reaches the destination. Thus, an adversary can initiate malicious route request flooding in the network aiming at increasing the resource consumption of the nodes. In order to protect against this, Ariadne is extended with a rate limiting mechanism based on one-way hash chains (for a detailed description of hash chains, see also Appendix A). The idea is the following: each node generates a hash chain and releases its element in reverse order, one element with each route request message that it originates. Since route requests are flooded in the entire network, all nodes learn the most recent hash chain element of all other nodes. When a node receives a route request message, it verifies if the hash chain element in the message is fresher than the most recent hash chain element that belongs to the source of the route request. This can be done by hashing the value received in the route request and comparing the result to the stored hash chain element. The route request is re-broadcast only if this verification is successful. Because of the one-way property of the hash chains, the adversary cannot predict the next element of the chain to be released, and therefore, cannot initiate the flooding of malicious route requests.

endairA

endairA is another secure on-demand source routing protocol proposed in [12]. The design of endairA has been inspired by Ariadne, hence, the two protocols are quite similar. The difference is that instead of signing the route request, in endairA, intermediate nodes sign the route reply. This explains the name endairA, which is the reverse of Ariadne. A remarkable feature of endairA is that it can be proven to be secure in a formal model. We will elaborate on this in the next section. Here, we describe the operation of the protocol.

The operation and the messages of endairA are illustrated in Figure 7.8. In endairA, the source generates a route request that contains the identifiers of the source and the destination, and a randomly generated request identifier. Each intermediate node that receives the request for the first time appends its identifier to the route accumulated so far in the request and re-broadcasts the request. When the request arrives at the destination, it generates a route reply. The route reply contains the identifiers of the source and the destination, the accumulated route obtained from the request, and a digital signature of the destination on these elements. The reply is sent back to the source on the reverse of the route found in the request. Each intermediate node F_i that receives the reply verifies that its identifier is in the node list carried by the reply, and that the preceding identifier F_{i-1} (or that of the source if there is no preceding identifier in the node list) and the following identifier F_{i+1} (or that of the destination if there is no following identifier in the node list) belong to neighboring nodes. Each intermediate node also verifies that the digital signatures in the reply are valid and that they correspond to the following identifiers in the node list and to the destination. If these verifications fail, then the reply is dropped. Otherwise, it is signed by the intermediate node and passed to the next node on the route (towards the source). When the source receives the route reply, it verifies if the first identifier in the route carried by the reply belongs to a neighbor. If so, then it verifies all the signatures in the reply. If all these verifications are successful, then the source accepts the route.

endairA has a significant advantage over Ariadne (and similar protocols): it is more efficient, because it requires less cryptographic computation overall. This is because in endairA only the processing of the route reply message involves cryptographic operations meaning that only those nodes need to perform cryptographic computations that are in the node list carried in the route reply. In contrast to this, in Ariadne, the route request messages need to be digitally signed by all intermediate nodes; however, owing to the way a route request is propagated, this means that each node in the network must sign each and every route request.

One problem with the basic endairA protocol is that it is vulnerable to malicious route request flooding attacks. This is because the route request messages

$$\begin{array}{lll}
S \rightarrow * & : & (rreq,\ S,\ D,\ id,\ [\,]) \\
F_1 \rightarrow * & : & (rreq,\ S,\ D,\ id,\ [F_1]) \\
F_2 \rightarrow * & : & (rreq,\ S,\ D,\ id,\ [F_1, F_2]) \\
D \rightarrow F_2 & : & (rrep,\ S,\ D,\ id,\ [F_1, F_2],\ [sig_D]) \\
F_2 \rightarrow F_1 & : & (rrep,\ S,\ D,\ id,\ [F_1, F_2],\ [sig_D,\ sig_{F_2}]) \\
F_1 \rightarrow S & : & (rrep,\ S,\ D,\ id,\ [F_1, F_2],\ [sig_D,\ sig_{F_2},\ sig_{F_1}])
\end{array}$$

Figure 7.8. An example of the operation and the messages of endairA. The source is S, the destination is D, and the intermediate nodes are F_1 and F_2. id is a randomly generated request identifier. The sig_D, sig_{F_1}, and sig_{F_2} denote the digital signature of D, F_1, and F_2, respectively. Each signature is computed over the message fields (including the signatures) that precede the signature.

are not authenticated, and hence, an adversary can initiate route discovery processes by spoofing route request messages. These spoofed route requests would be flooded in the network, because only the impersonated source can detect that they are spoofed. In order to prevent this, the route request can be digitally signed by the source, and rate limiting techniques similar to the one used by Ariadne can be applied to endairA too. Naturally, such extensions put more burden on the nodes, as now they also need to verify the signature of the source in each route request message and to maintain information that is required by the rate limiting mechanism.

Finally, we note that endairA can be optimized with respect to communication overhead by replacing the signature list in the route reply with a single aggregate signature (e.g., using the scheme described in [60]). This aggregate signature is computed by the intermediate nodes iteratively similarly to the iterated MAC computation in the optimized version of Ariadne.

SAODV

SAODV [391] is a secure variant of AODV. Its operation is similar to that of AODV, but it uses cryptographic extensions to provide authenticity and integrity of routing messages, and to prevent the manipulation of the hop count information.

Conceptually, SAODV routing messages (i.e., route requests and route replies) have a non-mutable and a mutable part. The non-mutable part includes, among other fields, the node sequence numbers, the addresses of the source and the destination, and a request identifier, whereas the mutable part contains the hop count information. Different mechanisms are used to protect the different parts.

The non-mutable part is protected by the digital signature of the originator of the routing message (i.e., the source, in the case of a route request, and the destination, in the case of a route reply). This ensures that the non-mutable fields cannot be changed by an adversary without the change being detected by non-compromised nodes.

SAODV uses hash chains in order to prevent the manipulation of the hop count information. There are four specific fields in the routing messages that are used by the hop count protection mechanism: HopCount, MaxHopCount, Hash, and TopHash. When a node originates a routing message (i.e., a route request or a route reply), it first sets the HopCount field to 0, and the MaxHopCount field to the TTL (Time-to-Live) value. Then, it initializes the Hash field of the routing message to a random value. After that, it calculates the TopHash field by hashing the seed iteratively MaxHopCount times. The MaxHopCount and the TopHash fields belong to the non-mutable part of the message, whereas the HopCount and the Hash fields are mutable. Every node receiving a routing message hashes the value of the Hash field (MaxHopCount − HopCount) times, and verifies whether the result matches the value of the TopHash field. Then, before re-broadcasting a route request or forwarding a route reply, the node increases the value of the HopCount field by one, and updates the Hash field by hashing its value once.

The rationale behind using the above hash chaining mechanism is that given the values of the Hash, the TopHash, and the MaxHopCount fields, anyone can verify the value of the HopCount field. Preceding hash values, however, cannot be computed starting from the value in the Hash field owing to the one-way property of the hash function. This is intended to ensure that an adversary cannot decrease the hop count, and thus, cannot make a route appear shorter than it really is. This is not true in general, because an adversarial node that happens to be in the route between the source and the destination can pass on the routing message without increasing the value of the HopCount field and updating the value of the Hash field. In this way, she can make the route seemingly shorter. In addition, as we have already pointed out, the adversary can always increase the hop count.

ARAN

ARAN (Authenticated Routing for Ad hoc Networks) is another secure, on-demand distance vector routing protocol for ad hoc networks proposed in [337]. Just like SAODV, ARAN uses public key cryptography to ensure the integrity of routing messages. Its operation and message format are illustrated in Figure 7.9.

The source node begins the route discovery process by broadcasting a route request message. This route request contains the identifier of the destination, the public key certificate of the source, a nonce generated by the source, the current timestamp, and the digital signature of the source on all these elements. The nonce, the timestamp, and the identifier of the source together uniquely identify the message, and they are used to detect and discard duplicates of the same request (and reply).

Later, when the request is propagated in the network, intermediate nodes also sign it. Hence, in general, the request contains two signatures: that of the source and that of the last intermediate node that processed it. Intermediate nodes process

$S \rightarrow *$	:	$(rreq, D, cert_S, n, t, sig_S)$
$F_1 \rightarrow *$	:	$(rreq, D, cert_S, n, t, sig_S, sig_{F_1}, cert_{F_1})$
$F_2 \rightarrow *$	:	$(rreq, D, cert_S, n, t, sig_S, sig_{F_2}, cert_{F_2})$
$D \rightarrow F_2$	:	$(rrep, S, cert_D, n, t, sig_D)$
$F_2 \rightarrow F_1$	:	$(rrep, S, cert_D, n, t, sig_D, sig_{F_2}, cert_{F_2})$
$F_1 \rightarrow S$	:	$(rrep, S, cert_D, n, t, sig_D, sig_{F_1}, cert_{F_1})$

Figure 7.9. Operation example of ARAN. S and D are the identifiers of the source and the destination, respectively, and F_1 and F_2 are the identifiers of the intermediate nodes; n is a nonce generated by S, and t is the current timestamp when generating the route request. The $cert_X$ and sig_X are the public key certificate and the digital signature of X, respectively. All signatures are generated on the message fields that precede the signature.

the request as follows: When an intermediate node F_{i+1} receives the request from another intermediate node F_i, it verifies the signatures of the source S and F_i, and the freshness of the nonce. If these verifications are successful, then F_{i+1} sets an entry in its routing table with S as destination, and F_i as next hop. Then, F_{i+1} removes the certificate and the signature of F_i, signs the request, appends its own certificate, and re-broadcasts the updated request.

When the destination receives the first route request that belongs to this route discovery, it performs the verifications and updates its routing table in a similar manner by the intermediate nodes. Then, it sends a route reply message to the source. The route reply is propagated back on the reverse of the discovered route as a unicast message. The route reply is signed by the destination; in addition, like in the case of the route request, it is also signed by the intermediate node that has just passed it on. The processing of the route reply by the intermediate nodes is analogous to the processing of the route request.

As can be seen from the description, ARAN does not use hop counts as a routing metric. Instead, the nodes update their routing tables using the information obtained from the routing messages that arrive first; any later message that belongs to the same route discovery is discarded. This means that ARAN does not necessarily discover the shortest paths in the network, but rather it discovers the quickest ones. In effect, ARAN uses the message propagation delay (i.e., physical time) as a routing metric. This results in a robust protocol: indeed, ARAN is proven to be secure in [11]. However, a major drawback of ARAN is that it needs extensive signature generation and verification during the route request flooding phase.

SEAD

SEAD (Secure Efficient Ad hoc Distance vector routing) is a proactive distance vector based routing protocol for mobile ad hoc networks proposed in [177]. It can be viewed as a secure variant of the DSDV protocol where the destination sequence numbers and the hop count values are protected using one-way hash chains. More precisely, SEAD tries to ensure that sequence numbers cannot be increased, and hop count values cannot be decreased by an adversary, but no attempt is made to prevent the modification of these values in the other direction (i.e., decreasing sequence numbers and increasing hop count values).

When using SEAD, each node generates a one-way hash chain of length $n + 1$ using a publicly known hash function H, where n is a multiple of the maximum diameter m of the network. Hence, the hash chain of a node can be denoted by $h_0, h_1, \ldots, h_n$, where $n = k \cdot m$, h_0 is a random value, and $h_i = H(h_{i-1})$. It is assumed that h_n is securely distributed to all other nodes in the network.

When a node S sends out a route update message about itself with sequence number i and hop count value 0, it reveals $h_{(k-i)m}$ in the same message. A neighboring node will update its routing table entry that belongs to S by recording sequence number i, hop count value 1, and hash value $H(h_{(k-i)m}) = h_{(k-i)m+1}$. Then, it sends out a route update message, and its neighboring nodes will record sequence number i, hop count value 2, and hash value $H(h_{(k-i)m+1}) = h_{(k-i)m+2}$ for destination S, and so on.

Each route update concerning S can be verified by the nodes using a previously known hash value from the hash chain of S; indeed, the nodes update their routing table entry for S only if this verification is successful. Let us assume, for instance, that D knows sequence number i, hop count value c, and hash value $h = h_{(k-i)m+c}$ for S, and now it receives a route update concerning S with sequence number $j > i$, hop count c', and hash value h'. Then, D accepts this update only if $H^{(j-i)m+c-c'}(h') = h$, where H^x means that we invoke H iteratively x times (see Figure 7.10 for illustration).

The main idea is that if an adversarial node knows a hash value belonging to a given sequence number and hop count, then, owing to the one-way property of the hash function, it cannot compute hash values that belong to larger sequence numbers, or the same sequence number and smaller hop count values. Thus, the adversary cannot advertise a fresher or a shorter route to S.

SMT

SMT (Secure Message Transmission) [302] is a secure data communication protocol for ad hoc networks that thwarts gray hole attacks. SMT simultaneously uses a set of diverse routes – preferably node disjoint – between a source and a destination. The source first invokes the route discovery function of some underlying routing protocol in order to discover a set of routes to the destination. Then, it splits the message to be sent to the destination into several pieces using a coding scheme [320] that ensures that the original packet can be reconstructed from at least a given

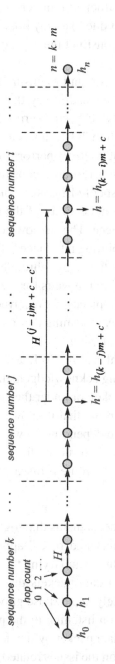

Figure 7.10. Illustration of the hash chain used in SEAD. Each route update concerning S can be verified by the nodes using a previously known hash value from the hash chain of S. For instance, assume that D knows sequence number i, hop count value c, and hash value $h = h_{(k-i)m+c}$ for S, and now it receives a route update concerning S with sequence number $j > i$, hop count c', and hash value h'. Then, D accepts this update only if $H^{(j-i)m+c-c'}(h') = h$, where H^x means that we invoke H iteratively x times.

number of pieces. After that, the pieces are sent to the destination through the set of routes established earlier (one piece per route). At the destination, the message can be reconstructed successfully if a sufficient number of pieces are received. In other words, even if some pieces are lost due to a gray hole attack, or corrupted by other means, the destination can still reconstruct the message if it receives enough number of pieces correctly.

SMT uses MACs to protect the integrity and to ensure the authenticity of the message pieces. The MACs are computed with a key that is shared between the source and the destination; hence, only they can verify the correctness of the pieces. As we have discussed in Subsection 7.3.1, this is a design trade-off, which ensures that intermediate nodes do not need to perform cryptographic computations at the cost of admitting increased resource consumption at intermediate nodes due to forwarding modified or spoofed message pieces. In any case, the destination can verify the integrity and authenticity of the received pieces, and it acknowledges each correctly received piece. The acknowledgement is sent back to the source using the same principle of splitting into pieces. The pieces for which no acknowledgement arrives are re-sent by the source through different routes in order to avoid repeating failures. The destination waits for the re-transmission of the missing pieces, and once enough correct pieces are received, it acknowledges the successful reception of the entire message, meaning that no more re-transmissions are needed even if some pieces are still missing.

Acknowledgements play an important role in SMT, as they allow the source to learn which routes are working: a missing acknowledgement is a strong indication that the corresponding route is either broken or under the control of the adversary. The source maintains a rating for each route that it knows to the destination. The rating of a route is increased or decreased depending on whether the message piece that was sent through that route was delivered successfully or not. When the rating of a route falls below a lower threshold, the route is discarded and not used again.

Watchdog and Pathrater

Watchdog and Pathrater [267] are two mechanisms that together implement a gray hole mitigation tool based on the "detect and react" approach. Watchdog is in charge of continuously monitoring neighbors and trying to identify gray holes (i.e., misbehaving nodes that do not forward data packets that they should forward). Pathrater is used to select routes that likely avoid those gray holes.

The operation of Watchdog is based on listening in the promiscuous mode and trying to catch the transmission of the data packet by the neighbor to which it was forwarded. We have already elaborated on the issues related to the operation of this approach in Subsection 7.3.1; the same can be said about Watchdog. Therefore, here we focus on the operation of Pathrater.

Pathrater assumes that each node maintains a rating in the interval [0, 1] for all the other nodes it knows in the network. Then, the reliability of a route is quantified by the source of a data packet by averaging the ratings of the nodes in that route. The nodes prefer routes with a higher average rating.

Nodes assign ratings to other nodes according to the following algorithm: when a node B becomes known to another node A, then A assigns a neutral rating of 0.5 to B. The ratings of the nodes in each active route are incremented by 0.01 at periodic intervals. The maximum value a rating can reach is 0.8. At the same time, the rating of a node is decreased by 0.05 when a link break is detected and that node becomes unreachable. In addition, the highly negative rating of -100 is assigned to nodes that are suspected of misbehaving by Watchdog. When the route metric is calculated, negative average ratings indicate that the route has a gray hole in it, and the route is not selected for data transmission. Of course, nodes that are incorrectly accused by Watchdog should not be excluded from routing forever, but they should be able to regain a normal rating. Therefore, all ratings are slowly increased in time or set back to a non-negative value after a long timeout.

The specific values of the parameters used by Pathrater seem a bit arbitrary, but the principle is clear, and the simulation results in [267] show that Watchdog and Pathrater can considerably increase the throughput of the network even if a large portion (e.g., 40%) of the nodes are misbehaving.

ODSBR

The ODSBR (On-Demand Source routing with Byzantine Robustness) protocol was proposed in [35] as a source routing protocol for wireless ad hoc networks that tries to detect gray holes (and other misbehavior causing packet delivery failure) with an adaptive acknowledgement scheme and to discover routes that avoid those gray holes.

The protocol consists of three components: (i) Byzantine fault detection, (ii) link weight management, and (iii) route discovery with fault avoidance. Component (i) is responsible for identifying faulty links in the network over which the packet loss ratio exceeds a pre-defined threshold. Each node uses component (ii) to maintain a weight for every link in the network that it knows about. The default weight of a link is 1, and the weight is increased if the link is detected to be faulty. Component (iii) is a route discovery mechanism that takes into account the link weights assigned by the source to the links of the network when selecting routes. In particular, component (iii) is responsible for finding the least weight route from the source to the destination. Since the link weights are related to the reliability of the links, the least weight route should be the most reliable one.

For the detection of faulty links, ODSBR uses the adaptive acknowledgement scheme that we described in Subsection 7.3.1 as a gray hole detection mechanism.

This means that when the packet loss ratio exceeds a given threshold on a route, the source specifies probe nodes on the route that should return acknowledgements for subsequent packets. The selection of the probe nodes implements a binary search on the route that results in the identification of the link where the packets are lost. Either end of this link can be a misbehaving node. ODSBR does not attempt to identify which node is misbehaving, instead, it tries to avoid links that are detected to be faulty.

The mechanism for specifying the list of the probe nodes in a packet is essential for the correct operation of the protocol. The list contains the probe nodes in the same order as they appear on the route, and it is encrypted in an onion-like, layered manner. Each layer corresponds to a probe node on the list, and besides information destined to that probe node, it contains all subsequent layers. The layers are encrypted in such a way that each layer can be decrypted by the probe node corresponding to the previous (outer) layer. This induces a processing order. The first probe node removes the information destined to it and then decrypts the rest of the list before forwarding the packet further. All subsequent probe nodes do the same. This layered encryption prevents the adversary from incriminating other links by removing specific nodes from the probe list.

Acknowledgements must also be handled with care. If the adversary can drop individual acknowledgements, then she can incriminate any arbitrary link along the route. In order to prevent this, each probe node does not send its acknowledgement immediately, but waits for the acknowledgement from the next probe node and combines them into one acknowledgement. If no acknowledgement is received within a timeout, then the probe node gives up waiting, and creates and sends its own acknowledgement only.

The route discovery mechanism of ODSBR floods the network with both route request and route reply messages. The flood of the route request is required to guarantee that it reaches the destination. However, route requests are digitally signed by the source in order to avoid malicious route request flooding by an adversary. The route reply must also be flooded because if it was unicast, a single adversary could prevent a route from being established. If an adversary was able to prevent routes from being established, the fault detection algorithm would be unable to detect and avoid the faulty link, since it requires a route as input in order to operate. The route reply messages are signed by the destination, in order to prevent malicious route reply flooding, and by the intermediate nodes that pass on the route reply, in order to authenticate themselves to the source.

As we mentioned above, the route discovery mechanism of ODSBR finds the least weight route between the source and the destination. This is done in the following way. The source creates and signs a route request that includes the destination, the source, a sequence number, and the weight list of the source (i.e., a list that contains the weights assigned to the links of the network by the source).[7] This route request is flooded in the network, until it reaches the destination. The destination generates a route reply that contains the source, the destination, a sequence number, and the weight list obtained from the route request. The route reply is also flooded in the network. When receiving a route reply, an intermediate node computes the total weight of the sub-route contained in the reply using the weight list included in the reply, and compares it to the minimum weight that it has computed for previously forwarded route replies. If the new route reply contains a route with a smaller weight, then the node appends its identifier to the route reply, signs it, and re-broadcasts it; otherwise, the route reply is dropped. In this way, the source obtains the least weight working route from itself to the destination.

7.4 Provable security for ad hoc network routing protocols

As we have seen in the previous section, many secure routing protocols have been proposed for mobile ad hoc networks. However, the security of those protocols has been analyzed by informal means only. It is well-known that informal arguments about security can be prone to errors, therefore, there is a strong need for a more rigorous analysis technique. In this section, we introduce such an analysis technique based on the simulation paradigm that has already been used in other contexts to prove the security of cryptographic algorithms and protocols.

7.4.1 Why do we need a more rigorous analysis technique?

Our main goal in this subsection is to demonstrate that attacks against ad hoc routing protocols can be very subtle, and therefore, difficult to discover. Consequently, it is also difficult to gain sufficient assurance that a protocol is free of flaws. The approach of verifying the protocol for a few specific configurations can never be exhaustive, thus it is far from satisfactory as a method for security analysis.

In order to support our claims above, we present an attack against Ariadne when used with MACs. We note that a similar attack can also be carried out when TESLA is used, or when digital signatures are used and, for efficiency reasons, intermediate nodes do not verify the signature list in the route request (which is an assumption that is compliant with the description of Ariadne in [182]).

Let us consider the network configuration illustrated in Figure 7.11. We assume that the adversary controls two adversarial nodes (represented by the black nodes in the figure), and it uses only a single compromised identifier A.

---

[7] Note that it is sufficient to include in the weight list only those weights that have non-default values.

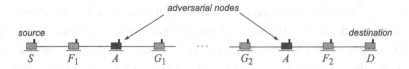

Figure 7.11. Part of a configuration where an attack against Ariadne is possible. The adversary controls two adversarial nodes, depicted in black, and it uses only a single compromised identifier A. Used with permission, from [12], © IEEE, 2006.

S initiates a route discovery process toward D. The first adversarial node receives the following route request:

$$msg_1 = (rreq,\ S,\ D,\ id,\ h_{F_1},\ [F_1],\ [mac_{F_1}]).$$

The adversary does not append the MAC of A to the request, instead, it puts h_{F_1} on the MAC list, and re-broadcasts the following request:

$$msg_2 = (rreq,\ S,\ D,\ id,\ h_{F_1},\ [F_1, A],\ [mac_{F_1}, h_{F_1}]).$$

Recall that the intermediate nodes cannot verify the MACs in the request. Note also that MAC functions based on cryptographic hash functions (e.g., HMAC [237]) output a hash value as the MAC, and therefore, h_{F_1} looks like a MAC. Hence, G_1 will not detect the attack, and the following request arrives at the second adversarial node:

$$msg_3 = (rreq,\ S,\ D,\ id,\ H(G_2, \ldots, H(G_1, h_{F_1})),$$
$$[F_1, A, G_1, \ldots, G_2],\ [mac_{F_1}, h_{F_1}, mac_{G_1}, \ldots, mac_{G_2}]).$$

The adversary removes $G_1, \ldots, G_2$ from the node list and the corresponding MACs from the MAC list. The adversary can do this in the following way: by recognizing identifier A in the accumulated route, the adversary knows that the request passed through the first adversarial node. By looking at the position of identifier A in the node list, the adversary will know where h_{F_1} is on the MAC list. From h_{F_1}, the adversary computes $h_A = H(A, h_{F_1})$ and a MAC on $(rreq, S, D, id, h_A, [F_1, A], mac_{F_1})$, and re-broadcasts the following request:

$$msg_4 = (rreq,\ S,\ D,\ id,\ h_A,\ [F_1, A],\ [mac_{F_1}, mac_A]).$$

Since the per-hop hash value and both MACs are correct in msg_4, D will receive a correct request, and returns the following reply:

$$msg_5 = (rrep,\ D,\ S,\ [F_1, A, F_2],\ mac_D).$$

When the reply reaches the second adversarial node, it will forward the following message to G_2:

$$msg_6 = (rrep,\ D,\ S,\ [F_1, A, G_1, \ldots, G_2, A, F_2],\ mac_D).$$

Note that $G_1, \ldots, G_2$ cannot verify the MAC in msg_6. In addition, their identifiers are in the route carried by the reply, and the preceding and following identifiers belong to their neighbors. Therefore, each of them forwards the reply. Finally, when the first adversarial node receives the reply, it removes $G_1, \ldots, G_2$ and one of the A's from the node list:

$$msg_7 = (rrep, \; D, \; S, \; [F_1, A, F_2], \; mac_D).$$

In this way, S receives the route reply that D sent. This means that the MAC verifies correctly and S accepts the route (S, F_1, A, F_2, D), which is non-existent.

It must be noted that in msg_6, the compromised identifier A appears twice in the node list. Note, however, that Ariadne does not specify that intermediate nodes should check the node list in the reply for repeating identifiers. If each honest node checks only that its own identifier is in the list and that the preceding and following identifiers belong to its neighbors, then the attack works. Moreover, a slightly modified version of the attack would work even if the intermediate nodes checked repeating identifiers in the reply. In that case, the second adversarial node would send the following reply towards S:

$$msg'_6 = (rrep, \; D, \; S, \; [F_1, X, G_1, \ldots, G_2, A, F_2], \; mac_D),$$

where X can be any identifier that is different from the other identifiers in the node list. With non-negligible probability,[8] X is a neighbor of G_1, and thus, G_1 will pass the reply on, so that the first adversarial node can overhear it. Then, the adversary can remove the identifiers $X, G_1, \ldots, G_2$, and send the reply containing the node list (F_1, A, F_2) to F_1. F_1 will process the reply, because it contains no repeating identifiers and A is its neighbor.

This is an attack aimed at creating an incorrect routing state in some nodes. In particular, the source will accept a non-existent route and cache it in its route cache. In addition, the attack is powerful, because despite the use of the per-hop hash mechanism, the adversary manages to shorten an existing route, and therefore, the source will probably prefer this short route over others (assuming there are other alternative routes between S and D that are not illustrated in Figure 7.11). As a consequence, the source will probably start sending data packets through a non-existent route.

At this point, it must be clear that proving that a routing protocol is free from this and similar kinds of attacks is virtually impossible by informal reasoning.

---

[8] In fact, the probability that X is a neighbor of G_1 is greater than n_{G_1}/n, where n is the number of nodes in the network and n_{G_1} is the number of G_1's neighbors.

7.4.2 A framework for security analysis

In this subsection, we shortly introduce a framework in which routing protocols can be analyzed in a rigorous manner. For a detailed description of this framework, we refer to [12]. The framework has been developed for topology-based protocols and, in particular, for on-demand source routing and on-demand distance vector based protocols. In addition, routing security is defined in terms of resistance against attacks aimed at creating an incorrect routing state in the network. Thus, route disruption, route diversion, generation of extra control traffic, and gray hole attacks are not considered within the framework. In this sense, the framework is somewhat restricted; but as we will see, dealing only with attacks aimed at creating an incorrect routing state is already sufficiently complicated if we want to be really precise.

The framework consists of a static and a dynamic model of the system and a formal definition of routing security based on these models. That is described in this subsection. In the next subsection, we illustrate the use of the framework by using it to prove the security of the endairA protocol.

Static representation of the system

Network model We model the ad hoc network as an undirected labeled graph $G(V, E)$, where V is the set of vertices and E is the set of edges. Each vertex represents a node, and there is an edge between two vertices if the corresponding nodes can hear each other (via either a radio link or perhaps a wormhole).

We assume that the nodes use authenticated identifiers (e.g., public keys) during neighbor discovery and in the routing protocol. We denote the set of identifiers by L, and we label each vertex v of G with the identifiers used by the node corresponding to v. We assume that honest (non-adversarial) nodes use a single identifier that is unique in the network, whereas adversarial nodes can use multiple compromised identifiers (see attacker model below).

For the purpose of modeling distance vector routing protocols, we also assign cost values to the nodes and to the radio links that can be interpreted as, respectively, processing and transmission costs and can be used to compute routing metrics.

Adversary model We assume that the adversary is not all powerful, but it launches its attacks from adversarial nodes that it controls and that have communication capabilities similar to regular nodes. We denote the vertices that correspond to adversarial nodes by V^*. In addition, we assume that the adversary compromises some identifiers, by which we mean that the adversary compromises the cryptographic keys that are used to authenticate those identifiers. We denote the set of compromised identifiers by L^*. We further assume that the adversary distributes all compromised identifiers to all adversarial nodes. Using the notation intro-duced in [182], the adversary described above is an Active-y–x adversary, where $x = |V^*|$ and $y = |L^*|$. In addition, we assume that the adversary is static in the sense that it does not corrupt more nodes and does not compromise more identifiers during the operation of the system.

As neighboring adversarial nodes can communicate with each other in an unre-stricted manner (e.g., by sending encrypted messages), they can appear as a single node (under all the compromised identifiers) to the other nodes. Hence, without loss of generality, we assume that adversarial nodes are not neighbors in G; if they were, we could merge them into a single adversarial node that would inherit all the neighbors of the original nodes.

Configuration A configuration is represented by the graph G, the set of adversar-ial nodes V^*, the labelling of the nodes with identifiers from L, and the assignment of costs to the nodes and to the links. We make the simplifying assumption that the configuration is static (at least during the time interval that is considered in the analysis).

Correctness criteria

Source routing From secure source routing protocols, we require that they return only "existing" routes. However, we must take into account that the adversary can always emulate the execution of the routing protocol using the compromised iden-tifiers locally within a single adversarial node. Hence, the adversary can always extend any route that passes through an adversarial node with any sequence of compromised identifiers. This is a fact that our definition of security must tolerate, otherwise we cannot hope that any routing protocol will satisfy it. This observation leads to the following definition of "existing" routes.

Definition 7.1 A sequence $\ell_1, \ell_2, \ldots, \ell_p$ of identifiers is a *plausible route* with respect to a configuration if each ℓ_i is different, and the sequence can be partitioned into k sub-sequences in such a way that each of the resulting partitions is a subset of the identifiers assigned to a vertex in V, and in addition, these vertices form a path in G.

Distance vector routing In distance vector routing, no explicit routes are returned by the route discovery procedure, but rather the state of the system is represented by the routing tables of the non-adversarial nodes. We assume that an entry of the routing table of every node contains the following three fields: the identifier of the target node, the identifier of the next hop towards the target, and the cost value that represents the believed cost of the route to the given target via the given next hop. Without loss of generality, we assume that the routing metric is such that routes with lower cost values are preferred.

Then, we define a correct state as follows.

Definition 7.2 The system is in a *correct state* if all the routing table entries of the non-adversarial nodes are correct. If non-adversarial node v has an entry for target ℓ_{tar} with next hop ℓ_{nxt} and cost c, then this entry is correct if there exists a route in the network that starts from node v, ends at a node that uses the identifier ℓ_{tar}, passes through a neighbor of v that uses identifier ℓ_{nxt}, and has a cost that is smaller than or equal to c.

Dynamic representation of the system

The dynamic behavior of the system is represented by two models, each consisting of a set of interactive probabilistic Turing machines. One of the models is called the *real-world model*, and it represents the behavior of the real system; the other is called the *ideal-world model*, and it describes how the system should work ideally. In both models, there is an adversary whose behavior is not constrained, apart from requiring it to run in time polynomial in the security parameter (e.g., size of the cryptographic keys used by the cryptographic primitives). This allows us to consider *any* feasible attacks and makes the approach very general. Although the adversary is not constrained, the construction of the ideal-world model ensures that all of its attacks are unsuccessful against the ideal-world system. In other words, the ideal-world system is secure by construction (e.g., non-plausible routes are never returned).

Formal definition of security

Once the models are defined, the goal is to prove that for any real-world adversary, there exists an ideal-world adversary that can achieve essentially the same effects in the ideal-world model as those achieved by the real-world adversary in the real-world model (i.e., the ideal-world adversary can simulate the real-world adversary). The existence of a proof means that no attacks can be successful in the real-world model (or more precisely, attacks can be successful only with negligible probability), otherwise an attack would be successful in the ideal-world model too, which is impossible by definition. This leads to the following definition of routing security.

Definition 7.3 A routing protocol is said to be (*statistically*) *secure* if, for any configuration and any real-world adversary, there exists an ideal-world adversary, such that the output of the real-world model is (statistically) indistinguishable from the output of the ideal-world model.

Hence, if a routing protocol is (statistically) secure, then any system using this routing protocol returns a non-plausible route or gets into an incorrect state only with negligible probability. This negligible probability is related to the fact that the adversary can always forge the cryptographic primitives (e.g., generate a valid digital signature) with a very small probability.

7.4.3 An example – the security proof of endairA

In this subsection, we want to illustrate the use of the framework introduced above. For this reason, we will prove the security of the endairA protocol. However, in order to do this, we need to give some more details about the dynamic representation of the system.

Recall that the dynamic representation of the system consists of an ideal-world model and a real-world model, where the ideal-world model is defined in such a way that attacks are not possible in it, whereas the real-world model allows all kinds of misdeeds by the adversary. Specifically, in source routing protocols, the ideal-world model is constructed in such a way that the Turing machine that is responsible for passing messages between the Turing machines that represent the nodes marks route reply messages that carry non-plausible routes. We can imagine this as route reply messages with a fictive plausibility flag attached to them, and the value of this flag is set by the Turing machine that handles the communication in the model. Then, in the ideal-world model, it is ensured that the routes received in marked route reply messages are not accepted by the honest nodes, whereas obviously, in the real-world model, the plausibility flags of the messages are ignored (since in reality there is no such flag attached to the messages). Thus, the ideal-world model is ideal in the sense that non-plausible routes are never returned to the honest nodes by definition.

Let us now prove the following theorem.

Theorem 7.1 *The endairA protocol described in Subsection 7.3.2 is (statistically) secure if the signature scheme used in the protocol is secure against chosen message attacks.*

Proof We provide only a sketch of the proof. We want to show that for any configuration and for any adversary, a route reply message in the ideal-world model is discarded owing to the value of its plausibility flag (meaning that the message carries a non-plausible route) with negligible probability. Discarding a message because of its plausibility flag leads to a simulation error, because the message is not discarded in the real-world model where plausibility flags are ignored. By showing that messages are discarded with negligible probability due to their plausibility flags in the ideal-world model, we show that simulation errors occur with negligible probability, or in other words, the effects of the real-world adversary can be simulated in the ideal-world model with overwhelming probability, and this is the basis of our security definition.

In what follows, we will refer to non-adversarial nodes with their identifiers. Let us suppose that the following route reply is received by a non-adversarial node ℓ_{src} in the ideal-world model:

$$msg = (rrep, \ell_{src}, \ell_{dst}, id, [\ell_1, \ldots, \ell_p], [sig_{\ell_{dst}}, sig_{\ell_p}, \ldots, sig_{\ell_1}]).$$

Let us suppose that msg passes all the verifications required by endairA at ℓ_{src}, which means that all signatures in msg are correct, and ℓ_{src} has a neighbor that uses the identifier ℓ_1. Let us further suppose that msg has been received with a plausibility flag indicating that the message contains a non-plausible route. This means that $(\ell_{src}, \ell_1, \ldots, \ell_p, \ell_{dst})$ is a non-plausible route in the given configuration. Hence, msg is dropped due to the value of its plausibility flag. We will show that this situation is possible only if the adversary forged the digital signatures of some non-adversarial nodes.

Recall that, by definition, adversarial nodes cannot be neighbors. In addition, each non-adversarial node has a single and unique non-compromised identifier assigned to it. It follows that every route, including $(\ell_{src}, \ell_1, \ldots, \ell_p, \ell_{dst})$, has a unique *meaningful* partitioning, which is the following: Each non-compromised identifier, as well as each sequence of consecutive compromised identifiers should form a partition.

Let $P_1, P_2, \ldots, P_k$ be the unique meaningful partitioning of $(\ell_{src}, \ell_1, \ldots, \ell_p, \ell_{dst})$. The fact that this route is non-plausible implies that at least one of the following two statements holds:

- *Case 1* There exist two partitions $P_i = \{\ell_j\}$ and $P_{i+1} = \{\ell_{j+1}\}$ such that both ℓ_j and ℓ_{j+1} are non-compromised identifiers, and the corresponding non-adversarial nodes are not neighbors.
- *Case 2* There exist three partitions $P_i = \{\ell_j\}$, $P_{i+1} = \{\ell_{j+1}, \ldots, \ell_{j+q}\}$, and $P_{i+2} = \{\ell_{j+q+1}\}$ such that ℓ_j and ℓ_{j+q+1} are non-compromised and $\ell_{j+1}, \ldots, \ell_{j+q}$ are compromised identifiers, and the non-adversarial nodes that correspond to ℓ_j and ℓ_{j+q+1} have no common adversarial neighbor.

As we will see, in both cases, the adversary must have forged the digital signature of a non-adversarial node.

In Case 1, node ℓ_{j+1} does not sign the route reply, because it is non-adversarial and it detects that the identifier that precedes its own identifer in the route does not belong to a neighboring node. Hence, the adversary must have forged $sig_{\ell_{j+1}}$ in msg.

In Case 2, the situation is more complicated. Let us assume that the adversary has not forged the signature of any of the non-adversarial nodes. Node ℓ_j must have received

$$msg' = (rrep,\ \ell_{\text{src}},\ \ell_{\text{dst}},\ id,\ [\ell_1, \ldots, \ell_p],\ [sig_{\ell_{\text{dst}}}, sig_{\ell_p}, \ldots, sig_{\ell_{j+1}}])$$

from an adversarial neighbor, say A, since ℓ_{j+1} is compromised, and thus, a non-adversarial node would not send out a route reply message with $sig_{\ell_{j+1}}$. In order to generate msg', A must have received

$$msg'' = (rrep,\ \ell_{\text{src}},\ \ell_{\text{dst}},\ id,\ [\ell_1, \ldots, \ell_p],\ [sig_{\ell_{\text{dst}}}, sig_{\ell_p}, \ldots, sig_{\ell_{j+q+1}}])$$

because, by assumption, the adversary has not forged the signature of ℓ_{j+q+1}, which is non-compromised. As A has no adversarial neighbor, it could have received msg'' only from a non-adversarial node. However, the only non-adversarial node that would send out msg'' is ℓ_{j+q+1}. This would mean that A is a common adversarial neighbor of ℓ_j and ℓ_{j+q+1}, which contradicts the assumption of Case 2. This means that our assumption that the adversary has not forged the signature of any of the non-adversarial nodes cannot be true.

It should be intuitively clear that if the signature scheme is secure, then the adversary can forge a signature only with negligible probability, and thus, a route reply message in the ideal-world model is dropped because of the value of its plausibility flag only with negligible probability. Nevertheless, we sketch how this could be proven formally. The proof is indirect. We assume that there exists a configuration and an adversary such that a route reply message in the ideal-world model is dropped due to its plausibility flag with probability ϵ, and then, based on that, we construct a forger F that can break the signature scheme with probability ϵ/n, where n is the number of non-adversarial nodes in the network. If ϵ is non-negligible, then so is ϵ/n, and thus the existence of F contradicts the assumption about the security of the signature scheme.

The construction of F is the following. Let puk be an arbitrary public key of the signature scheme. Let us assume that the corresponding private key prk is not known to F, but F has access to a signing oracle that produces signatures on submitted messages using prk. F runs a simulation of the ideal-world model where all nodes are initialized with the keys of the corresponding nodes, except that the public key of a randomly selected non-adversarial node ℓ_i is replaced with puk. During the simulation, whenever ℓ_i signs a message m, F submits m to the oracle, and replaces the signature of ℓ_i on m with the one produced by the oracle. This signature verifies correctly at other nodes later, as the public verification key of ℓ_i is replaced with puk. By assumption, with probability ϵ, the simulation of the ideal-world model will result in a route reply message msg such that all signatures in msg are correct and msg contains a non-plausible route. As we saw above, this means that there exists a non-adversarial node ℓ_j such that *msg* contains the signature sig_{ℓ_j} of ℓ_j, but ℓ_j has never signed (the corresponding part of) *msg*. Let us assume that $i = j$. In this case, sig_{ℓ_j} is a signature that verifies correctly with the public key *puk*. Since ℓ_j did not sign (the corresponding part of) *msg*, F did not call the oracle to generate sig_{ℓ_j}. This means that F managed to produce a signature on a message that verifies correctly with *puk*. Since F selected ℓ_i randomly, the probability of $i = j$ is $1/n$, and hence, the success probability of F is ϵ/n. $\qquad\square$

7.5 Secure routing in sensor networks

Wireless sensor networks are envisaged to use multi-hop communications in order to reduce the interference between the nodes and the overall energy consumption of the network. Ultimately, using multi-hop communications is expected to result in an increased network lifetime, which is crucial in many sensor network applications, because recharging the batteries of the nodes may be impossible, or at least very impractical. However, using multi-hop communications raises the problem of secure routing in sensor networks too.

The types of attacks that an adversary can mount against the routing protocol in a wireless sensor network are similar to those listed for mobile ad hoc networks with a somewhat stronger emphasis on increased resource consumption as their objectives, because sensor nodes are highly resource constrained. This similarity stems from the similar assumptions that can be made about the capabilities and the attack mechanisms of the adversary in both cases. Consequently, the applicable countermeasures are similar too. One may even think that secure routing protocols developed for mobile ad hoc networks could directly be used in wireless sensor networks. However, this is not the case in general owing to the following important differences between mobile ad hoc networks and wireless sensor networks.

- First of all, in sensor networks, node-to-node communications are usually not required. Rather, the sensor nodes must be able to communicate with the base station (e.g., to send sensor readings), and vice versa (e.g., to send control information or specific queries). Thus, in sensor networks, the prevailing communication types are many-to-one and one-to-many, in contrast to mobile ad hoc networks, where most of the communications are one-to-one. As a consequence, secure ad hoc network routing protocols designed to support one-to-one communications may not be efficient for many-to-one and one-to-many communications.
- In the majority of the envisaged sensor network applications, the nodes are static, and therefore, the topology changes are less dynamic in sensor networks than they are in ad hoc networks. Some topology changes may still occur in sensor networks, as sensor nodes may disappear temporarily owing to some failure,

or permanently owing to battery depletion, but the resulting dynamicity of the network is much lower than that in ad hoc networks where the nodes are mobile. This means that secure ad hoc network routing protocols designed to cope with the dynamic nature of the network may contain features that are unnecessary in sensor networks, or at least, that can be implemented in a more efficient way.

- Sensor nodes are assumed to be much more resource constrained than the nodes in mobile ad hoc networks. Typically, in mobile ad hoc networks, the nodes are hand-held devices, such as PDAs or mobile phones, or even laptop class computers. These have orders of magnitude more resources than a typical sensor node. For instance, a contemporary PDA is equipped with a memory of several hundreds of megabytes, while a typical sensor node has only a few kilobytes of memory. In terms of CPU speed, the differences are similarly large. We note that these differences will likely persist in the future owing to the objective of keeping the price of sensor nodes at a very low level. In addition to the differences in memory size and CPU speed, reduced energy consumption is even more critical in sensor networks than it is in ad hoc networks. For all these reasons, secure ad hoc network routing protocols that rely on public key cryptography (e.g., ARAN, SAODV, and endairA) cannot be directly used in sensor networks.

In the rest of this section, we sketch three approaches to secure routing in wireless sensor networks. The first approach extends a known topology-based sensor network routing protocol, TinyOS beaconing, with cryptographic protection to defend against spoofing routing control messages. In this way, a reasonable level of security can be achieved against an outsider adversary, but the protocol can still be successfully attacked by an insider adversary. The second approach uses the principles of link-state routing. The advantage of link-state routing is that the routing control messages do not need to be modified by intermediate nodes, and hence, they can be protected in an end-to-end manner, which simplifies the protocol. Finally, the third approach is based on extending a position-based routing protocol with security measures. The rationale of starting from a position-based routing protocol is that such protocols usually do not maintain routing state in the nodes and, hence, the routing state cannot be corrupted. For the same reason, position-based routing protocols also incur less overhead, and hence, they are more energy efficient than the topology-based protocols.

7.5.1 Authenticated TinyOS beaconing

TinyOS beaconing is a simple (but insecure) sensor network routing protocol. This protocol establishes a routing tree rooted at the base station. Once the tree is established, sensor nodes forward data packets towards the base station by sending them to their parent in the tree. The establishment of the tree is based on a network wide flooding. The base station generates a route update message and broadcasts it to its neighbors. Each node that receives the route update message for the first time sets the node from which the route update was received as its parent, and then re-broadcasts the route update message. Any copy of the route update that is received later by the node is discarded.

Since route update messages are not authenticated, an adversary can spoof them. As a result, the adversary can initiate the routing tree establishment process, and she can become the root of the established tree. Then, every sensor node will send data packets to the adversary, who can inspect and drop them. Thus, the adversary can easily obtain information from the entire system, which means increased control over the communications. Moreover, the quality of the service provided by the system is decreased, as the base station receives no sensor readings anymore, and the resource consumption of the nodes is increased.

To protect against this attack, the base station can authenticate the route update message. Since every node must be able to verify the authenticity of the route update, the base station should use a broadcast authentication scheme. Taking into account the resource constraints of the sensor nodes, a good candidate for this would be the TESLA broadcast authentication protocol, which uses only symmetric key cryptography. This would prevent the adversary from initiating the routing tree establishment process, but there are still other attacks that she can perform. For instance, she can spoof the node identifier of a far-away node when re-broadcasting the route update message. All nodes that hear the spoofed route update message will set that far-away node as their parent. Later, when these nodes want to forward data packets towards the base station, they will send the packets into void. This results in decreased quality of service and increased resource consumption by some nodes.

The above described attack will be discovered quickly if data packets are acknowledged at the link layer. However, an even more subtle attack is possible against the protocol and can only be detected in an end-to-end manner (meaning that much more resources are wasted before successful detection). The attack is illustrated in Figure 7.12. Let us assume that the adversary resides near a node u, and u has a neighbor v, which is further away from the base station than u itself. When the adversary receives the authenticated route update message, it re-broadcasts it in the name of v, and therefore, u sets v as its parent. When u re-broadcasts the route update message, v sets u as its parent. Thus, the adversary creates a routing loop between u and v by arranging that they both set each other as parent. The result is decreased quality of service, because some data packets will never reach the base station, and increased resource consumption for nodes u and v, and for all nodes downstream from them.

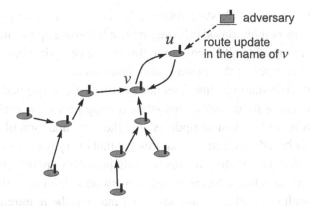

Figure 7.12. Illustration of an attack against the TinyOS beaconing protocol, where the adversary creates a routing loop between two nodes u and v by arranging that they both set the other as parent in the routing tree. In order to achieve this, the adversary re-broadcasts the route update message in the name of v. The result of the attack is decreased quality of service, because some data packets will never reach the base station, and increased resource consumption for nodes u and v, and for all nodes downstream from them.

To protect against the spoofing of node identifiers, the route update message should also be authenticated in a hop-by-hop manner. This requires pairwise keys between neighboring nodes, which can be set up as described in Section 5.1. While hop-by-hop authentication results in a larger overhead, this can still be bearable because only symmetric key cryptography is used, and the routing tree establishment procedure is run rather infrequently due to the static nature of the network.

The authenticated TinyOS beaconing protocol provides a reasonable level of protection against an outsider adversary, but an insider adversary (i.e., one that compromised the cryptographic keys of some sensor nodes) can still mount some attacks. Another disadvantage of the authenticated TinyOS beaconing protocol is that it supports only tree topologies, only node to base station communication, and only single path forwarding.

7.5.2 *Centralized link-state routing*

In the TinyOS beaconing protocol, the routing state of every node is determined by the identity of the neighbor from which it receives the routing update message. Therefore, in order to prevent the creation of incorrect routing states in the nodes, this identity information needs to be protected from spoofing. That is the reason for hop-by-hop authentication of routing update messages.

Another approach could be to avoid that the routing state depends on information contributed by intermediate nodes to routing control messages. One way to achieve this is to use link-state routing. In link-state routing protocols, each node distributes its neighborhood information in a link-state update message. Since other nodes do not need to add anything to this link-state update message, it is sufficient if only the source of the message authenticates it.

At first sight, link-state routing does not seem to be a very good idea for sensor networks, because in traditional link-state routing protocols, each node floods the entire network with link-state updates, and the authentication of link-state update messages is based on digital signatures (so that every node can verify them). Note, however, that traditional link-state routing protocols are designed to support one-to-one communications. Since in sensor networks, the nodes do not need to communicate with each other in an end-to-end manner, the requirement of flooding the entire network with link-state update messages can be relaxed. In addition, if link-state update messages are not flooded in the entire network, then there is no need to ensure that every node can verify them, and hence, digital signatures can be replaced with more efficient primitives.

Link-state routing in wireless sensor networks could work in the following way. Each node collects its neighborhood information locally, and sends the list of its neighbors to the base station in a link-state update message. Based on the received link-state update messages, the base station constructs the routing table for every node, and distributes the computed routing tables to the nodes. A protocol based on these principles would have many advantages.

- Each node needs a single symmetric key that it shares with the base station. This key can be used to protect routing control messages (link-state updates originating from the node and routing tables destined to the node) using an efficient symmetric key MAC.
- The computation of the routing topology is performed by the base station, which is much more powerful than the sensor nodes, and it has no resource constraints.
- The routing tables are computed centrally using information obtained from the entire network. This allows the construction of highly optimized routing topologies, which may greatly reduce the overall energy consumption of the network.
- Related to the previous point, the base station can run centralized wormhole detection algorithms such as those described in Subsection 6.2.1.
- There is no restriction on the form of the routing tables that are distributed to the nodes. Hence various kinds of routing schemes can be easily supported in the link-state approach ranging from simple routing trees to more complex multipath routing.

One remaining problem is that the sensor nodes do not necessarily know how to route link-state update messages to the base station, since at that stage of the protocol, they have not received their routing tables yet. Indeed, the very purpose of sending the link-state update messages to the base station is to let it compute the routing tables.

This problem can be solved by a mechanism that is similar to TinyOS beaconing: the base station floods the network with a link-state request message, which informs the nodes that the base station collects link-state information, and helps to establish an initial routing state in every sensor node that receives the request. In particular, each node can set the node from which it received the link-state request for the first time as the next hop towards the base station. Later, the node can forward the link-state update messages to the base station via this next hop node.

Note that the above described mechanism assumes that the links are bi-directional. Note also that the same kind of attacks may be possible here as in the case of TinyOS beaconing. However, the difference is that flooding the network with the link-state request establishes only an initial routing state in the nodes, which is used during the transfer of the link-state update messages (i.e., for a very limited time), and later, this initial routing state is replaced by the routing tables computed by the base station. Therefore, the attacks on this initial phase have a limited impact on the resource consumption of the nodes. In addition, as link-state update messages cannot be spoofed and modified by the adversary due to their MAC, the only possible attack that the adversary can effectively mount against the protocol is to arrange that some link-state updates do not arrive to the base station. However, this can always be achieved by jamming as well.

It is also important to prevent that the adversary can initiate the flooding of the network with a spoofed link-state request, because flooding consumes a lot of energy, and also because that would trigger the transfer of the link-state update messages, which also consumes energy. A simple approach to achieve this is to let the base station create a one-way hash chain and release the elements of this hash chain in the link-state request messages in such a way that a released element can be checked against the previously released, already authenticated elements, but the next element of the chain cannot be computed. This ensures that the adversary cannot generate a link-state request message before the base station releases the next element, and therefore, malicious flooding is prevented.

The distribution of the routing tables can use source routing, because the base station has a full view of the network topology, and hence, it can easily determine source routes to every node in the network. Alternatively, each node u may record the identifier of those nodes whose link-state update messages it forwarded towards the base station. Later, when node u overhears a transmission of the routing table destined to one of the recorded nodes, it re-broadcasts that routing table to its neighbors. In this way, the nodes that forwarded the link-state update of a node

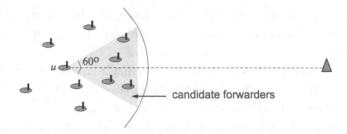

Figure 7.13. Illustration of the selection of the next forwarder node in the IGF protocol. The current forwarder u broadcasts an RTS containing its own position and the position of the destination. The neighboring nodes that reside in the 60° sextant centered on the direct line from u to the destination are the candidate next forwarders. The first that sends a CTS will be selected as the next hop. CTS timers are set so that nodes that are closer to the destination and have more remaining energy are favored.

also forward the routing table of the same node, and therefore, the routing tables are delivered to their destinations.

Obviously, sending the link-state update messages to the base station and distributing the routing tables to the nodes result in a considerable overhead. However, this is counterbalanced by the fact that the nodes' routing tables are highly optimized and the bulk of the communications (i.e., sending the sensor readings to the base station) will use these routing tables. Moreover, sensor networks are rather static networks, meaning that link-state changes occur rarely. Thus, the overhead can be reduced considerably by requiring the nodes to send only the changes with respect to their previous link-states to the base station. Similarly, the base station can send to each node only the changes made in the routing table of the node.

7.5.3 Secured position-based routing

In general, position-based routing protocols do not require the nodes to maintain any routing state (apart from their own geographical position). For this reason, position-based routing protocols incur less overhead than topology-based protocols. In addition, this property also makes them inherently resistant against attacks aiming at creating incorrect routing states in the nodes. Therefore, a position-based routing protocol can be a good starting point to construct a secure sensor network routing protocol.

Let us consider, for instance, the Implicit Geographic Forwarding (IGF) protocol [57]. In IGF, routing is integrated with the RTS/CTS handshake of the MAC layer. A node u that wants to forward a data packet broadcasts an RTS containing its own position and the position of the destination. The neighboring nodes that reside in the 60° sextant centered on the direct line from u to the destination are eligible to forward the packet further (see Figure 7.13 for illustration). These candidate forwarder nodes set their CTS response timer inversely proportional to a weighted sum of their distance from u, remaining energy, and perpendicular distance to the line from u to the destination. Hence, the node that is the most desirable candidate to forward the packet sends its CTS first. The other candidate nodes hear this CTS (due to the fact that they are in the sextant) and cancel their timers. Finally, u sends the data packet to the candidate neighbor that sent the CTS. Voids are dealt with by shifting the forwarding sextant to the side and repeating the above described procedure.

The problem of IGF is that an adversarial neighbor of u can always arrange that it is selected as the next forwarder by not respecting the protocol and sending a CTS immediately after receiving the RTS. Then, the adversary can drop the data packets, by which she can effectively disable the communication between the sensor nodes downstream from u and the base station.

Fortunately, IGF can be modified to make it more resistant to this attack. The idea is that instead of sending the data packet to the first neighbor that responds with a CTS, u can wait in a specified time window and collect multiple CTS responses. Clearly, this also requires that the candidate next forwarders do not cancel their CTS timers when they hear a CTS response to the RTS of u. Then, u can select the next forwarder randomly, thereby reducing the chances of the adversary to be selected. It is also possible to select multiple next forwarders, which reduces the impact of adversarial nodes met along the way to the base station. This extension of IGF still results in a stateless protocol, which is advantageous with respect to security.

However, there is still one problem: the adversary can spoof node identifiers, and hence, a single adversarial node can appear as several different nodes at different locations in the neighborhood of u. In other words, the adversary can mount a Sybil attack (see Chapter 4 for the description of this attack), and this has a negative impact on the extended IGF protocol, because it increases the chances that the adversary is selected as the next forwarder.

In order to prevent this attack, we must give up the fully stateless property of the protocol. A straightforward way to mitigate this attack is to let each sensor node authenticate its neighbors and restrict the selection of the next forwarder to the set of authenticated neighbors. This, however, requires pairwise keys between the neighboring nodes.

Neighbor authentication excludes attacks from outsider adversaries completely, but an insider adversary that can use the identities of some compromised nodes may still try to introduce virtual nodes in the network under the compromised identities. To mitigate this attack as well, the nodes can monitor the behavior of their neighbors locally and assign trust values to them. The trust value of a neighbor that often fails to forward data packets, introduces long delays, or reports itself at different positions frequently can be progressively decreased until it reaches a threshold below which that neighbor is not selected as next forwarder anymore. Note that the trust values are computed by the nodes based on their local observations, and no reputation reports need to be disseminated in the network. This makes sense because sensor networks are static, and therefore, the local neighborhood of the nodes does not change considerably in time. This makes it possible to progressively determine the trust values. In addition, there is no need for recommendations for new neighbors, as new nodes appear very rarely in the neighborhood.

7.6 Summary

In this chapter, we studied the problem of securing routing in multi-hop wireless networks, notably in mobile ad hoc networks and in wireless sensor networks. First, we gave an overview on the different approaches for routing in mobile ad hoc networks. Then, we described why and how ad hoc network routing protocols can be attacked. This involved the presentation of the different objectives an adversary could have (such as increasing adversarial control over the communications between some nodes, degrading the quality of service provided by the network, and increasing the resource consumption of some nodes) and the detailed description of the different attack types (such as route disruption, route diversion, creation of incorrect routing state, generation of extra control traffic, and creation of gray holes).

Following this, we presented general countermeasures against the attacks (such as origin authentication of control packets, protection of mutable information in control packets, detection of tunnels, and techniques to mitigate the effect of gray holes), and we illustrated how these countermeasures can be applied by describing a set of ad hoc network routing protocol designed with security in mind.

We argued that informal reasoning about the security of ad hoc network routing protocols is prone to errors and we demonstrated, via an example, how subtle the attacks against (secured) routing protocols can be. Then, we briefly presented a formal framework in which routing security can be precisely defined and rigorous proofs about the security of routing protocols can be carried out. We demonstrated the usage of this framework by sketching the security proof of the endairA protocol.

Finally, we studied the problem of secure routing in wireless sensor networks. We identified the main differences between mobile ad hoc networks and wireless sensor networks, which render the application of secure ad hoc network routing protocols largely inappropriate in sensor networks. Then, we described three approaches to secure routing in wireless sensor networks: authenticated TinyOS beaconing, centralized link-state routing, and secured position-based routing. We argued that link-state routing and position-based routing are advantageous with respect to security, because in link-state routing, the routing control messages do not need to be modified by intermediate nodes (hence, there is no added information that needs to be authenticated), and in position-based routing, the nodes do not need to maintain routing state (hence, the routing state cannot be corrupted).

7.7 To probe further

Surveys on ad hoc network routing protocols and routing protocols for wireless sensor networks can be found in [333, 16].

An overview of position-based routing protocols is presented in [269]. Location services for geographic routing protocols are proposed in various papers. For instance, the location service of the DREAM protocol is presented in [42], a quorum-based distributed location service is proposed in [158], a grid-based hierarchical location service is presented in [258], and the concept of the virtual home region (a geographical area determined by the ID of the node) is introduced in [151]. Another interesting approach for position-based routing that does not require a location service is called Last Encounter Routing and described in [123]. The effect of falsified location data on position-based routing protocols is studied in [254].

Several groups of researchers have studied the problem of securing ad hoc network routing protocols, including Papadimitratos and Haas [299, 301, 302, 303, 304]; Hu, Perrig, and Johnson [177, 181, 182]; Zapata and Asokan [391]; and Sanzgiri, Dahill, Levine, Shields, and Belding-Royer [337]. An overview of the various approaches for securing routing in ad hoc networks can be found in [178].

The tunneling attack was first introduced by Papadimitratos and Haas in [299]. Later, it was further analyzed by Kruus *et al.* in [238] (where it is called *in-band wormhole*).

A survey of two signature aggregation techniques is published by Boneh, Gentry, Lynn, and Shacham in [59]. It would an interesting research direction to study how these techniques can be used for replacing signature lists in the messages of ad hoc network routing protocols.

The formal framework for the security analysis of ad hoc network routing protocols described in Section 7.4 was proposed by Ács, Buttyán, and Vajda in [78, 11, 12]. Another approach based on the simulation paradigm has been proposed by Kong, Hong, and Gerla in [228]. Yet another formal model for analyzing the security of ad hoc network routing protocols was proposed by Yang and Baras in [385]. Finally, in [305], Papadimitratos *et al.* introduced adversary models and definitions of routing security, and formally proved some properties of the SRP protocol. Besides routing, formal analysis techniques have also been used to analyze other types of wireless security protocols, such as 802.11i [166].

The problem of securing routing in wireless sensor networks is studied by Karlof and Wagner in [217]. That paper lists possible attacks and illustrates them on mainstream sensor network routing protocols proposed in the literature. The authenticated TinyOS beaconing protocol was inspired by the authenticated sensor routing protocol proposed by Perrig *et al.* in [316]. The centralized link-state routing protocol for sensor networks that we described in Subsection 7.5.2 is based on the basic INSENS protocol proposed by Deng, Han, and Mishra in [111]. Finally, the secured position-based routing protocol that we described in Subsection 7.5.3 was originally proposed by Wood *et al.* in [377].

7.8 Questions

(1) In Subsection 7.2.3, we presented some examples of attacks against on-demand ad hoc network routing protocols. Try to construct similar example attacks for pro-active and position-based routing protocols.
(2) Try to construct attacks mounted by colluding adversarial nodes against the secure ad hoc network routing protocols described in this chapter.
(3) What is the role of the *id* field in the route reply of the endairA protocol?
(4) Try to construct an insider attack against the authenticated TinyOS beaconing protocol.
(5) Could you propose a mechanism to prevent the replay of old link-state update messages in the centralized link-state routing protocol described in Subsection 7.5.2?
(6) Why is measuring the round trip time to the two-hop neighbors insufficient to detect a tunnel?

8
Privacy protection

Privacy means that one can control when, where, and how information about one-self is used and by whom. Privacy is *not* about hiding one's personal information *from everybody else* in the world. In fact, revealing personal information to authorized parties under well-defined circumstances can be very useful and should be made possible. For instance, assume that someone suffers an accident and she is transferred to the hospital in an unconscious state. In this case, it is quite useful if the medical doctors can look up her medical record in a database.

It is clear, however, that we do not want that everyone can access our medical record. In particular, employers and insurance companies should not have access to it. However, the problem is that once personal information has fallen into the wrong hands, we cannot control its use anymore. Therefore, hiding personal information *from unauthorized parties* is indeed very important. This is one of the most powerful (and sometimes the only) means to retain control.

Privacy is a problem that cannot be solved solely by technical means. For instance, there is no technical solution that guarantees that a misbehaving doctor cannot reveal sensitive medical data to an unauthorized party (e.g., an employer). Therefore, the problem is usually addressed with a combination of technical and legal means. In fact, the technical and the legal approaches can nicely complement each other. Problems that cannot be solved technically are often related to human behavior. These problems can usually be tackled by legal means (as in other fields) by defining laws and putting serious punishments in prospect for violators of those laws. The enforcement of laws, however, can be very expensive and difficult (mainly because it needs proofs of misbehavior that may be difficult to collect). Technical approaches can be used to make law enforcement more efficient and/or less expensive, for instance, by providing digital evidence. Although it is important to understand that privacy is not an exclusively technical issue, this book being a technical one, we will focus on technical approaches to privacy protection.

In communication networks, quite naturally, privacy is about controlling information related to communications. First of all, we do not want unauthorized parties to access the content of our messages. This can be easily solved by encrypting them. But there are other problems as well. Even if the adversary cannot access the content of the messages, she may still be able to observe with whom we are communicating by inspecting the headers of our messages and the frequency and duration of our communication sessions (this is usually called *traffic analysis*, as already mentioned in Chapter 2). Note that header information is used by the networks to route messages to their destinations, thus it is not encrypted by default. The identity of our communication partners and the frequency and duration of our communication sessions can reveal personal information about ourselves. For instance, the kind of Web sites that we visit indicates our areas of interest. Hence, *communication privacy* is about controlling access to the content of the communication, as well as to the meta-information related to the communication (e.g., who is communicating with whom, how often, how long, etc.).

In addition, in mobile communication systems, unauthorized parties may learn our location by observing our communications. Our whereabouts are personal data that should not be revealed to everyone. Hence, besides communication privacy, there is also a need for ensuring *location privacy*.

Communication and location privacy problems have already arisen in existing wireless networks. Yet there are strong reasons to believe that in the upcoming wireless networks described in Chapter 2, privacy is even more endangered, for two main reasons. First, the density of deployment of wireless devices will be much higher, and many of them will be embedded in real objects with which people interact in their everyday life. This makes monitoring activities easier to carry out continuously in space and in time. In addition, monitoring our Web transactions allows the adversary to determine our activities in the virtual world of the World Wide Web, and monitoring our interactions with real objects will allow her to track our activities *in the real world*.

Second, in many of the upcoming systems, wireless devices associated with the users will operate autonomously without conscious actions of the users. For instance, in vehicular networks, it is envisioned that vehicles communicate with other vehicles and with the roadside infrastructure without requiring any action from the drivers. In Radio Frequency Identification (RFID) systems, RFID tags can be read silently by nearby readers without the tag bearer noticing this. The autonomous operation of devices means that the level of user control is reduced, making access to personal information by unauthorized parties easier.

In this chapter, we will study three representative examples of privacy issues in upcoming wireless systems.

- *Privacy in RFID systems* Wireless devices, for various purposes, often need to identify themselves explicitly. For instance, when a device wants to access some services, it usually needs to authenticate itself. Many authentication protocols require that the device first identifies itself and then proves its identity by some cryptographic means. Identification is done by sending the device identifier through a wireless channel that can be easily eavesdropped. If identification is required frequently, then it is easy for an eavesdropping adversary to track the movement of the device, hence its user. RFID is a wonderful framework to study this problem, because the very purpose of RFID is wireless identification.
- *Location privacy in vehicular networks* In addition to explicit identification, wireless devices reveal their identities to eavesdroppers simply by sending messages. The reason is that the MAC protocol usually requires the device to include its MAC address in the messages that it sends. In addition, MAC addresses are usually static. This provides a convenient means for an eavesdropping adversary to track the movement of a communicating device, hence its user. A suitable context in which this problem can be studied is vehicular networks, because in those networks vehicles move and communicate more or less continuously. For instance, it is envisioned that vehicles will emit their current position and speed with a frequency high enough (several times per second, typically) to enable crash avoidance by other nearby vehicles.
- *Privacy preserving routing in ad hoc networks* The operating principles of upcoming wireless networks are very different from those of existing networks. One important difference is that some of the upcoming networks are based on multi-hop wireless communications. This represents some risk with respect to privacy; in particular, there is a privacy problem related to the way routes are discovered by on-demand routing protocols proposed for wireless ad hoc networks. These protocols flood the entire network with route request messages when discovering new routes, where the route request contains the identifiers of the source and the destination of the intended communication. Thus, an eavesdropping adversary can easily observe (independently of where she resides in the network!) who wants to communicate with whom.

We must note that the levels of maturity of the three areas described above are *very* different. The research community has carried out a lot of work in the domain of RFID privacy, whereas the other two domains are still in their infancy. The number of papers published in the area of privacy preserving routing in ad hoc networks is very limited, perhaps because it is unlikely that this kind of networks will be deployed and used on a large scale for personal communications. The problem of location privacy in vehicular networks has only been identified recently, therefore, the published papers in this area are even more limited.

8.1 Important privacy related notions and metrics

Before discussing the examples described in the previous section, we first introduce some fundamental notions related to privacy and some approaches for the quantification of privacy that we will use in the analysis of the examples later.

As we said before, an effective way to retain control over the use of one's personal information is to hide it as much as possible from unauthorized parties. Depending on what kind of information is being hidden, we can distinguish between the following notions.

- *Anonymity* Anonymity is concerned with hiding who performed a given action. Actions are usually communication actions, such as sending and receiving messages. When the identity of the message sender is hidden, we speak about sender anonymity. Similarly, when the identity of the receiver is hidden, we speak about receiver anonymity. In addition, both types of anonymity can be provided with respect to external eavesdroppers or with respect to communicating parties. For instance, sender anonymity can be provided with respect to the receiver, meaning that the receiver of a message cannot identify its sender.

  It is clear why anonymity is relevant for privacy. If one can carry out an action anonymously, then an adversary cannot link that action to anybody. She may observe all the details related to the action, except that she does not know who performed it. Hence, she obtains no personal information.

  Anonymity can be achieved only if there exists a set of subjects with similar attributes such that all of those subjects could potentially have performed the action in question. If the adversary knows that a certain action can be performed only by a single entity, and she observes that action, then no matter how effectively we hide the identity of the entity, the adversary can trivially identify it as the source of the action. This observation is so fundamental that it serves as the basis of the commonly known definition of anonymity [317]: "Anonymity is the state of being not identifiable within a set of subjects, the *anonymity set*."

- *Untraceability* Untraceability aims at making it difficult for the adversary to identify that a given set of actions were performed by the same subject. Anonymity is useful for ensuring untraceability, but it is not sufficient: the adversary may not know who exactly performed the actions, but if she knows that all of them were performed by the same subject, then untraceability is not provided.

- *Unlinkability* The notions of anonymity and untraceability are generalized by the notion of unlinkability. Unlinkability means hiding information about the relationships between *any* items (e.g., subjects, messages, actions, etc.). In the case of anonymity, the items that cannot be related by the adversary are the observed action and the identifier of the subject who performed that action. In the case of untraceability, the items that cannot be related to each other are the actions of the same subject.

Unlinkability allows us to describe requirements that are weaker than anonymity but still meaningful with respect to privacy. For instance, it may be the case that the adversary can determine who sends messages and who receives messages (i.e., neither sender nor receiver anonymity hold), and yet the adversary may still be unable to relate senders to receivers (i.e., unlinkability of senders and receivers is provided). This means that although the adversary can see who is communicating, she does not know who communicates with whom.

- *Unobservability* In contrast to unlinkability, which is concerned with hiding the relationships between items, unobservability is concerned with hiding the items themselves. For instance, assume that instead of hiding the identity of the sender of a message, we want to hide the fact that a message was sent at all. In this case, we need sender unobservability instead of sender anonymity.
- *Pseudonymity* Pseudonymity means that, instead of one's real identifier, one uses a pseudonym to identify oneself. This makes sense if the pseudonym and the real identifier are unlinkable for the adversary. In this case, pseudonyms can be used to refer to the subject that performed a given action without jeopardizing the privacy of that subject. This is important, because as we already mentioned, communication protocols often require references to message senders and receivers.

In certain cases, unlinkability of pseudonyms and real identifiers may not be enough, but we also need unlinkability of different pseudonyms of the same subject. In particular, if the subject performs several actions under different pseudonyms, then in order to ensure untraceability those pseudonyms must be unlinkable.

There are different techniques to achieve unlinkability (anonymity and untraceability), and unobservability. Sender anonymity can be achieved with the help of so-called DC (Dining Cryptographers) networks [99], and receiver anonymity can be achieved by broadcast communications. Unlinkability between senders and receivers can be achieved by so-called MIX networks [98]. Each of these mechanisms can be extended, with the injection of dummy traffic, to achieve unobservability.

It makes sense to talk about unlinkability only with respect to some observations by the adversary. However, different observations may yield different amounts of information to the adversary, and thus they may result in different levels of unlinkability. The question is how these levels can be determined. Or in other words, how can unlinkability be quantified and measured?

As far as anonymity is concerned, a popular metric is the size of the anonymity set (i.e., the set of subjects that might have performed the observed action). Intuitively, the larger the anonymity set is, the more uncertainty the adversary has about who might have performed the observed action. If the anonymity set is a singleton, then obviously no anonymity is provided.

The advantage of the anonymity set size as a metric for anonymity is that it is usually easy to compute. Its disadvantage is that it is not really a precise measure. Assume, for instance, that the adversary observes a message and she knows that the message might have been sent by someone among one million potential senders, but one of these potential senders is the real sender with probability 0.9, and each of the other potential senders is the real sender with probability around 10^{-7}. In this case, the anonymity set size is one million, which seems like a large number, still the adversary can almost be sure who is the real sender. It is easy to see what is wrong with the anonymity set size as a metric for anonymity: it is a good measure only if all the members of the set are equally likely to have performed the observed action. In general, however, this is rarely the case.

Thus, a better metric can be obtained by taking into account the probability distribution over the members of the anonymity set. Since intuitively, anonymity means that the adversary is uncertain about the real identity of the subject that performed the observed action, it is natural to use the standard metric of uncertainty, namely the *entropy* of the probability distribution, to quantify the level of anonymity. For this reason, let us denote the anonymity set by A, and let us denote by p_x the probability (for the adversary) that the observed action has been performed by subject $x \in A$. Then, the entropy based measure of anonymity is defined as

$$- \sum_{\forall x \in A} p_x \cdot \log p_x. \tag{8.1}$$

One can also normalize this metric by dividing by $\log |A|$, and obtain a number between 0 and 1. Normalization makes it easier to compare the level of anonymity in cases when the anonymity sets have different sizes.

The same approach can be used to define an entropy based metric for unlinkability in general. Let us assume that the adversary has made some observations and that she wants to relate two sets of items, I_1 and I_2, to each other. For instance, I_1 can contain message senders and I_2 may contain message receivers. By definition, a relation is a subset of $I_1 \times I_2$. Let p_R be the probability (for the adversary) that the real relationship between the items in I_1 and I_2 is captured by relation $R \subseteq I_1 \times I_2$. Then, the measure of unlinkability can be defined as

$$- \sum_{\forall R \subseteq I_1 \times I_2} p_R \cdot \log p_R. \tag{8.2}$$

Although these entropy based metrics precisely capture the idea behind measuring the level of anonymity and unlinkability, their main disadvantage is that, in order to compute them, one needs to know the probability distribution, and this is usually difficult to determine.

8.2 Privacy in RFID systems

The main problem with low-cost RFID tags is that they are not capable of enforcing any kind of access control to the data that they store. This means that low-cost tags respond to the query of any reader without authenticating it, and therefore clandestine scanning of low-cost tags is a plausible threat. At the same time, these low-cost tags are expected to be deployed on a large scale. Hence, there is a potential danger of obtaining private information about people by stealthily scanning the tags that they carry.

More precisely, there are two problems that arise here: *inventorying* and *tracking*. Inventorying means that a reader can stealthily determine what objects a person is carrying, including books, medicaments, banknotes, and clothes. Most of the information obtained in this way is highly personal. For instance, medicaments convey information about the diseases that a person has, and books may reveal their personal interests or religion. Moreover, no one wants to reveal the number and the denomination of the banknotes in their wallet, especially when walking in a deserted street at night.

The other problem is that, based on the identifiers emitted by the tags carried by a person, the location of that person can be tracked. Even if the tags do not emit unique identifiers, a specific constellation of object types may be unique to a person, and can be the basis for tracking. For instance, at a given period of time, there may be a single person in a city wearing a specific type of shoes and wrist watch, and carrying a specific book in a specific type of suitcase.

When talking about clandestine scanning, it is useful to remember that in practice, various read ranges exist. First of all, there is a *nominal read range* associated with a tag. This is the value that one can find in the official specifications, and it refers to the maximum distance from which a normally operating reader can reliably read the tag. Depending on the application, the nominal read range can vary form a few centimeters to a few meters.

The *rogue read range* is the distance from which the tag can be read by a rogue reader that uses a stronger signal than that used by normally operating readers. Clearly, the rogue read range is larger than the nominal read range. For instance, in the case of proximity smart cards, the nominal range is around 10 cm, but it has been demonstrated [221] that those cards can be read by a rogue reader from as far as 1 m, which means 10 times farther than the nominal range.

In addition to directly reading the tag, an adversary may try to eavesdrop the communication (including identifying information) between tags and readers. As we have mentioned above, because of the way in which passive tags operate, the tag-to-reader communication is more difficult to eavesdrop on than the reader-to-tag communication. Still, the *tag-to-reader eavesdropping range* can be much larger than the rogue read range. For instance, there have been some rumors [394] that the US electronic passport can be eavesdropped on from a few meters' distance, perhaps a hundred times farther than its nominal range. Compared with that, the *reader-to-tag eavesdropping range* can even be two orders of magnitude larger. Although eavesdropping the reader-to-tag communications seems to be less dangerous with respect to the privacy of the tag bearers, in some protocols the reader may reveal tag specific information that could be exploited by an adversary in an attack against privacy.

8.2.1 Solutions for low-cost tags

There have been many proposals for privacy protecting protocols for RFID tags. They can be broadly classified into two groups: protocols for low-cost tags that cannot perform any computations and protocols for higher-tier tags that can perform certain cryptographic operations. In this subsection, we give an overview of the first group of solutions; the second group will be presented in the next subsection.

Killing and sleeping

One of the simplest solutions for protecting the privacy of consumers carrying tagged objects is to "kill" the tags. For instance, upon the purchase of a product, an RFID reader at the Point Of Sale (POS) terminal can send a kill command to the RFID tag attached to the product that changes its state to permanently disabled. Of course, such a command should be somehow authenticated, otherwise tags can be killed by unauthorized readers. One option for this is to require the kill command to be PIN protected. In other words, the kill command must be accompanied by a PIN in order to be executed by the tag. This is the approach that is supported by the Electronic Product Code (EPC) standards.

Killing tags is a very efficient solution to the privacy problems described above: neither inventorying nor tracking of killed tags is possible. However, the drawback of this approach is that all the post-purchase benefits of RFID for the consumers, such as returning an item without a receipt, are lost too. In addition, in some applications, tags simply cannot be killed. An example is the library application, where the tags in the books must be operational after check-out in order to be able to read them when the books are returned.

Another similar approach is to put the tags in disabled state only temporarily. This means that instead of sending a kill command to the tags, the reader at the POS terminal would send a sleep command to them. The advantage of this approach is that just like dead tags, sleeping tags cannot be stealthily read, whereas unlike dead tags, sleeping tags can be woken up if needed. Similarly to the kill command, the sleep and the wake up commands must be authenticated, for instance, by an accompanying PIN. Note, however, that whereas the PINs for the kill and the sleep commands are managed by the retailer, the PINs for the wake up commands must be managed by the consumer. This may be a problem as people already have difficulties with properly managing their PINs and passwords for their various accounts, and it can be expected that they would have many more tags in the future than accounts today. Therefore, this solution requires a careful design and a proper management of PINs and keys that balances the effort for all stakeholders, and in particular, reduces the burden on the consumer.

Renaming

Let us assume now that the tags are neither killed nor put in a sleeping state. This means that they can be read at any time. Recall that there are two problems that we want to solve: preventing clandestine inventorying and preventing clandestine tracking.

The easiest problem is the prevention of clandestine inventorying. It can be solved by using pseudonyms such that the pseudonyms can be decoded into real identifiers only by the authorized readers. This prevents an adversary who does not have an authorized reader to identify the tags, and hence, the objects to which they are attached. However, using static pseudonyms does not solve the problem of clandestine tracking: if the tag always responds with the same pseudonym, then the pseudonym serves as a meta-identifier, which makes tracking possible.

In order to prevent tracking, the tag must change its pseudonym from time to time. One approach to achieve this is to store a list of pseudonyms in the tag and to require that the tag rotates these pseudonyms and uses a new pseudonym each time it is scanned. Authorized readers would know all the pseudonyms of the tag and they would be able to identify the tag based on the emitted pseudonym; and an unauthorized reader that does not know the pseudonym list cannot correlate the different responses of the same tag. In addition, authorized readers would also be able to refresh the list of pseudonyms in the tag. As the tag is assumed to be a low-cost tag that does not support any cryptography, the new pseudonym list would be transmitted to the tag in clear. Hence, it must be assumed that the adversary cannot continuously follow the tag, and pseudonym list refresh operations cannot be eavesdropped by the adversary. Another problem is that an adversary could rapidly query the tag several times so that it would emit all its pseudonyms in the list. Then, the tag would continue to use the same list of pseudonyms until the next refresh operation, consequently the adversary could still track the tag for some time. In order to prevent this, the bandwidth of the tags must be restricted (e.g., by hardware means), so that they can emit their pseudonyms only at a relatively low rate. Then the attack requires the adversary to have access to the tag continuously for an extended period of time, which is impractical.

Another approach for renaming uses some special properties of the ElGamal public-key cryptosystem. In such a system, the public key of an entity, say Alice, is a triplet (p, g, A), where p is a large prime, g is a generator of the multiplicative group Z_p^*, $A = g^a \pmod{p}$, and a is a secret value known only to Alice. In order to encrypt a message m with the public key (p, g, A), one has to generate a random number r, and compute $R = g^r \pmod{p}$ and $C = m \cdot A^r \pmod{p}$. The ciphertext is the pair (C, R). An interesting feature of the ElGamal cryptosystem is that a ciphertext can be re-encrypted without first being decrypted (i.e., without the knowledge of the private key). Re-encryption works in the following way. One generates a random number r', and computes $R' = R \cdot g^{r'} \pmod{p}$ and $C' = C \cdot A^{r'} \pmod{p}$. It is easy to verify that $R' = g^{r+r'} \pmod{p}$ and $C' = m \cdot A^{r+r'} \pmod{p}$, and therefore, (C', R') is a valid ciphertext for m.

We can use this re-encryption property of the ElGamal cryptosystem for changing pseudonyms of RFID tags in the following way. The pseudonym of each tag is computed by encrypting its real identifier using the ElGamal cryptosystem. Authorized readers know the public key with which they can re-encrypt and rewrite the pseudonym of the tags. Thus, after an interaction with such a reader, the pseudonym of the tag is changed, which makes tracking difficult for the adversary. Some special readers (who also need to know the real identifier of the tag) possess the private key with which the pseudonyms can be decrypted into real identifiers.

This re-encryption scheme is quite elegant, but it has two disadvantages. First, between two subsequent re-encryptions, the tag can still be tracked. Second, it must be assumed that the adversary cannot eavesdrop on the communication between the re-encrypting reader and the tag, otherwise she can link the new pseudonym to the old one. This is a strong assumption because the reader-to-tag eavesdropping range is usually quite large. Still, it is reasonable to assume that the adversary cannot continuously follow the tag, and thus cannot eavesdrop each and every new pseudonym. This makes tracking difficult.

Blocking

An interesting approach to prevent tracking takes advantage of the very properties of the low level anti-collision protocols used in RFID systems. In order to avoid collisions, the reader first determines which tags are in its vicinity, and then it addresses each of those tags individually. The procedure of determining which tags are nearby is called *singulation*.

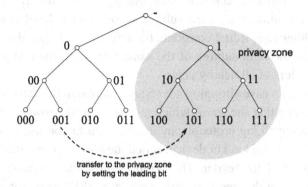

Figure 8.1. Illustration of a labeled binary tree used in the binary tree walking protocol. Tag identifiers are assigned to the leaves of the tree. The internal vertices of the tree are labeled with the largest common prefix of their descendant vertices. The root has an empty label. The reader performs a depth-first search in this tree to determine which tags are nearby. The figure also illustrates the concept of the privacy zone. The privacy zone is a specific part of the identifier space. Tags can be transferred from the non-protected zone into the privacy zone by changing their identifiers (e.g., setting the leading bits). When the reader recurses in the privacy protected part of the tree, a special blocking device simulates collisions thereby preventing the identification of any nearby tags in the privacy zone.

One example of a singulation procedure is the *binary tree walking* protocol.[1] In this protocol, it is assumed that the tag identifiers are initially assigned to the leaves of a binary tree. The internal vertices of the tree are labeled with binary strings in such a way that the left child of a vertex that has label ℓ is labeled with $\ell 0$, and the right child of the same vertex is labeled with $\ell 1$. The root has an empty label. Such a labeled binary tree is illustrated in Figure 8.1.

The reader performs a depth-first search in this tree. First, it requests all nearby tags to emit the first bit of their identifiers. If every tag responds with a 0 (or a 1), then no collision occurs at the physical layer, and the reader knows that all nearby tags are in the left half (or in the right half) of the tree. In this case, the reader recurses in the left (or in the right) subtree below the root. If, instead, some tags respond with 0 and some tags respond with 1, then the reader detects a collision and it knows that there are tags nearby from both halves of the tree. In this case, the reader recurses in both subtrees below the root.

The same procedure is applied in each step of the recursion. Let us assume that the root of the currently considered subtree is labeled with ℓ. The reader requests all nearby tags whose identifier starts with prefix ℓ to emit the next bit of their

---

[1] The binary tree walking protocol is not frequently used in practice. Nevertheless, we use it here as an example of a singulation procedure in order to explain the idea of blocking as a privacy enhancing technique.

identifiers. If no collision occurs, then depending on whether the response was 0 or 1, the reader continues with the subtree whose root is labeled with $\ell 0$ or $\ell 1$, respectively. Otherwise, it first considers the left subtree below the vertex labeled with ℓ, and then the right subtree of the same vertex. At the end, this procedure yields the identifiers of all nearby tags.

As we can see, an eavesdropping adversary can obtain the same information as the reader, meaning that she learns which tags are nearby. However, the properties of the binary tree walking protocol can also be used to confuse an eavesdropper. The idea is that a special kind of device, called the *blocker*, can simulate a collision upon each request of the reader. This would coerce the reader to walk through the whole tree. Since the tree is usually very large (EPC tags, for instance, have 96-bit identifiers, meaning that the tree has 2^{96} leaves!), the reader would stall. More importantly, the adversary would not learn any useful information either, as it would seem to her that all tags of the system are nearby.

The nice thing about blocking is that the user can decide when she wants to activate the blocking device. For instance, she can carry the blocker with her most of the time, preventing clandestine scanning and tracking of her tags. However, if she wants to return an item to a merchant, she can simply do so by de-activating the blocker device for the time of this specific transaction. In addition, the blocker device can be manufactured almost as cheaply as an ordinary tag.

The blocking approach described above has a problem: if there is a blocker device nearby, then it would prevent the reading of *all* tags in the vicinity, even those that are legitimately being attempted to be read. For instance, the blocker device of one consumer may prevent the reader from reading the tags of another nearby consumer. One way to solve this problem is to divide the identifier space into two zones, one of which is called the *privacy zone* (see Figure 8.1 for illustration). Then, each attempt to read tags in the privacy zone is considered to be an adversarial action and it should be blocked, whereas reading tags in the non-protected zone should be made possible. Therefore, the blocking devices would start to simulate collisions only when the reader recurses in the privacy zone. For instance, the privacy zone can be defined such that all identifiers whose leading bit is 1 are considered to be in the privacy zone. Then, each time the reader sends a query with a prefix starting with 1, the blocker device would simulate a collision, but it would stay quiet when the prefix begins with 0. In addition, tags could be transferred from the non-protected zone into the privacy zone and back by appropriately setting the leading bit of their identifiers. This can even be done automatically. For instance, upon purchase of a product, its tag can automatically be transferred into the privacy zone at the POS terminal, thus providing privacy to the consumer when she leaves the shop. Clearly, the reader that transfers the tag from one zone into the other must be authenticated (e.g., using a PIN), in order to avoid that tags are stealthily transferred back into the unprotected zone.

8.2.2 Solutions for crypto-enabled tags

In this subsection, we will assume that the tags are capable of cryptographic operations. We must note, however, that though advances in technology make it possible to integrate more and more functions into microchips, public key cryptography still seems to be prohibitively expensive for RFID tags. As the price of RFID tags is a critical issue owing to their potentially very large number, it is unlikely that low-cost RFID tags will be capable of public key cryptographic operations in the near future. Therefore, we make the realistic assumption that tags can use only symmetric key cryptography.

The advantage of crypto-enabled tags over the simple tags that we considered in the previous subsection is that these tags can generate their pseudonyms themselves. One simple approach for pseudonym generation is that the tag encrypts its real identifier ID_i and some random number R (freshly generated by the tag itself) with a key K_i that it shares with the authorized readers. Let us denote the resulting ciphertext by $E_{K_i}(ID_i||R)$. The random number R is needed to ensure that the pseudonyms are changing. Otherwise, the encrypted identifier would be static, and it would serve as a meta-identifier for the tag. When the reader receives $E_{K_i}(ID_i||R)$, it must try all possible tag keys, until it finds K_i that properly decrypts the pseudonym.[2] Once the pseudonym is decrypted, the reader learns the real identifier of the tag. An eavesdropping adversary, in contrast, learns no practically useful information regarding the tag's identity. The only thing that she can do is to mount a ciphertext-only attack against the cipher E.

Clearly, the main disadvantage of this approach is that the reader must perform $O(n)$ operations for the identification of the tag, where n is the total number of tags in the system. As n can be large, the delay needed to identify the tag may be too long. Note that the tag cannot help the reader by providing information about which key it should use, because that would also help the adversary to identify the tag. Therefore, we need some clever way to improve the above scheme by reducing the complexity of the identification procedure at the reader's side, while preserving the anonymity of the tag with respect to an eavesdropping adversary.

Time–memory trade-off

Observe that the problem of determining which tag key was used to produce a pseudonym is almost identical to the problem of breaking the key of a symmetric

---

[2] This requires some redundancy in the encrypted message that allows the reader to recognize that it used the right key.

cipher. There is a well-known technique, called Hellman's time–memory trade-off [167], that can be used for breaking symmetric keys with less effort than that of the brute force key search. The brute force approach requires either $O(n)$ computation and no storage, or $O(n)$ storage and no computation. In contrast to this, Hellman's technique needs only $O(n^{\frac{2}{3}})$ computation, but it also requires $O(n^{\frac{2}{3}})$ storage; hence the trade-off. Thus, one can use Hellman's technique to speed up the identification of the tags at the cost of an increased storage requirement in the reader.

Maintaining state

Another approach to avoid the brute force key search is for the reader to main-tain some state information for each tag. For instance, each tag can generate its pseudonym by encrypting a counter c with the tag key K_i. Assuming that the reader knows the current counter value for each tag, it can pre-compute the cur-rent pseudonym of each tag. When the tag presents $E_{K_i}(c)$, the reader only needs to look up the received value in its database and in this way identify the tag to which this pseudonym belongs. After this transaction, both the tag and the reader increment the counter and compute the next pseudonym. The same approach can also be implemented by using a one-way hash function, instead of a symmetric key cipher, to compute the pseudonym from the counter. In addition, if there are multiple authorized readers, then the state can be stored at a central place where all readers can access it on-line.

There is a problem, however, with this simple solution. It may happen that the tag and the reader become de-synchronized, meaning that the current counter val-ues at the tag and at the reader are not the same. For instance, if a rogue reader queries the tag, then the tag increments its counter, while the state associated with that tag at the reader remains unchanged. In order to solve this problem, the reader may pre-compute several future pseudonyms for each tag. Let us assume, for in-stance, that the current counter value of a tag is c, and the tag successfully identified itself to the authorized reader using its current pseudonym $E_{K_i}(c)$. Then, the reader can pre-compute and store $E_{K_i}(c + j)$ for all $1 \leq j \leq d$, where d is a parameter chosen by the reader. This will increase the storage requirement at the reader, but at the same time, this scheme can tolerate de-synchronization up to a difference of d between the counter values at the tag and at the reader. Unfortunately, if d is known to the adversary, then she can still enforce de-synchronization between the tag and the reader by querying the tag more than d times.

Another problem with the above solution is that if a tag is compromised, mean-ing that its tag key K_i and its current counter c become known to the adversary, then she can compute all pseudonyms that the tag used in the past. Hence, if the adversary has a log of all past transactions, it can easily identify the trans-actions of the compromised tag in that log. In other words, the adversary can do

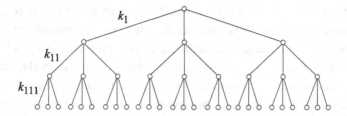

Figure 8.2. Illustration of a key-tree. Each leaf represents a tag. There is a unique key assigned to each edge. Each tag possesses the keys assigned to the edges of the path starting from the root and ending in the leaf that corresponds to the given tag. For instance, the tag that belongs to the leftmost leaf in the figure possesses the keys k_1, k_{11}, and k_{111}. Derived from [74], with kind permission of Springer Science and Business Media.

delayed tracking, which may still be dangerous concerning the privacy of the tag bearer.

This problem can be solved by replacing the counter with a hash value. This hash value is re-hashed by the tag and the reader after each transaction using a one-way hash function. If a tag is compromised and its current state becomes known to the adversary, then she still cannot compute the previous states owing to the unidirectionality of the hash function. This property is often called *forward secrecy* or *break backward protection*.

Using key-trees

Another idea to reduce the complexity of tag identification while preserving privacy is to use key-trees. A key-tree is a tree where a unique key is assigned to each edge (see Figure 8.2 for illustration). The leaves of the tree represent the tags. Each tag possesses the keys assigned to the edges of the path starting from the root and ending in the leaf that corresponds to the given tag. The reader knows all keys in the tree (or it can generate them from a master key). In order to identify itself to the reader, a tag uses all of its keys, one after the other, starting from the first level of the tree and proceeding towards lower levels. The reader first determines which first-level key has been used; for this, it needs to search through the first-level keys only. Once the first key is identified, the reader continues by determining which second-level key has been used; for this, it needs to search only through those second-level keys that reside below the already identified first-level key in the tree. This process is continued until all keys are identified, which at the end, identifies the tag. The crucial point is that the reader can reduce the search space considerably each time a key is identified, because it should consider only the subtree below the recently identified key. Hence, the complexity of the procedure is $O(\log n)$, where n is the total number of tags in the system.

The problem of the above described tree-based approach is that upper-level keys in the tree are used by many tags, and therefore if a tag is compromised and its keys become known to the adversary, then the adversary gains partial knowledge of the keys of other tags too. This obviously reduces the privacy provided by the system, because by observing the transaction of a non-compromised tag, the adversary can recognize the use of some compromised keys, and therefore its uncertainty regarding the identity of the tag is reduced (it may be able to determine which subtree the tag belongs to).

One interesting observation is that the basic scheme described at the beginning of this subsection can be viewed as a special case of the key-tree based approach, where the key-tree has a single level and each tag has a single key. Regarding the above described problem of compromised tags, the basic scheme is in fact optimal, because compromising a tag does not reveal any key information of other tags. At the same time, as we have seen above, the identification delay is the worst in this case. In contrast, in a binary key-tree, the compromise of a single tag strongly affects the privacy of the other tags (as we will see below), yet the binary tree is very advantageous in terms of identification delay. Thus, there seems to be a trade-off between the level of privacy and the identification delay in the key-tree based scheme that depends on the parameters of the key-tree. A detailed analysis of this trade-off can be found in [74].

We will now illustrate how to determine the level of privacy provided by the key-tree based scheme when a single tag is compromised. For this, we will use the concept of anonymity sets. Recall that the anonymity set of a tag v is the set of tags that are indistinguishable from v from the adversary's point of view. The size of the anonymity set is a good measure of the level of privacy provided for v, because it is related to the level of uncertainty of the adversary. Clearly, the larger the anonymity set is, the higher the level of privacy is. The minimum size of the anonymity set is 1, and its maximum size is equal to the number of all tags in the system. In order to make the privacy measure independent of the number of tags, we can divide the anonymity set size by the total number of tags, and obtain a normalized privacy measure between 0 and 1. Such normalization makes the comparison of different systems easier.

Now, let us consider a key-tree with ℓ levels and branching factor b at each level, and let us assume that exactly one tag is compromised (see Figure 8.3 for illustration). Knowledge of the compromised keys allows the adversary to partition the tags into partitions $P_0, P_1, P_2, \ldots, P_\ell$, where

- P_0 contains the compromised tag only,
- P_1 contains the tags whose parent is the same as that of the compromised tag, and that are not in P_0,

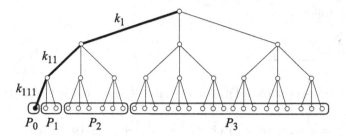

Figure 8.3. Illustration of what happens when a single tag is compromised. Without loss of generality, we assume that the tag corresponding to the leftmost leaf in the figure is compromised. This means that the keys k_1, k_{11}, and k_{111} become known to the adversary. This knowledge of the adversary partitions the set of tags into anonymity sets P_0, P_1, ..., P_ℓ of different sizes. Tags that belong to the same partition are indistinguishable to the adversary, though it can distinguish between tags that belong to different partitions. For instance, the adversary can recognize a tag in partition P_1 by observing the use of k_1 and k_{11} but not that of k_{111}, where each of these keys are known to the adversary. Tags in P_3 are recognized by not being able to observe the use of any of the keys known to the adversary. Clearly, tags that belong to larger partitions enjoy more privacy, because the uncertainty of the adversary is greater. The level of privacy provided by the system can be characterized by the level of privacy provided to a randomly selected tag, or in other words, by the expected size of the anonymity set of a randomly selected tag. From [74], with kind permission of Springer Science and Business Media.

- P_2 contains the tags whose grandparent is the same as that of the compromised tag, and that are not in $P_0 \cup P_1$,

- ...

- P_ℓ contains all the remaining tags that are not in $P_0 \cup \ldots \cup P_{\ell-1}$ (these are the tags that have no compromised keys).

Tags of a given partition are indistinguishable for the adversary, though it can distinguish between tags that belong to different partitions. Hence, each partition is the anonymity set of its member tags.

The level of privacy provided by the system can be characterized by the level of privacy provided to a randomly selected tag, or in other words, by the expected size of the anonymity set of a randomly selected tag. By definition, the expected anonymity set size is:

$$\bar{S} = \sum_{i=0}^{\ell} \frac{|P_i|}{n} |P_i| = \sum_{i=0}^{\ell} \frac{|P_i|^2}{n}, \tag{8.3}$$

where n is the total number of tags, and $|P_i|/n$ is the probability of selecting a tag from partition P_i. Then, the normalized expected anonymity set size can be computed as follows:

$$\frac{\bar{S}}{n} = \sum_{i=0}^{\ell} \frac{|P_i|^2}{n^2} \tag{8.4}$$

$$= \frac{1}{n^2}(1 + (b-1)^2 + ((b-1)b)^2 + \cdots + ((b-1)b^{\ell-1})^2)$$

$$= \frac{1}{n^2}(1 + (b-1)^2(1 + b^2 + (b^2)^2 + \cdots + (b^2)^{\ell-1}))$$

$$= \frac{1}{n^2}\left(1 + (b-1)^2 \cdot \frac{b^{2\ell} - 1}{b^2 - 1}\right)$$

$$= \frac{b-1}{b+1} + \frac{2}{(b+1)n^2} \tag{8.5}$$

where we used that $n = b^\ell$ and

$$|P_0| = 1$$
$$|P_1| = b - 1$$
$$|P_2| = (b-1)b$$
$$|P_3| = (b-1)b^2$$
$$\cdots \quad \cdots$$
$$|P_\ell| = (b-1)b^{\ell-1}.$$

Using this result, for a binary tree (i.e., when $b = 2$), we achieve that the normalized expected anonymity set size of the system, when only one tag is compromised, drops from 1 to around $1/3$ (assuming that n is large). This reinforces our previous statement that in a binary key-tree, the compromise of a single tag strongly affects the privacy of the non-compromised tags.

The above calculations can be generalized for key-trees that have different branching factors at different levels, and for the case when more than one tags are compromised. For the details, the interested reader is referred to [74].

8.3 Location privacy in vehicular networks

In this section, we study the location privacy problem in mobile wireless networks and, in particular, in vehicular networks. The location privacy problem in vehicular networks stems from their envisioned operational principles, specifically from the requirement that each vehicle must broadcast its current location and speed with a relatively high frequency. The idea is that nearby vehicles and roadside units can pick up this information and they use it in many applications, such as assisting the drivers in making a left turn at a crossroad, changing lanes, and warning the drivers when a collision is predicted or an emergency vehicle is approaching. When a vehicle always broadcasts its position and speed under the same identity (e.g., using a static MAC address), it becomes possible for an eavesdropping adversary to track the movement of that vehicle. In addition, it is relatively easy to find out who is the owner of a given vehicle, meaning that not only the vehicle, but its owner, can be tracked too.

8.3.1 Changing pseudonyms

One appealing approach to solve this problem is that the vehicles broadcast their messages under pseudonyms that they change with some frequency. These pseudonyms should be generated in such a way that a new pseudonym cannot be directly linked to previously used pseudonyms of the same vehicle. In most of the applications related to road safety, it is important to let other vehicles know that there is a vehicle at a given position moving with a given speed, but it is not important which particular vehicle it is. Thus, using pseudonyms is just as good as using the real identifiers, as far as the functionality of the applications is concerned.

But what do we really gain with using ever-changing pseudonyms? In fact, we gain practically nothing against a global eavesdropper that can hear all communications in the network. The reason is that such an adversary can link different pseudonyms of the same vehicle together with very high probability, based on the position and speed information in the messages that she eavesdrops. Let us assume, for instance, that the adversary eavesdrops the message "at time 13h:34m:52s vehicle v is at position (45 m, 102 m) heading East with a speed of 50 km/h," and the message "at time 13h:34m:53s vehicle u is at position (59 m, 102 m) heading East with a speed of 50 km/h." There is a very strong correlation between these two messages. Indeed, by looking at the first message, the adversary can compute that at time 13h:34:53s, vehicle v should most probably be at position (58.8 m, 102 m), because it goes to East with a speed of 50 km/h, which is around 13.8 m/s. Moreover, it is unlikely that another vehicle will be at the same position at the same time, because that would mean that the two vehicles crashed. Thus, it is extremely likely that the second message came from the same vehicle as the first message, meaning that pseudonym u refers to the same vehicle as pseudonym v. Figure 8.4 illustrates how an adversary who is able to hear all transmitted messages can track vehicles based on the information obtained from those messages despite the fact that the vehicles may change their pseudonyms.

8.3.2 The mix zone

Actually, the assumption that the adversary is a global eavesdropper is a very strong one, as we discussed in Chapter 3. It is more reasonable to assume that the adversary can monitor the ether only at a limited number of places and only in

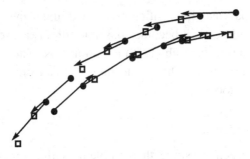

Figure 8.4. Illustration of how the adversary can track vehicles based on the messages that she eavesdrops despite the fact that the vehicles may change their pseudonyms. The dots correspond to messages that were broadcast at some time t, and the rectangles correspond to messages that were broadcast one second later. Each dot and rectangle is placed at the position indicated in the corresponding message. The arrows show the direction of the movement obtained from the corresponding message, and the length of each arrow corresponds to the distance that the given vehicle would cover in one second starting from the given position and moving straight with a speed indicated in the message. We could have also labeled the dots and the rectangles with the pseudonyms obtained from the messages, but we deliberately did not do that in order to show that it is quite easy to determine which rectangles belong to which dots, and thus, which messages belong to the same vehicle, without any identity information. This clearly shows that when vehicles broadcast their position and speed with a relatively high frequency (as it is envisioned in future vehicular networks), changing pseudonyms is not an effective mechanism to protect location privacy with respect to a global eavesdropper.

a limited range. This means that vehicles that change pseudonyms in the unmonitored region are mixed and the adversary is confused.

In essence, the unmonitored area functions much in the same way as a mix node in a mix network [98]. A mix network is a store-and-forward network that offers anonymous communication facilities. It contains some special routers called mix nodes. A mix node processes messages in batches of a given size n. It collects n messages and forwards them all at the same time. Then, it collects the next n messages and forwards them again simultaneously, and so on. In each batch, it changes the encoding and the order of the messages, so an eavesdropper cannot easily determine which outgoing message belongs to which incoming message. Similarly, in our case, the adversary cannot easily tell which of the vehicles exiting the unmonitored area belong to which of the vehicles entering it. For this reason, one can view the entire unmonitored area as a *mix zone*.

Privacy level provided by the mix zone

The concept of mix zone is very appealing, but one must be careful because the geometry of the mix zone influences the level of protection that it provides. As an

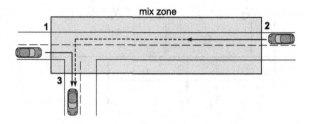

Figure 8.5. A mix zone with three gates.

example, let us consider Figure 8.5, which illustrates a mix zone with three gates (where vehicles can enter and exit the mix zone). The geometry of this mix zone is such that gates 1 and 3 are close to each other, and gate 2 is remote from both gates 1 and 3. Assume that the adversary observes two vehicles entering the mix zone almost at the same time at gates 1 and 2, respectively. Let us further suppose that the mix zone was empty before, meaning that now there are only these two vehicles in it. After some short time, the adversary observes a vehicle exiting the mix zone at gate 3. It may be that it is impossible or very unlikely to go from gate 2 to gate 3 in this short amount of time, and therefore the adversary can link the vehicle exiting at gate 3 to the one that entered at gate 1. In this example, the mix zone does not provide any protection.

In order to quantify the level of location privacy provided by the mix zone, we assume that the adversary has a model of the mix zone. This model consists of a matrix $P = [p_{ij}]$ of size $n \times n$, where n is the number of gates of the mix zone, and of n^2 discrete probability density functions $d_{ij}(t)$ $(1 \le i, j \le n)$. Here p_{ij} is the conditional probability of exiting the mix zone at gate j given that the entry point was gate i; $d_{ij}(t)$ describes the probability distribution of the delay when crossing the mix zone between gate i and gate j. We assume that time is slotted, that is why $d_{ij}(t)$ is a discrete function.

Equipped with this model, for each observation consisting of some entering events and some exiting events, the adversary can compute the probability of a particular mapping between those events. We can use these probabilities to compute the entropy corresponding to the given observation, which provides the means for quantifying the level of unlinkability between the observed events, and hence, the level of location privacy offered by the mix zone (in the case of this particular observation). Below, we explain how to do this.

Let an event consist of two elements: a gate number and a timestamp. There are two kinds of events: entering events and exiting events. For an entering event $N = (n, \tau)$, n denotes the gate where the vehicle entered the mix zone, and τ denotes the time when this happened. Similarly, for an exiting event $X = (x, t)$, x denotes the gate where the vehicle exited the mix zone, and t denotes the time

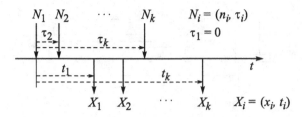

Figure 8.6. Illustration of an observation. The horizontal axis represents time, the arrows above the axis represent entering events, and the arrows below the axis represent exiting events. The placement of an arrow indicates the timing of the corresponding event. We assume that the first entering event happens at time $\tau_1 = 0$, and the times of all other events are measured from this first entering event. We further assume that the mix zone was empty before time 0 and it becomes empty again after the kth exiting event.

when this event happened. An observation is a set of entering and exiting events, which we will shortly denote by $(\bar{N}, \bar{X})$.

Let us consider Figure 8.6, which illustrates an observation made by the adversary. In the figure, entering and exiting events are represented by vertical arrows above and below the time axis, respectively. We assume that the first entering event happens at time $\tau_1 = 0$, and the times of all other events are measured from this first entering event. For simplicity, we further assume that the mix zone was empty before the first entering event and it becomes empty again after the kth exiting event.

As the vehicles change pseudonyms while in the mix zone, the adversary cannot directly map exiting events to entering events, but she considers many mappings possible. Each possible mapping between the exiting and entering events can be represented by a permutation π on $\{1, 2, \ldots, k\}$. We will denote by m_π the mapping described by the permutation π. Thus, we have:

$$m_\pi = \left(N_1 \to X_{\pi(1)}, N_2 \to X_{\pi(2)}, \ldots, N_k \to X_{\pi(k)}\right),$$

where $N \to X$ means that the exiting event X belongs to the same vehicle as the entering event N, and $\pi(i)$ is the ith element of the permutation π.

Now, we would like to determine the probability that mapping m_π holds given the observation $(\bar{N}, \bar{X})$, which we will shortly denote by $\Pr\{m_\pi | \bar{N}, \bar{X}\}$. In order to do this, we first observe that

$$\Pr\{m_\pi | \bar{N}, \bar{X}\} = \frac{\Pr\{m_\pi, \bar{X} | \bar{N}\}}{\Pr\{\bar{X} | \bar{N}\}}, \tag{8.6}$$

and we compute the numerator and the denominator separately as follows:

$$\Pr\{m_\pi, \bar{X} | \bar{N}\} = \prod_{i=1}^{k} p_{n_i x_{\pi(i)}} d_{n_i x_{\pi(i)}} \left(t_{\pi(i)} - \tau_i\right) = q_\pi \tag{8.7}$$

Table 8.1. *Discrete probability density functions of the delay between gates 1 and 3, and the delay between gates 2 and 3 in the example,* $d_{13}(t) = d_{23}(t) = 0$ *for* $t < 0.1$ *and* $t > 1.2$

t	0.1	0.2	0.3	0.4	0.5	0.6	0.7	0.8	0.9	1.0	1.1	1.2
$d_{13}(t)$	0.1	0.2	0.3	0.2	0.1	0.04	0.03	0.02	0.01	0	0	0
$d_{23}(t)$	0	0	0	0.1	0.2	0.3	0.2	0.1	0.04	0.03	0.02	0.01

and

$$\Pr\{\bar{X}|\bar{N}\} = \sum_{\pi'} \Pr\{m_{\pi'}, \bar{X}|\bar{N}\} = \sum_{\pi'} q_{\pi'}, \tag{8.8}$$

where in (8.7) we made the simplifying assumption that each vehicle chooses its exit gate independently from the other vehicles, and the experienced delays of the vehicles are independent too.

The entropy corresponding to the observation $(\bar{N}, \bar{X})$ is then computed as

$$H(\bar{N}, \bar{X}) = -\sum_{\pi} \frac{q_{\pi}}{\sum_{\pi'} q_{\pi'}} \log \left(\frac{q_{\pi}}{\sum_{\pi'} q_{\pi'}} \right), \tag{8.9}$$

where q_{π} is defined in (8.7).

One problem with (8.9) is that it considers mappings on the entire set of observed events. It may be the case that each of the possible mappings has an equally low probability (meaning that the entropy is high), but still a particular exiting event can be related to a particular entering event with a much higher probability than to the rest of the entering events. In other words, the adversary may be clueless regarding the full mapping, but she can still track a particular vehicle. In order to see whether this is the case, we can determine the probability $\Pr\{N_i \to X_j|\bar{N}, \bar{X}\}$ for all pairs of i and j separately as follows:

$$\Pr\{N_i \to X_j|\bar{N}, \bar{X}\} = \sum_{\pi:\, \pi(i)=j} \Pr\{m_\pi|\bar{N}, \bar{X}\}, \tag{8.10}$$

where $\Pr\{m_\pi|\bar{N}, \bar{X}\}$ is defined in (8.6–8.8).

An example

Although (8.9) does not give an immediate insight into the problem, it does provide the means for computing the entropy for any observation given the model of the mix zone. We illustrate this by a quantitative example. Let us consider a mix zone with three gates. Let the time-slot size be 0.1 unit. The discrete probability density functions of the delay between gates 1 and 3, and the delay between gates 2 and 3 are given in Table 8.1.

Let us assume that the adversary observes two vehicles entering the mix zone at gates 1 and 2 at time 0, and two vehicles exiting the mix zone at gate 3 at time 0.4 and at time 0.7. Thus, we have two entering events $N_1 = (1, 0)$ and $N_2 = (2, 0)$, and two exiting events $X_1 = (3, 0.4)$ and $X_2 = (3, 0.7)$.

There are two possible mappings between these events: $m_{12} = (N_1 \to X_1, N_2 \to X_2)$ and $m_{21} = (N_1 \to X_2, N_2 \to X_1)$. Let us compute q_{12} and q_{21}:

$$q_{12} = p_{13}d_{13}(0.4 - 0)p_{23}d_{23}(0.7 - 0) = 0.2 \cdot 0.2 \cdot p_{13}p_{23},$$
$$q_{21} = p_{13}d_{13}(0.7 - 0)p_{23}d_{23}(0.4 - 0) = 0.03 \cdot 0.1 \cdot p_{13}p_{23}.$$

From this, we get that

$$\Pr\{m_{12}|\bar{N}, \bar{X}\} = \frac{0.2 \cdot 0.2 \cdot p_{13}p_{23}}{0.2 \cdot 0.2 \cdot p_{13}p_{23} + 0.03 \cdot 0.1 \cdot p_{13}p_{23}} = \frac{0.04}{0.043} = 0.93$$

$$\Pr\{m_{21}|\bar{N}, \bar{X}\} = \frac{0.03 \cdot 0.1 \cdot p_{13}p_{23}}{0.2 \cdot 0.2 \cdot p_{13}p_{23} + 0.03 \cdot 0.1 \cdot p_{13}p_{23}} = \frac{0.003}{0.043} = 0.07$$

and

$$H(\bar{N}, \bar{X}) = -0.93 \cdot \log 0.93 - 0.07 \cdot \log 0.07 = 0.366.$$

This means that the first mapping is much more likely than the second one (given the observation), and consequently the entropy of the observation is small. In other words, the uncertainty of the adversary is not really increased by the mix zone, hence the mix zone is ineffective in this example.

8.4 Privacy preserving routing in ad hoc networks

In Chapter 7, we presented some routing protocols proposed for wireless ad hoc networks. In that chapter, we were concerned with the security of those protocols against an adversary that wants to increase her control over the communications in the network, to degrade the quality of service provided by the network, or to increase the resource consumption of the nodes. Now, we will extend the above list, and consider an adversary that wants to obtain information about who is communicating with whom in the network. Note that the secure routing protocols discussed in Chapter 7 do not prevent the adversary from learning this information: it is quite trivial for the adversary to obtain this information by inspecting the headers of the control and the data packets.

The privacy problem is most apparent in on-demand routing protocols where route discovery is based on flooding the entire network with a route request that contains the identifiers of the source and the destination of the intended communication. Hence, anybody can observe who is communicating (or about to communicate) with whom with virtually no effort, even eavesdroppers who are not on the routes between the communicating nodes. Therefore, we will focus on on-demand ad hoc network routing protocols in this section.

Before beginning to develop solutions, it is worth refining our objectives and specifying what kind of adversary we are dealing with. Our primary goal is to hide the relationships between communicating sources and destinations; thus, we are aiming at unlinkability. We assume that the adversary is a global eavesdropper, who may also compromise some nodes. Therefore, we want to hide the relationship between every pair of communicating source and destination from global eaves-droppers and *from all nodes other than the source and the destination*, including the forwarding nodes between them. This latter requirement immediately rules out approaches that are based on encrypting every control and data packet with a com-mon key shared by all nodes in the network. Such a solution would work against an external eavesdropper, but not against an insider node.

In addition, we want to hide the identity of the forwarding nodes from each other and from a global eavesdropper because the knowledge of the identity of the forwarding nodes reduces the uncertainty of the adversary regarding the identity of the source and the destination. For instance, the adversary would know that the source is one of the neighbors of the first forwarding node and the destination is one of the neighbors of the last forwarding node.

One may also consider unobservability as an objective (i.e., making it difficult for the adversary to determine who is communicating at all), but in order to achieve this, we would need to introduce a considerable amount of dummy traffic in the network, which may not be desirable in the common case when the nodes have battery constraints. Hence, we do not consider unobservability as an objective.

8.4.1 An effective but inefficient solution

As mentioned before, flooding route requests in the entire network is disadvanta-geous with respect to privacy. In fact, the problem does not stem from the flooding, but from the fact that the route request contains the identifier of the source and the destination in clear. Indeed, flooding is a broadcast operation, and broadcast com-munication is known to be an effective mechanism to achieve recipient anonymity (if used correctly). So we keep the flooding of route requests, but we want to hide the identity of the source and the destination in such a way that only the destination can determine that it is the target of the route request, and all other nodes can only determine that they are not the target.

One way to achieve this is to encrypt the source and the destination identifiers in the route request with the public key of the destination. Any node can attempt to decrypt the encrypted identifiers with its own private key. If decryption fails, the node will only know that it is not the target of the request, but it will not know who is the target. If the decryption succeeds, then the node will know that it is the target, and it will also learn the identity of the source.

The next thing we want to ensure is that the destination can return a route reply that creates the necessary routing state in the forwarding nodes. However, we must do this in a way that reveals neither the identity of the destination nor that of the forwarding nodes to other forwarding nodes and to an external eavesdropper. We can achieve this by using the route request to set up a secret key between each intermediate node that relays the route request and the destination. These secret keys can then be used by the destination to establish the necessary routing state in the forwarding nodes. The question is how an intermediate node can set up a secret key with the destination, when it does not know who the destination is.

In order to solve this problem, the source can generate an asymmetric key-pair and include the public key in the route request. This public key can be used by every intermediate node to set up a secret key with the destination. Because the destination needs the corresponding private key, the source encrypts that key with the public key of the destination together with the identifiers of the destination and the source. The private key also serves as a salt that makes dictionary type attacks on the encrypted identifiers inefficient.[3]

We are almost finished, but there are still two questions to answer. First, what should the routing state be in the forwarding nodes? Second, how should data packets be handled? These two questions are related, because data packets are handled according to the routing state in the forwarding nodes.

Note that the encoding of data packets must be changed at each hop, otherwise a global eavesdropper can track a data packet from its source to its destination based on its static appearance. Hence, data packets must be re-encrypted hop-by-hop. This suggests the idea of representing the routing state in each forwarding node as a pair of link keys (k^{in}, k^{out}) with the following operating principle: when the node receives an encrypted data packet, it tries to decrypt it with k^{in}, and if this succeeds, then it re-encrypts the data packet with k^{out} and broadcasts the encrypted packet. It is convenient that the node does not need to know with which neighbor it shares k^{in} and k^{out}. Hence, this solution can satisfy the requirement that the forwarding nodes should not know the identity of the other forwarding nodes.

The link keys can be established with the help of the route reply message. This message can have an onion-like structure where each layer is encrypted with the key shared by the destination and the corresponding forwarding node, and it contains the link keys destined for that forwarding node. Each forwarding node peels

---

[3] In a dictionary attack, the adversary builds a dictionary containing plaintext–ciphertext pairs by encrypting a set of known plaintexts with a known public key. Later, she can use the dictionary to look up the plaintext that belongs to an observed ciphertext. This attack can be very efficient, if the set of possible plaintexts is small.

off one layer of the route reply, which ensures that the encoding of the route reply changes at each hop, so it cannot be tracked.

Putting all these ingredients together, we arrive at the following protocol: assume that a source S wants to communicate with a destination D. For this reason, it must first establish a route to D, and then send data packets to the destination via that route. Route establishment is based on flooding the network with a route request, and then returning a route reply, which creates the necessary routing state in the forwarding nodes.

In order to create the route request message, the source generates an asymmetric key-pair (K, K^{-1}), a secret key k_0, and a nonce n_0. It then encrypts D, S, and K^{-1} with the public key K_D of the destination. In addition, it encrypts k_0 and n_0 with the public key K. Then, it broadcasts the route request:

$$E_{K_D}(D||S||K^{-1}) || K || E_K(k_0||n_0).$$

Assume that node F_1 receives this route request. It first checks, if it has received this request before. The public key K can serve as a request identifier, thus, F_1 checks if it has seen K before. If so, then the request is dropped, otherwise K is stored (until some time), and the processing continues.

Next, F_1 verifies if it is the target of the request. For this reason, it attempts to decrypt $E_{K_D}(D||S||K^{-1})$ with its private key $K_{F_1}^{-1}$. If this is successful, then F_1 is the target, otherwise it is not. Let us assume for the moment that F_1 is not the target; the processing in the other case will be discussed later.

If F_1 is not the target, then F_1 generates a secret key k_1 and a nonce n_1, concatenates them to $E_K(k_0||n_0)$, and encrypts the result with K. Then, it broadcasts the following route request:

$$E_{K_D}(D||S||K^{-1}) || K || E_K(k_1||n_1||E_K(k_0||n_0)).$$

In addition, k_1 and n_1 are stored for later use.

Each intermediate node does the same, and hence, the general format of the route request message is the following:

$$E_{K_D}(D||S||K^{-1}) || K || E_K(k_i||n_i|| \ldots E_K(k_0||n_0) \ldots).$$

After receiving the route request, D attempts to decrypt $E_{K_D}(D||S||K^{-1})$ with its private key, and it succeeds. Thus, D realizes that it is the target of the request. Nevertheless, it broadcasts a dummy request:

$$E_{K_D}(D||S||K^{-1}) || K || garbage.$$

This is needed, since otherwise the adversary may identify the destination as the node that does not forward the request.

As D successfully decrypted $E_{K_D}(D||S||K^{-1})$, it learns the private key K^{-1}. With that it decrypts $E_K(k_\ell||n_\ell|| \ldots E_K(k_0||n_0) \ldots)$, and obtains the secret keys and the nonces of the forwarding nodes. Then it generates a link key for each link of the discovered route, and constructs a route reply, which has the following layered structure:

$$E_{k_\ell}(n_\ell||k_\ell^{in}||k_\ell^{out} \; || \; E_{k_{\ell-1}}(n_{\ell-1}||k_{\ell-1}^{in}||k_{\ell-1}^{out} \; || \; \ldots E_{k_0}(n_0|| - ||k_0^{out}) \ldots)).$$

Each layer (except the most inner one) contains the nonce of the node corresponding to that layer and two link keys intended for the given node. The most inner layer corresponds to the source S that needs only one link key k_0^{out} with which it can encrypt data packets to be sent. In addition, as a link key must be shared by the two ends of the link, we have that $k_i^{out} = k_{i+1}^{in}$ for all $0 \leq i < \ell$. Finally, k_ℓ^{out} is stored by the destination together with the identifier S.

When a route reply of this kind is received by an intermediate node F_i, it tries to decrypt it with k_i that it stored before. In fact, F_i may have several such keys if it participated in multiple route discoveries, and it must check each of those keys. If none of the keys work,[4] then F_i is not involved in the route, and the route reply is discarded. If k_i works, then the node must check if it received back its nonce n_i corresponding to key k_i in the reply. If this is the case, then F_i peels the outer layer off the route reply, applies some padding to retain its original length and re-broadcasts the updated route reply. In addition, F_i stores k_i^{in} and k_i^{out} in its routing table.

Data packets are handled as follows. The source encrypts the packet with k_0^{out} and broadcasts it. Each node that receives the transmission tries to decrypt it with its incoming link keys. If none of the keys work, then the packet is discarded. Otherwise, if F_i manages to decrypt the packet with k_i^{in}, then it re-encrypts it with k_i^{out}, and re-broadcasts it. This process is repeated, until the packet arrives to the destination.

Brief informal analysis

Route request messages do not contain explicit identifiers. However, a global eavesdropper can observe where a particular route request is originated, hence she may guess who the source is; whereas the way in which the route requests are handled, and especially the fact that the destination also re-broadcasts the route request targeted to it, ensures that even a global eavesdropper cannot determine (solely by observing the route request) who the destination is. Note that route request messages contain a static part (the encrypted identifiers and the public key), but this is not a problem because these messages are flooded in the network, hence tracking them is futile.

---

[4] A key works if after decryption, the node can see one of its nonces in the decrypted message.

Route reply messages do not contain explicit identifiers either. In addition, their entire encoding is changed (while preserving their original length) at each hop, which makes it non-trivial to track them from the destination to the source. Still, a global eavesdropper can observe the timing of the transmissions of the route reply messages, which may allow her to track them: two route reply messages eavesdropped upon at the same place right after each other are very likely two encodings of the same reply. This problem is somewhat mitigated by the fact that each node may be involved in many route discoveries simultaneously, thus nodes act as mixes for the route reply messages. If this is deemed insufficient (e.g., the average number of simultaneous route discoveries is low in the network), then the only way to overcome the problem is to require that each node periodically injects dummy route reply messages in the network, which are forwarded up to a few hops and then discarded.

Another problem is that a node transmitting a route reply immediately after receiving a route request is very likely the target of that request. Again, periodic transmission of dummy route requests may help to solve this problem.

Data packets are similar to route request messages in the sense that they are unicast messages and they change their entire encoding at each hop. Thus, the same analysis applies to them as to route reply messages. One additional point that is worth mentioning is that data packets are decrypted and re-encrypted by forwarding nodes. Thus, their content should be protected so that forwarding nodes cannot determine who are their sources and destinations. This can be achieved easily by end-to-end encryption.

Notice that no attempt is made to hide who is source and who is destination. However, owing to the fact that data packets are re-encrypted at each hop, and that forwarding nodes act as mixes (with additional dummy traffic if needed), it becomes difficult even for global eavesdroppers to determine which destinations belong to which sources.

8.4.2 Improving efficiency

The main problem with the protocol described above is that it requires much computation from the nodes. Firstly, for each and every route request, each node in the network must determine if it is the target of that route request by attempting to decrypt the encrypted identifiers in the request. This involves asymmetric key operations, which are very resource consuming. Secondly, when a node receives a route reply, it must figure out if it is involved in the route or not by attempting to decrypt the route reply with all its secret keys that belong to the pending route requests processed by that node earlier. The number of those secret keys may be large, which is advantageous with respect to privacy (because it means that many simultaneous route discoveries are taking place), but it is disadvantageous with respect to efficiency. Finally, when a node receives a data packet, it attempts to decrypt it with all its incoming link keys. Again, the number of those keys may be large, which is good for privacy, but it is bad for efficiency.

Note that we are facing a problem that is similar to the one that we encountered in the case of private identification of RFID tags. In other words, we want to use symmetric key cryptography for efficiency reasons, but we cannot provide information on which key should be used to process the messages, because that would also help the adversary to break privacy. So we may try to borrow ideas from proposals for private identification of RFID tags. The approach that we will follow below is based on synchronized state maintained between the communicating source–destination pairs and between neighboring nodes.

First of all, we replace the public key encryption of the source and destination identifiers with symmetric key encryption. For this, we assume that the source and the destination already share a secret key k_{SD}. In addition, they also share a counter c_{SD} from which the source can compute a one-time hint for the destination, for instance, by computing the keyed hash value $h(k_{SD}, c_{SD})$. Each node can pre-compute and store the current hint of each possible source for which it is a destination, and therefore the nodes need to do a table lookup only when processing route request messages. Once the route request to D has been sent, S increments the counter c_{SD} and, similarly when the route request is received by D, it increments its own counter c_{SD}. In this way, the counters and the hints remain synchronized (assuming that no route request is lost).

The modified route request has the following form:

$$h(k_{SD}, c_{SD}) \,||\, E_{k_{SD}}(D||S||K^{-1}) \,||\, K \,||\, E_K(k_i||n_i|| \ldots E_K(k_0||n_0) \ldots).$$

Next, we include hints in the route reply too. Here, we take advantage of the common secret state between the destination and each of the intermediate nodes, which was established by the route request message. In particular, we derive a hint for node F_i from the nonce n_i, which is a secret value shared by D and F_i, by hashing it with a one-way hash function g. The modified route reply has the following form:

$$g(n_\ell) \,||\, E_{k_\ell}\left(n_\ell \big\| k_\ell^{in} \big\| k_\ell^{out} \big\| g(n_{\ell-1}) \,\big\|\right.$$
$$\left. E_{k_{\ell-1}}\left(n_{\ell-1} \big\| k_{\ell-1}^{in} \big\| k_{\ell-1}^{out} \big\| \ldots g(n_0) \,||\, E_{k_0}(n_0 \big\| - \big\| k_0^{out}) \ldots\right).\right)$$

Each forwarding node F_i stores $g(n_i)$ besides n_i and k_i when processing a route request. Thus, when processing route reply messages, the node has to do only a table lookup to determine which key it should use to decrypt the route reply. In addition $g(n_i)$ just looks as a random string to anyone who does not know n_i, so it reveals no information to eavesdroppers and other forwarding nodes that see it.

Finally, we use again synchronized counters to generate hints for forwarding nodes that help them processing encrypted data packets. Here, we take advantage of the link keys established by the route reply message. Let us consider two consecutive forwarding nodes F_i and F_{i+1}. They share a link key k, which is referred to as k_i^{out} at F_i and as k_{i+1}^{in} at F_{i+1}. They can both initialize a shared counter c by computing $c = g(k)$. Then F_i can generate a one-time hint $h(k_i^{out}, c)$ as a keyed hash of the counter. F_{i+1} can pre-compute this hint and store it in its routing table together with k_{i+1}^{in}. When sending the next data packet encrypted with k_i^{out}, F_i also sends the hint $h(k_i^{out}, c)$ and then increments c. When receiving a data packet, F_{i+1} uses the hint to do a table lookup and determine on which link the packet arrived, and which key it should use to decrypt it. If the link is identified, then F_{i+1} also increments c and pre-computes the next hint.

8.5 Summary

In this chapter, we studied the very important issue of privacy in upcoming wireless networks through three specific examples: privacy in RFID systems, location privacy in vehicular networks, and privacy preserving routing in wireless ad hoc networks.

Before the discussion of these examples, in Section 8.1, we introduced some fundamental notions related to privacy, such as *anonymity*, *unlinkability*, *unobservability*, and *pseudonymity*. We also presented metrics to quantify the level of privacy based on the *anonymity set size*, as well as on the *entropy* of the probability distribution over the members of the anonymity set. We used these notions and metrics extensively in the subsequent sections.

In Section 8.2, we investigated privacy problems and possible solutions in RFID systems. RFID is an interesting example for at least two reasons. First, it is expected that low-cost RFID tags will be deployed massively in the future, which raises serious concerns about the privacy of the tag bearers. Second, low-cost RFID tags are extremely resource constrained, meaning that any privacy protecting solution must be very carefully designed and optimized. In particular, low-cost tags will not support public key cryptography in the foreseeable future.

In terms of privacy problems, we identified clandestine *reading* and *eavesdropping* of low-cost tags as a plausible threat that makes stealth *inventorying* and *tracking* of tags possible. We also elaborated on the various read ranges and eavesdropping ranges of passive RFID tags. Then, we presented privacy protecting solutions for tags that cannot perform any computations, and also for tags that can perform symmetric key cryptographic operations. We illustrated how one of the presented solutions, the key-tree based scheme, can be analyzed using the notion of anonymity set. In particular, we characterized the level of privacy provided by the scheme with the (normalized) expected anonymity set size of a randomly selected tag, and we computed this measure in the case when a single tag is compromised.

In Section 8.3, we were concerned with the problem of location privacy in vehicular networks. We showed that it is nearly impossible to provide location privacy with respect to a global eavesdropper even if the vehicles *use pseudonyms* that they change frequently. This is mainly the result of the envisioned operational principles of vehicular networks that require that each vehicle broadcasts its current position and speed with a high frequency. We proposed a more realistic adversary model, where we distinguished between monitored zones and unmonitored zones. The latter was called a *mix zone*, because it provides similar services for moving vehicles as a mix node for messages. Then, we studied how the level of location privacy provided by the mix zone can be quantified using an entropy based metric, and illustrated the computation of the privacy level through an example.

Finally, in Section 8.4, we investigated the problem of privacy in ad hoc network routing protocols. Most of the protocols that we discussed in Chapter 7 trivially allow for an adversary to learn who is communicating with whom in the network. In order to counter this problem, we proposed a routing protocol that prevents global eavesdroppers and forwarding nodes to learn who the communicating pairs are. We first introduced a basic protocol that was effective but inefficient, and then we proposed modifications to the basic protocol to improve its efficiency. The building blocks that we used in our design were *broadcast communications*, *hop-by-hop re-encryption*, *padding to a fixed length*, and *dummy messages*.

8.6 To probe further

Privacy in general Precise (but informal) definitions of anonymity, unlinkability, unobservability, and pseudonymity are given by Pfitzmann and Köhntopp in [317]. Entropy based metrics for anonymity and for unlinkability are proposed by Diaz *et al.* in [113], by Serjantov and Danezis in [342], and by Steinbrecher and Köpsellin in [359]. In [326], Reiter and Rubin propose another approach, based on probabilities, to measure the level of anonymity. They introduce an "informal continuum" with six degrees of anonymity, and they illustrate how this measure can be used via the analysis of a practical system called Crowds, designed for anonymous Web transactions.

Chaum pioneered techniques to protect privacy by proposing various mechanisms including DC (Dining Cryptographers) networks [99] for anonymous communications and MIX networks [98] for unlinkability of message senders and receivers. Later on, many researchers made various extensions and improvements to Chaum's original ideas.

RFID privacy Weis, Sarma, Rivest, and Engels were among the first researchers who began to investigate privacy (and security) problems in RFID systems [375, 338]. Later on, many privacy protecting schemes were proposed for RFID tags. A comprehensive survey of the state of the art can be found in [213] written by Juels. This paper also identifies several research questions in the field of RFID security and privacy.

Concerns about the possibility of eavesdropping on the US electronic passports are raised by Zetter in [394]. The feasibility of remote reading of RFID based contactless smart cards is reported in [221] by Kfir and Wool.

The privacy protecting scheme based on pseudonym lists stored in the tags was proposed in [212] by Juels. In the same paper, the author also proposes a formal model for analyzing the security and privacy of RFID systems, which might be of independent interest for the interested readers. The re-encryption scheme based on the ElGamal cryptosystem is proposed in [214] by Juels and Pappu. One disadvantage of this scheme is that to perform the re-encryption operation, the readers must know the system public key. Later, this limitation was overcome by the universal re-encryption scheme proposed by Golle, Jakobsson, Juels, and Syverson in [154]. The approach based on the blocker device is proposed by Juels, Rivest, and Szydlo in [215]. Many papers explore the approach of maintaining synchronized state in the tag and in the reader [212, 294, 116]. Among them, Ohkubo, Suzuki, and Kinoshita were the first to identify forward secrecy as an important requirement in [294].

The key-tree based approach for privacy protecting tag identification is proposed by Molnar and Wagner in [278]. The approach of using Hellman's time–memory trade-off [167] is proposed by Avoine, Dysli, and Oechslin in [34]. In the same paper, the authors identify the problem of the decreasing level of privacy provided by the key-tree based scheme when some tags are compromised. However, instead of anonymity sets, the authors use a cryptographic approach to quantify the level of privacy. In their model, the adversary is first allowed to compromise some tags, then she chooses a target tag that she wants to track, and she is allowed to interact with the chosen tag. Later, the adversary is given two tags such that one of them is the target tag chosen by the adversary earlier. The adversary can interact with the given tags, and she must decide which one is her target. The level of privacy provided by the system is quantified by the success probability of the adversary. An entropy based approach for quantifying the level of privacy provided by the key-tree based scheme is proposed by Nohara *et al.* in [293]. The expected anonymity set size as a measure of privacy provided by the key-tree based scheme is proposed by Buttyán, Holczer, and Vajda in [74]. In the same paper, the authors also show that careful design of the key-tree can minimize the decrease of privacy caused by the compromised tags. They also propose an algorithm that finds the optimal key-tree that provides the maximum level of privacy given some constraints on the identification delay.

Location privacy in vehicular networks The envisioned operational principles, as well as many applications of vehicular communications, are described in [365]. El Zarki *et al.* were among the first researchers to identify security and privacy problems in vehicular networks [392]. Later, Hubaux, Čapkun, Luo, and Raya, as well as Parno and Perrig published papers on the same topic [186, 323, 307]. The concept of the mix zone is proposed by Beresfrod and Stajano in [51, 52] in the context of location based services in ubiquitous computing systems. We adopted that model for the vehicular setting. Another approach for location privacy in vehicular networks is proposed by Sampigethaya *et al.* in [335].

At the time of this writing, the problem of location privacy in vehicular networks is an on-going research topic. For instance, the quantification of the level of privacy provided by mix zones needs further investigation. Questions are still open about what characteristics of the mix zone have an effect on the level of privacy, and what exactly that effect is. Another question is whether the adversary can choose its monitoring area in an optimal way. Yet another open question is what effect the change of pseudonyms has on the performance of the system, and in particular, on the routing protocol.

Privacy preserving routing in ad hoc networks The routing protocol that we have described in Section 8.4 was inspired by the protocol proposed in [344] by Seys and Preneel. A few other papers consider the problem of privacy preserving routing in ad hoc networks [227, 397]. Seemingly, this problem has not yet received as much attention from the research community as, for instance, RFID privacy.

8.7 Questions

(1) Let us consider the key-tree based approach to privacy preserving identification of RFID tags described in Section 8.2. Let us assume that the reader sends a challenge to the tag, and the tag uses its keys to encrypt that challenge. Does the key-tree based approach in this case ensure forward secrecy? If so, why? If not, could the scheme be extended to ensure it?

(2) Consider again the key-tree based approach to private authentication of RFID tags. Generalize the computation of the expected anonymity set size for key-trees that have different branching factors at different levels of the tree. How would you estimate the expected anonymity set size in the case when more than one tags are compromised?

(3) In Section 8.3, we computed the level of privacy provided by the mix zone in the case of a given observation of the adversary (see expression (8.9)). How would you make the privacy metric independent of the observation?

(4) In the calculations of Section 8.3, we assumed that the mix zone was empty before the first entering event, and it becomes empty again after the k-th exiting event. In practice, however, the mix zone may not become empty at all. How would you re-define the notion of observation in this case? What effect does this modification have on the calculations?

(5) In the anonymous routing protocol that we described in Section 8.4, when sending a route request, the source generates an asymmetric key-pair (K, K^{-1}), and encrypts the identifiers of the destination and the source, together with K^{-1} as salt, with the public key of the destination. Explain why it is important here to include some salt in the encryption.

(6) In the anonymous routing protocol that we described in Section 8.4, when an intermediate node F_i receives the route reply, it peels one layer of encryption off, removes the information that is destined to it, and applies some random padding before re-broadcasting the message in order to retain its original length. How can the next intermediate node F_{i-1} on the path of the route reply identify and remove this padding?

(7) When we improved the efficiency of the anonymous routing protocol that we described in Section 8.4, we introduced a counter c_{SD} whose value is synchronously maintained by the source and the destination. This assumes that the destination receives the route request packet. How realistic is this assumption? What happens when the source and the destination are de-synchronized? Can you propose a mechanism that helps the source and the destination to maintain the synchrony?

Part III

Thwarting selfish behavior

So far, we have focused exclusively on malicious behavior, the traditional area of interest of security. But it is very important to keep in mind that security has been dominated for many years by military considerations, meaning that the "attacker" was indeed an attacker, equipped also with lethal weapons. Consequently, the goal was to defeat it at any cost, and it was pointless to try to quantify its motivation.

Commercial applications are very different from warfare, as we pointed out in Chapter 3. It is our profound belief that focusing exclusively on (malicious) attacks, as frequently done in current practice, captures only one aspect of the problem. Part III will explain how to model the *motivation* of a given party to depart from its nominal behavior.

The first two chapters of this part take the point of view of a selfish wireless station, considering first the MAC layer and then a fundamental mechanism of the network layer, namely packet forwarding. In both cases, we explain how to model the behavior of the stations by means of game theory. The reader unfamiliar with this discipline is strongly encouraged to first read Appendix B.

In the third chapter, we take the point of view of wireless operators having to cope with each other's presence in the same spectrum. Again, we explain how this problem can be modeled by means of game theory.

The fourth and final chapter of Part III (and therefore of the book) proposes techniques to enforce appropriate (non-selfish) behavior of the players in the specific case of a mobile ad hoc network. Through this example, we show that security protocols can be very helpful for this purpose.

As we have mentioned in Part I, the Internet community was taken by surprise by selfish behavior (spam in particular). At the time of this writing, it is still unclear how selfishness will materialize in practice in upcoming wireless networks. Hence the treatment provided in this Part III must be understood as *examples* of possible solutions, potentially to be adapted to slightly different networks, rather than anticipations of future protocol standards.

These examples have been carefully chosen: each addresses a fundamental question in its field.

9

Selfish behavior at the MAC layer of CSMA/CA

For our first analysis of selfish (or greedy) behavior in wireless networks, we will focus on the MAC layer.[1] More specifically, we will focus on IEEE 802.11, (the security of which we have already discussed in Chapter 1) because (i) this protocol is by far the most popular in existing wireless LANs, (ii) the devices (access points and wireless adapters) are very inexpensive, (iii) simulation code is widely available, and (iv) consequently the interested reader can relatively easily experiment the ideas developed in this chapter. In other words, the prominence of this protocol in wireless networks is such that it is the optimal candidate to illustrate the concepts of Part III of this book with a concrete example. The subsequent chapters will be (as all of the ones of Part II) standard-independent, with the exception of Section 11.2 in which we will refer to UMTS.

In IEEE 802.11, the stations (including the access point) in a given radio domain can transmit only one at a time. The temptation for selfish behavior is obvious, as the radio link is shared between all stations in power range: by departing from the protocol, a cheating station can substantially increase its bandwidth, at the expense of the other stations.

In the first section of this chapter, we briefly describe the operating principles of IEEE 802.11, which need to be known for a proper understanding of this chapter. In the second section, we address the problem of greedy behavior of a mobile station in the presence of an access point; this kind of misbehavior is typically to be expected against WiFi and mesh networks. The focus is on the system and engineering aspects of the problem. Finally, in the third section we consider greedy behavior in the absence of access points. This case typically corresponds to self-organized mobile ad hoc networks. The approach of that third section is more theoretic.

The reader interested only in the theoretic aspects can skip Section 9.2; this would not jeopardize the understanding of the rest of the chapter, nor of the other

---

[1] In this and in the following chapters, unless otherwise stated, MAC will mean "Medium Access Control."

chapters. More generally, the subsequent chapters can be understood also without reading the present chapter.

It is important to notice that in the whole chapter (and in all the other chapters of Part III), we consider exclusively **selfish** (and not malicious) behavior; if needed, refer back to Chapter 3 for the definitions of these terms.

9.1 Operating principles of IEEE 802.11

Currently, IEEE 802.11 is the *de facto* standard for WLANs [104]. It specifies both the *medium access control* and the *physical* layers for WLANs. The scope of IEEE 802.11 *working groups* (WGs) is to propose and develop MAC and PHY layer specifications for WLAN to handle mobile and portable stations. In this standard, the MAC layer operates on top of one of several possible physical layers. Medium access is performed using *carrier sense multiple access with collision avoidance* (CSMA/CA). Concerning the physical layer, three IEEE 802.11 standards are available at the time of this writing: a, b, and g. The first IEEE 802.11 compliant products were based on 11b. Since the end of 2001, higher data rate products based on the IEEE 802.11a standard have appeared on the market [192]. More recently, the IEEE 802.11 working group has approved the 802.11g standard, which extends the data rate of the IEEE 802.11b to 54 Mbps [195]. The IEEE 802.11g specification offers transmission over relatively short distances at up to 54 Mbps. The 802.11g PHY layer employs all available modulations specified for 802.11a/b.

Figure 9.1 shows the reference model of the IEEE 802.11 architecture [104]. All PHY layers consist of two sublayers and two management entities. The *physical medium dependent* (PMD) sublayer defines characteristics of the wireless medium and performs data encoding and modulation as well. The *physical layer convergence procedure* (PLCP) sublayer allows the MAC to operate with minimum dependence on the physical characteristics of the wireless medium. The PLCP sublayer also sets up the frame called *PHY protocol data unit* (PPDU) using the information provided by the MAC layer. The payload part of the PPDU frame is called *MAC protocol data unit* (MPDU).

The MAC layer communicates with the PHY layer using PLCP via specific primitives through a PHY service access point. Management entities (for each layer) perform the management of the PHY and MAC layer. Generally, the IEEE 802.11 WGs address the following issues [104].

- Functions required for an 802.11 compliant device to operate either in a peer-to-peer (ad hoc) fashion or integrated with an existing wired LAN.
- Operations of the 802.11 device within possibly overlapping 802.11 wireless LANs and the mobility of this device between multiple wireless LANs.

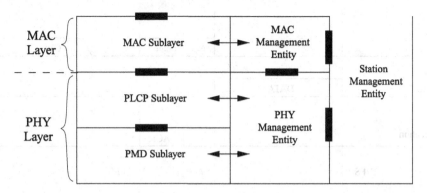

Figure 9.1. Reference model of PHY and MAC layer covered by IEEE 802.11 standards (used with permission, from [104]).

- MAC level access control and data delivery services.
- Several physical layer signaling techniques and interfaces.
- Privacy and security of user data being transferred over the wireless media.

In general, the wireless networking can be implemented in two significantly different operating modes: the *ad hoc* and *infrastructure* modes. The infrastructure mode consists of an *access point* (AP) acting as a hub for the network with each client communicating through it. This mode is the one used in virtually all of today's operational WLANs.

The ad hoc mode essentially eliminates the need for an access point. In this mode, the mobile nodes can be connected dynamically in an arbitrary manner. This mode is typically used when this MAC layer is used in mobile ad hoc networks, described in Chapter 2.

The *distributed coordination function* (DCF) is the basic medium access mechanism of IEEE 802.11, and uses a *carrier sense multiple access with collision avoidance* (CSMA/CA) algorithm to mediate the access to the shared medium.[2]

The DCF protocol in the IEEE 802.11 standard defines how the medium is shared among stations. DCF is based on CSMA/CA [104]. It includes a basic access method and an optional channel access method with *request-to-send* (RTS) and *clear-to-send* (CTS) exchanged as shown in Figures 9.2 and 9.3, respectively. First, we explain the basic access method.

If the channel is busy for the source, a backoff time (measured in slot times) is chosen randomly in the interval $[0, CW)$, where CW stands for the *contention*

---

[2] The standard describes centralized, polling-based access mechanism, the Point Coordination Function (PCF). It is very rarely used in practice and is not used at all in this book.

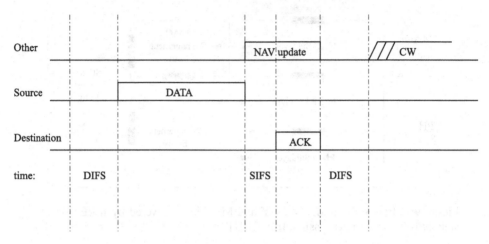

Figure 9.2. Basic access CSMA/CA protocol.

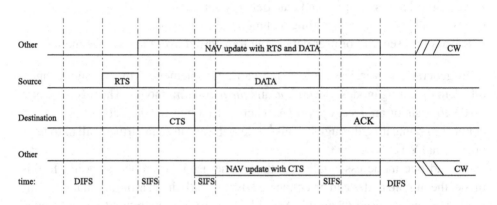

Figure 9.3. RTS/CTS exchange in the CSMA/CA protocol.

window.[3] This timer is decreased by one as long as the channel is sensed idle for a DIFS, i.e., *distributed inter-frame space* time. DIFS is equal to $SIFS + 2 \times SlotTime$, where SIFS stands for *short inter-frame space* (see values in Table 9.1). The timer stops when the channel is busy and resumes when the channel is idle again for at least a DIFS period. *CW* is an integer whose range is determined by the PHY layer characteristics: CW_{min} and CW_{max}. *CW* is doubled after each unsuccessful transmission, up to the maximum value (equal to $CW_{max} + 1$).

When the backoff timer reaches zero, the source transmits the data packet. The ACK is transmitted by the receiver immediately after a period of duration equal to SIFS. When a data packet is transmitted, all other stations hearing this transmission

---

[3] The slot time is the sum of the RX-to-TX turnaround time, MAC processing delay, and CCA detect time [104]. The value of slot time for different PHY layer protocols is shown in Table 9.1.

Table 9.1. *Inter frame space and CW time for different PHY layers*

Parameters	802.11a	802.11b (FH)	802.11b (DS)	802.11b (IR)	802.11b (High Rate)
Slot Time (μs)	9	50	20	8	20
SIFS (μs)	16	28	10	10	10
DIFS (μs)	34	128	50	26	50
EIFS (μs)	92.6	396	364	205 or 193	268 or 364
CW_{min} (Slot Time)	15	15	31	63	31
CW_{max} (Slot Time)	1023	1023	1023	1023	1023

adjust their *net allocation vector* (NAV). The NAV maintains a prediction of future traffic on the medium based on the duration information that is announced in DATA frames (or RTS/CTS frames as will be explained in the following) prior to the actual exchange of data. In addition, whenever a node detects an erroneous frame, the node defers its transmission by a fixed duration indicated by EIFS, i.e., *extended inter-frame space* time. This time is equal to the $SIFS + ACK_{time} + DIFS$ time.

The contention window is initially set to the minimum value of CW_{min}, equal for example to 15 (see Table 9.1). Every time a collision occurs, this is interpreted as a high load of the network, and each station involved in the collision throttles down its transmission rate by doubling the size of its contention window. In this way, the contention window can take values equal for example to 31, 63, 127, 255, 511, up to $CW_{max} = 1023$. Larger contention windows slow down the transmission of packets and reduce the probability of collisions. In case of a successful (i.e. collision-free) transmission, the transmitting station brings the value of its contention window back to CW_{min}. The mechanism we have just described is called *exponential backoff* or *binary exponential backoff*.

If the optional access method is used, an RTS frame should be transmitted by the source and the destination should accept the data transmission by sending a CTS frame prior to the transmission of the actual data packet. Note that stations in the sender's range that hear the RTS packet should update their NAVs and defer their transmissions for the duration specified by the RTS. Nodes that overhear the CTS packet update their NAVs and refrain from transmitting. In this way, the transmission of the data packet and its corresponding ACK can proceed without interference from other nodes. Table 9.1 shows the important time interval between frames in different standard specification called *inter-frame space* (IFS) [192, 193, 195]. IEEE 802.11g uses the IFS corresponding to its operating mode.

The increasing number of wireless users and the demand for high-bandwidth multimedia applications over WLANs led the IEEE working groups to extend the

MAC layer to provide QoS support (IEEE 802.11e [191]). We do not describe it here, as it is not used in this book.

9.2 Detecting selfish behavior in hotspots

IEEE 802.11 [104] wireless LANs were originally meant to be deployed in (relatively) protected locations such as corporate offices; as a result, security, billing, and guarantee of fair access received limited attention. But over the last few years, IEEE 802.11 has also become the standard solution for hotspots that provide public wireless access to the Internet.

In this framework MAC-layer greedy behavior is particularly tempting: a station can deliberately misuse the MAC protocol to gain bandwidth at the expense of other stations. The benefits of this misuse are the following.

- It can result in significant bandwidth gains as it directly deals with the wireless medium. Therefore, it is more efficient than misbehavior at the network and transport layers.
- It is hidden from upper layers and hence cannot be detected by any mechanism designed for those layers. Thus, it can be combined with upper layer misbehavior to enhance it.
- It is always usable in WiFi settings, because all the wireless stations use the same IEEE 802.11 MAC protocol. In contrast, for example, cheating with TCP yields no benefits against UDP competing sources.

As we will see, MAC-layer selfish behavior in IEEE 802.11 networks can lead to severe unfairness in bandwidth distribution. This can become a serious problem in public Internet access hot-spots where individual users have to pay for network access and hence can be motivated to cheat in order to increase their share of the medium. Once a hacker has implemented an attack, she can make it available on a Web site, thus jeopardizing the proper operation of many wireless networks around the globe.

It is important to note that the scope of this presentation goes beyond IEEE 802.11 networks; indeed, we provide a framework that can be adapted to the study of cheating and detection techniques in any network based on a shared medium.

In this section, we explore this space of MAC-layer greedy behavior. We consider that only a single mobile station can cheat (the Access Point is trusted, and complies to the protocol).[4] We propose a *classification* of the different MAC misbehavior

---

[4] If there are several cheaters, their greedy behavior will lead to a high proportion of packet collisions, which is a natural disincentive to cheat in such conditions; we will discuss the interactions between multiple cheaters in detail in Section 9.3.

techniques and illustrate them with representative examples. Then, we present a system, called DOMINO (Detection Of greedy behavior in the MAC layer of IEEE 802.11 public NetwOrks) for detecting MAC misbehavior in a way that is transparent to the operation of the network. The key features of DOMINO are its seamless integration in or near the AP,[5] its full compliance with existing standards, and its ability to identify the cheater.

Based on the output of the detection system, the WISP (Wireless ISP) can decide how to react to cheating users. For example, the operator can invite the suspected user to bring her mobile station for a technical scrutiny, charge a penalty bill, reduce the service quality, or even completely stop the service, depending on the extent of the observed cheating and the responsiveness of the cheater.

In order to make things very concrete, we deliberately go into the engineering details of the solution by presenting the results of real experiments that demonstrate the ease of cheating and the efficiency of DOMINO. It is indeed possible, by means of minor changes to a driver for IEEE 802.11 compliant cards, to obtain much higher throughput at the expense of stations equipped with unmodified drivers.

We will now explore possible misbehaving techniques. We will then present a detection system and illustrate its performance by means of simulation results; we will also describe its implementation.

9.2.1 Misbehavior techniques

In this section we present a taxonomy of MAC-layer greedy behavior in IEEE 802.11 hotspots. As mentioned, we deliberately do not address the techniques making use of the *security* weaknesses of the standard (we already addressed them in Chapter 1). We rather focus on *MAC greedy behavior*, which consists in modifying the operation of the IEEE 802.11 protocol by departing from the communication procedures or changing parameters defined in the standard. In the rest of the chapter, "misbehavior" (or "attack") means *greedy* behavior and does not relate to the security aspects of wireless networks.

Several studies have shown that around 90% of the traffic flowing over deployed wireless LANs is TCP-based and is mainly downlink, namely from the AP to the user stations (admittedly, this might change in case of massive adoption of Internet telephony). Hence it is important to distinguish misbehavior techniques according to the type of traffic they target. In the following, we describe greedy attacks on uplink traffic (both TCP and UDP) and downlink TCP traffic (greedy attacks against

---

[5] The actual component in which DOMINO has to be installed is the *hot-spot controller*, which provides access control and can control several APs [124]. Nevertheless we assume in the following, without loss of generality, that the hotspot controller is incorporated in a single AP and thus we refer, for simplicity, to both components as AP.

downlink UDP traffic are much more difficult to perpetrate and will not be described here).

Uplink traffic

- A greedy station can selectively scramble frames sent by other stations in order to increase their contention windows. The frames to be targeted can be the following:

  (1)   **CTS frames**. In this case the cheater hears an RTS frame sent by another station to the Access Point and intentionally causes collision and loss of the corresponding CTS frame in order to prevent the subsequent long frame exchange sequence (RTS/CTS handshake is generally used for large frames). As a result, the channel becomes idle after the corrupted CTS, the station whose CTS was jammed doubles its contention window, and the cheater has a better chance to send her data.

  (2)   **ACK and DATA frames**. Although jamming these frames does not result in saving the data frame transmission time, it causes the contention window of the ACK destination (i.e., the DATA source) station to be doubled and consequently makes the latter select larger backoffs. As before, the cheater increases her chances to access the channel.

- A greedy station can manipulate protocol parameters to increase bandwidth share.

  (1)   When the channel is idle, transmit after SIFS but before DIFS.

  (2)   When sending RTS or DATA frames, set the *duration* field (in the frame headers) to a high value; in this way, as the stations in range set their NAVs with this value, they will refrain from contending with the channel during all this time.

  (3)   Reduce the backoff time. This can be done by choosing a small fixed contention window; thus the backoff is always chosen from this small window.

A cheater can also combine several of the above techniques or adaptively change her misbehavior to avoid being detected. We will address this type of cheating in Section 9.2.5.

Downlink traffic

- In the case of the downlink traffic, the cheater will attempt to increase the share of traffic sent to her through the AP, thus increasing the number of packets destined to it in the AP's queue (usually there is indeed a single, FIFO queue); to achieve this goal, she will target the protocols responsible for filling this queue. We can distinguish two types of sources (e.g., Web servers) sending traffic to wireless stations through the AP.

(1) *UDP source*: Attacking UDP traffic is pointless because UDP requires no acknowledgements from the receiver and hence cannot be affected by channel conditions.[6]

(2) *TCP source*: In contrast with UDP, the TCP traffic rate reacts to the channel conditions by using congestion windows and acknowledgements from the receiver. Hence an attack can be mounted on the TCP traffic by exploiting the congestion avoidance mechanism and reducing the source rate, and eventually shutting down the flow.

Downlink attacks are relatively less intuitive and require more "effort" from the cheater's side to increase her share of the bandwidth, and from the AP's side to detect the misbehavior. Leveraging on the closed-loop nature of TCP flows, their impact goes beyond the local area (the hot-spot and associated nodes) to reach remote servers. Consider the topology in Figure 9.4 and the typical following scenario: two mobile nodes M and Mc are connected to the Internet via the AP. M and Mc download large files from two remote servers, S and Sc, respectively. Both downloads use FTP/TCP. To increase her download data rate, the cheater (Mc) can use the following two techniques to reduce S's data rate, thus freeing more bandwidth for himself at the AP (or at any common bottleneck between the servers and the AP).

- Mc jams the TCP-ACKs from M to the AP, so they never reach the server S. As TCP-ACKs get lost (jammed), S decreases its sending data rate, using TCP congestion control, and ends up killing the connection. At the AP, M's share of the bandwidth decreases, leading to an increase of the data rate from Sc to Mc.

- In the previous technique, the AP can still hear the collisions/jamming and can end up detecting Mc based on the number of retransmissions of M. Another option for Mc consists in jamming the AP's frames destined for M, therefore reducing S's data rate without being heard by the AP. However, Mc's packets share the same queue as M's packets at the AP. While jammed frames get repeatedly retransmitted by the AP, Mc's packets get delayed in the queue, and her data rate (from Sc) decreases as well. To prevent the AP's retransmissions and the queueing delays, Mc sends forged MAC-ACKs on behalf of M for the jammed packets.[7] This avoids retransmissions at the AP, and reduces the data rate from S. Furthermore, as we will show in the simulation results, Mc can jam only part of the AP's frames to M, saving her battery power and making detection even harder.

The effects of these misbehavior techniques can be devastating, given the extensive use of TCP in the Internet, especially for Web browsing and file transfers,

---

[6] To simplify the discussion, we do not address the case of applications with feedback loops running over UDP.

[7] In IEEE 802.11, ACK frames contain no source address fields, therefore making Mc's task easier.

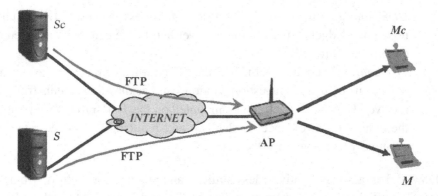

Figure 9.4. Generic scenario where *Mc* jams the AP's TCP packets destined to *M* (or the corresponding ACKs) in order to reduce the flow from server *S*. Used with permission, from [322], © IEEE, 2006.

which are major Internet applications nowadays. It should be noted that using IPsec to encrypt TCP packets will not prevent the cheater from mounting the above attacks because she can simply jam all the packets related to other stations without distinguishing whether they are TCP or not.

From the implementation point of view, the above attacks are feasible. In fact, the amount of time that a station has at its disposal to revert its radio (called Rx-TxTurnAroundTime) is smaller than or equal to 5 μs in IEEE 802.11b (which was the first deployed, and therefore the slowest version of IEEE 802.11); the MAC frame header is 30 bytes; the IP frame header is 20 bytes. Hence, assuming the highest rate of IEEE 802.11b (11 Mbps), which implies the shortest time available for jamming, the transmission time of the MAC header is around 22 μs and of the IP header is around 15 μs. Once the cheater knows the source and destination addresses from the MAC header, the short RxTxTurnAroundTime will allow her to jam the TCP frame or the TCP-ACK even before the whole IP header is transmitted.

Note that the use of TCP-splitting techniques will increase the tolerance of the TCP connection to the cheater's jamming, but only for a while if the latter is persistent. Hence, the connection will still be dropped, but after a longer delay.

9.2.2 A possible solution: DOMINO

Several approaches can be envisioned to counter the misbehavior techniques presented in Section 9.2.1. The one we describe here (DOMINO) is one of them. Given the number of possible attacks and their independence, DOMINO has a modular architecture, depicted in Figure 9.5. As mentioned, DOMINO is entirely implemented at the AP.

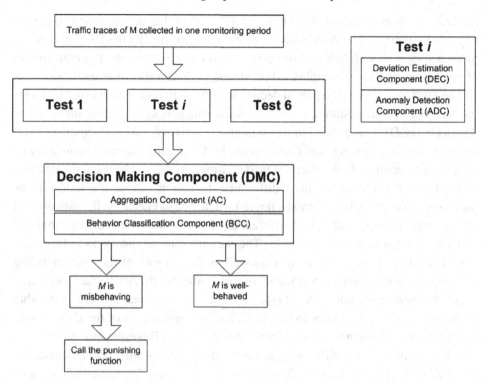

Figure 9.5. The modular architecture of DOMINO. Used with permission, from [322], © IEEE, 2006.

DOMINO periodically collects traffic traces of active user stations during short intervals of time called *monitoring periods* (the choice of their length is discussed in Section 9.2.5). A series of *tests*, each aimed at detecting a particular misbehavior technique, determines if the analyzed traffic presents behavior anomalies; these anomalies can be considered the *symptoms* of the corresponding misbehavior. The outputs of these tests are then fed into a *Decision Making Component* (DMC) that decides whether a given station is cheating. If so, the control is passed to the misbehavior handling mechanism that, as mentioned before, is dependent on the WISP policy.

The modular architecture presents several advantages. First, the tests as well as the decision making component can be implemented using several algorithms depending on the required accuracy and the tolerable complexity. Second, new tests for potential and yet undiscovered misbehavior techniques can be easily added.

In the following, we present the tests designed to detect the previously presented misbehavior techniques. Each test consists of two components: a *Deviation Estimation Component* (DEC) and an *Anomaly Detection Component* (ADC). The DEC is typically a statistical test that determines the amount of deviation of a station's behavior, inferred from its traffic trace, from a model of the expected behavior (derived by observing the behavior of the AP or the other active stations during a monitoring period). The ADC uses the deviation measured by the DEC in order to judge a station as well-behaved or suspected. It can be as simple as a comparison of two values or a more complicated technique such as a Bayesian inference.[8]

The DMC aggregates the partial decisions of the different tests in order to assess the behavior of a given station in the last monitoring period. Following the modular approach, the DMC is divided into two modules: an *Aggregation Component* (AC) and a *Behavior Classification Component* (BCC). Again, the implementation of either can be flexible. In the simplest instantiation, the AC can be implemented as a simple OR of the boolean outputs of different tests. This means that if a station cheats using any of the described methods, it will be detected as cheater. Alternatively, the AC can output a weighted sum of different test outputs; this sum is then normalized to 1 and compared with a threshold. The weights can be chosen to indicate the confidence in a given detection test, as well as the severity of the corresponding misbehavior. For example, a test whose output cannot be affected by factors such as channel conditions would have a higher weight than a test that is more vulnerable to these conditions. Similarly to the ADC, the implementation of the BCC can be based on a simple misbehavior tolerance threshold, or a Bayesian inference.

Whereas the ADCs in different tests can use different or similar implementations, the DEC is specific to each test. Therefore, in the following description of each test, we will focus on the algorithm behind the corresponding DEC. The tests described below use the following structure, where x indicates the test number.

It should be noted that all the tests described below are performed on each data sample successfully collected for a station M_i during the last monitoring period; if misbehavior is detected, the checking on M_i is interrupted as no further analysis is needed. For clarity, we present the operation of the tests on a single data sample.

"Scrambled frames"

This test aims at detecting misbehavior techniques that rely on frame scrambling; they correspond to the first attacks described in the uplink and downlink parts of Section 9.2.1.

In order to gain a significant share of the common wireless bandwidth using CTS/ACK/DATA scrambling, the cheater has to scramble a relatively large percentage of CTS, ACK, or DATA frames sent by other stations. As a result, its average number of retransmissions will be less than that of other stations, and it can be detected using Test 1 (for this test as for the following ones, if the inequality holds, it means that a greedy attack is probably taking place).

---

[8] Bayesian inference is a statistical technique in which evidence or observations are used to update or to newly infer the probability that a given hypothesis could be true.

Test 1 Scrambled frames

$$num\_rtx(M_i) < \phi \times E_{j \neq i}[num\_rtx(M_j)]$$

In this test, $num\_rtx(M)$ is the number of times station M retransmitted its last frame successfully received by the AP. ϕ is a tolerance parameter with a value between 0 and 1; it is applied to the average number of retransmissions of all "other" stations, $E_{j \neq i}$.

DOMINO can detect a retransmission by observing a repeated sequence number in the header of RTS or DATA frames when the corresponding CTS or ACK frames are scrambled, respectively. In the case of DATA frames, we might argue that the AP would not be able to distinguish retransmissions because the DATA frames are scrambled. However, the cheater cannot scramble the headers of these frames, otherwise it cannot know whether a given frame is destined to herself.

As we assume a rational attacker who jams other frames only when she needs to, her identity can be derived from the number of retransmissions. She cannot change this number to cheat because the sender (the AP in this case) will react to a wrong sequence number by discarding the frame (if the number is not larger than the last recorded one) or by sending a frame out of order (if the number is larger than the last recorded value) depending on the specific wireless card implementation. We also assume that the attacker cannot change the MAC address of her station because an authentication mechanism (e.g., WPA or IEEE 802.11i) is in place that prevents the arbitrary use of MAC addresses.

A potential cause of false positives for this test could be the bad channel conditions that lead to frame loss and retransmission. To avoid this pitfall, the AP can take the Received Signal Strength Indicator (RSSI) of stations into consideration when detecting misbehavior.

Detection of manipulated protocol parameters

In the following paragraphs we address misbehavior techniques that alter protocol parameters. We focus mainly on backoff manipulation because it is the easiest to implement (as we will show in Section 9.2.4) and the hardest to detect.

"Shorter than DIFS"

The AP can monitor the idle period after the last ACK and distinguish any station that transmits before the required DIFS period. After having observed this misbehavior repeatedly for several frames from the same station, the AP can make a reliable decision (Test 2).

Test 2 Shorter than DIFS

$$idle\_time\_after\_ACK(M_i) < DIFS$$

"Oversized NAV"

By measuring the actual duration of a transmission (including the DATA, ACK, and optional RTS/CTS) and comparing it with the *duration* field value in the RTS or DATA frame headers, the AP can detect a station that regularly sets the *duration* field (and therefore the NAV of listening stations) to very large values. In Test 3, the tolerance parameter A (greater than 1) ensures that the AP does not mistakenly incriminate well-behaved stations.

Test 3 Oversized NAV

$$A \times actual\_duration(M_i) < duration(M_i)$$

Backoff manipulation

Backoff manipulation detection is comprised of three tests described hereafter, namely "Actual backoff," "Consecutive backoff," and "Maximum backoff."

"Actual backoff"

This test (Test 4) consists in measuring the actual backoff, as shown in Figure 9.6. The main procedures of the test can be summarized as follows:

- If between two transmissions from a station M there are no collisions, we assume that M spent all its idle time backing off (although it could be just part of the M's inter-frame delay, if it is transmitting at low data rates). Then we estimate this backoff by computing the sum as illustrated in Figure 9.6.
- If a collision happens, it could be more difficult to know the identities of the senders of the colliding frames and which stations whose measured *actual backoff* should be updated. To avoid complexity, collisions are simply not taken into account; in case of collisions, neither the current backoff nor the next one are measured for any station.[9]

Test 4 Actual backoff

$$B_{ac}[M_i] < \alpha_{ac} \times B_{acnom}$$

In Test 4, $B_{ac}[M_i]$ denotes the average *actual backoff* (observed by the AP) of station M_i. B_{acnom} is the nominal backoff value, which is equal to the average backoff of the AP, assuming it has enough traffic to compute this value. The α_{ac} ($0 < \alpha_{ac} \leq 1$) parameter is configurable according to the desired true positive (correct detection) and false positive (wrong detection) percentages (for example, a value of $\alpha_{ac} = 90\%$ is used in the simulations).

---

[9] Stations that hear frame headers with a wrong CRC, caused by a collision, will defer their transmissions by EIFS (Extended InterFrame Spacing). This deferral does not interfere with the measurements because all deferrals of all nodes are not taken into account after a collision.

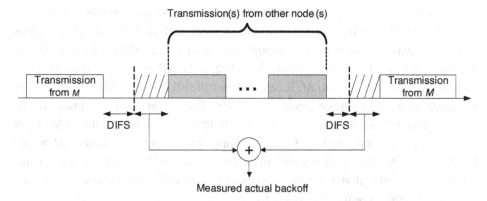

Figure 9.6. Measurement of the *actual backoff*. Transmissions from *M* are interleaved with one or more transmissions from other nodes (including the AP). The transmission includes, in addition to the DATA frame, all the control frames such as RTS, CTS, and ACK, as well as the interleaving idle periods of SIFS and DIFS. The measured value is the sum of all idle intervals (not including inter-frame spaces) between two transmissions from *M*. Used with permission, from [322], © IEEE, 2006.

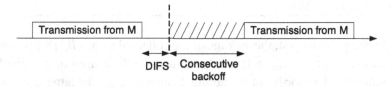

Figure 9.7. Measurement of the *consecutive backoff*. Backoff values are taken only between consecutive non-interleaved transmissions from *M*. Used with permission, from [322], © IEEE, 2006.

As it collects no data during collisions, the actual backoff test measures backoffs that are selected from only the $[0, CW_{min} - 1]$ range. Owing to its mechanism, this test fails to detect a misbehavior case if the cheater has inter-frame delays (e.g., a TCP source using congestion control). In fact, the test measures these delays instead of backoffs because it adds up the idle periods between transmissions from the same source (Figure 9.6). The solution to this problem is provided by the *consecutive backoff* test.

"Consecutive backoff"

Figure 9.7 illustrates this test (Test 5), which works in the case of sources with inter-frame delays. In practice, this is mainly the case of TCP sources (in this case the delay is typically due to the congestion control of TCP). The *actual backoff* test for these sources does not yield the correct values (as explained in the previous paragraph), and consequently cannot detect potential cheating.

Let us consider a station M sending TCP traffic. We assume that there is enough traffic from other sources on the common channel such that, between two frames sent by M and separated by a transport layer delay, there should be at least one interleaving frame from another station. Hence, if the AP observes two consecutive non-interleaved frames from M, it can consider the idle time between them as only a backoff in addition to the mandatory DIFS. These consecutive frames are the result of channel contention that can force M to queue packets at the MAC layer even if they were separated by a delay at upper layers. In this situation, M would benefit from cheating with backoff in order to free its MAC layer queue. Thus, DOMINO can collect significant samples of the backoff values chosen by M; we call these samples *consecutive backoffs*.

The above assumption of traffic level is realistic. In fact, if the traffic on the channel is low enough to invalidate this assumption, i.e., if M can send consecutive non-interleaved frames separated by a delay in addition to the backoff and DIFS, cheating would be pointless because reducing the backoff does not affect the upper layer delay. Misbehavior detection would not be needed in this case.

Test 5 Consecutive backoff

$\quad B_{co}[M_i] < \alpha_{co} \times B_{conom}$

As with the previous test, the average of the collected values $B_{co}[S_i]$ is compared with a fraction α_{co} ($0 < \alpha_{co} \leq 1$; a value of $\alpha_{co} = 90\%$ is used in the simulations that we will describe shortly) of the nominal value B_{conom}. The latter is the average consecutive backoff of the AP if enough data are available. Otherwise, it is an analytical value $E[B_{co}]$ (see [322]).

"Maximum backoff"

As the IEEE 802.11 protocol selects backoffs randomly from the range $[0, CW - 1]$ (where CW depends on the number of retransmissions), the maximum selected backoff over a set of frames sent by a given station should be greater or equal to $CW_{min} - 1$, if the number of samples is large enough. DOMINO uses this property to identify stations whose maximum backoff over a set of samples is smaller than a threshold value $threshold_{maxbkf}$. Clearly, a trade-off exists between the number of samples and the threshold; if we increase the threshold (its largest value is CW_{min}), we have to increase the number of sampled backoffs to obtain more distinct values and thus avoid false positives. In the simulations (Section 9.2.3), a threshold equal to $CW_{min}/2$ is used, thus the test works if the reduced contention window is in $[0, CW_{min}/2 - 1]$.

Unfortunately, this check can be easily tricked by a clever cheater who succeeds at making the monitor observe in every sample at least one backoff value larger than or equal to the threshold; channel conditions can also yield a similar result and thus make the check fail. Thus, the *maximum backoff* check is only auxiliary to the above two tests.

Scrambled TCP packets with forged MAC ACKs

Out of the two downlink attacks described in Section 9.2.1, the second one is the more sophisticated, so we will focus on it. This second attack is also the more difficult to detect, because the AP cannot hear collisions and, as the cheater forges the MAC ACKs corresponding to scrambled frames, DOMINO cannot rely on the number of retransmissions to detect misbehavior. To cope with this technique, DOMINO has two complementary mechanisms that implement the DEC and ADC components of the test in the system architecture (we call "Test 6" the combination of these two mechanisms). First, DOMINO measures the throughputs of the downlink flows (this is the DEC). Then, if there is a receiver that draws most of the traffic, DOMINO suspects it as a potential cheater. As will be explained in Section 9.2.5, throughput is not a reliable detection metric because of the different needs of users. Hence, DOMINO uses *Dummy Frame Probing* (DFP) to confirm the suspicion or reject it. DFP consists in sending dummy frames to virtual (nonexisting) stations. If any of these frames are followed by a MAC ACK, this is an indication of an existing cheater in the network. Longer throughput observation is then needed to determine the identity of the cheater. DFP combined with throughput comparison constitute the ADC.

A clever cheater can construct the list of virtual stations (by recording stations that do not reply with a MAC ACK) in order not to respond to the dummy frames. To detect this cheater, the AP should also generate fake ACKs: in this way, the cheater cannot easily distinguish the dummy frames from the others. But for the cheaters it is beneficial to attack only the connections with high throughput. Thus, in order to be effective, the dummy frames must be generated from time to time at high throughput as well, as a trap for the cheater. The advantage of dummy frames is that they represent a highly discriminating test: a simple sample is enough to raise a very high suspicion, even if they are generated during a small amount of the time (e.g., 5% to 10%). Thus, the resulting overhead is small.

9.2.3 Simulation results

In order to study the performance of the proposed solution, *ns*-2 [131] has been used to simulate DOMINO. As the frame scrambling misbehavior is fairly easy to detect using the number of retransmissions, this section examines in detail only the backoff manipulation tests and the complete detection mechanism.

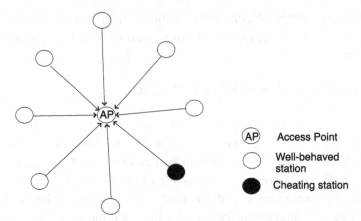

Figure 9.8. Simulation scenarios: 8 stations send UDP or TCP data to the AP, which also generates traffic similarly to one station and sends it to an additional receiver station (not shown in the figure). The distance between each station and the AP is 50m. All stations are within radio range of each other. Used with permission, from [322], © IEEE, 2006.

Uplink traffic

Simulation topology Further to the discussion in the previous section about the effect of traffic on Test 5, here we will study two cases (Figure 9.8) that are representative of common traffic types.

1. UDP traffic
   Besides the cheater, there are seven stations sending CBR traffic (the nominal rate is 500 bytes/packet, 200 packets/s); the cheater is also a CBR source. The cheating technique consists in decreasing the contention window.
   In any idle slot, there is at least one packet ready for transmission by any of the competing stations. The time elapsed between two transmissions from the same station (interleaved with transmissions from other stations) is therefore the result only at the backoff chosen by the IEEE 802.11 protocol.
2. TCP traffic
   Each of the eight stations runs an FTP application; one station is cheating by jamming TCP packets and forging the corresponding MAC ACKs.
   This case illustrates the effect of inter-frame delays (owings to TCP congestion control) on backoff measurement. This is the most realistic scenario.

In both cases the AP generates traffic similarly to one station, i.e., CBR in the first case and FTP in the second.

To take into account the fading effects present in real channels, the simulations are carried out with the shadowing channel model, represented by the following

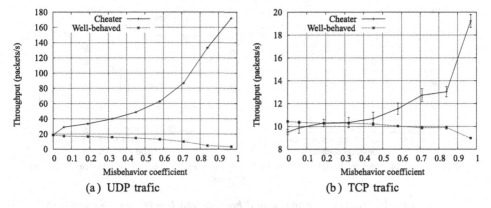

Figure 9.9. Throughput comparison between misbehaving and well-behaved stations. Used with permission, from [322], © IEEE, 2006.

equation:

$$\left[\frac{P_r(d)}{P_r(d_0)} \right]_{dB} = -10\beta \log \left(\frac{d}{d_0} \right) + X_{dB},$$

where $P_r(d)$ is the mean received power at distance d, d_0 is a reference distance, β is the path loss exponent, and X_{dB} is a Gaussian random variable with zero mean and standard deviation σ_{dB}. The following values have been used: $\beta = 2$ (free space propagation) and $\sigma_{dB} = 4$.

Results are averaged over 10 simulations, each of a duration of 110 s. The monitoring period is set to 10 s, which also corresponds to one decision (cheater or well-behaved) by the AP regarding each station. Thus, each point on the following graphs is averaged over 100 samples with a 95% confidence interval; the first 10 s of each simulation is an initialization period, where measurements are not taken into account in the results.

In the following, the *misbehavior coefficient* represents the amount of misbehavior. A misbehavior coefficient equal to m means that the corresponding station uses a fixed contention window equal to $(1 - m) \times CW_{min}$ and then chooses its backoff from this new window. Thus, $m = 0$ means no misbehavior, and $m = 1$ means that the station transmits without any backoff.

Impact of misbehavior on throughput Before presenting the performance of DOMINO, we compare the throughput values of cheating and well-behaved stations in both simulation scenarios.

Figure 9.9 shows that MAC misbehavior results in throughput benefits that are obtained at the expense of well-behaved stations and that increase with the amount of misbehavior.[10] We can also notice that this increase is less significant in the case

---

[10] The graphs also display 95% confidence intervals that are very small in some cases.

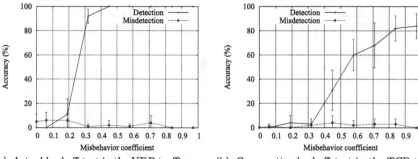

(a) *Actual backoff* test in the UDP traffic case. (b) *Consecutive backoff* test in the TCP traffic case.

Figure 9.10. Performance of the *Actual backoff* and *Consecutive backoff* tests. Used with permission, from [322], © IEEE, 2006.

of TCP sources. This is due to the TCP congestion control mechanisms and the dependence of the TCP throughput, including the cheater's, on the rate of the TCP ACKs, which are sent by the (well-behaved) AP.

Actual backoff From the simulation graphs we can draw the following observations.

- In the UDP traffic case, the test performs well, as shown in Figure 9.10a, because there is always at least one frame ready for transmission by each station. Hence the channel idle time between two transmissions from a station is the result of only the backoff mechanism (in addition to DIFS).
- In the TCP traffic case, the numbers of both correct and wrong detections are very small (the curves are practically superimposed with the x-axis; thus, the corresponding figure does not provide any important information and hence they have been omitted). The low correct-detection accuracy can be explained by the fact that the measured actual backoff is actually the idle period (not including transmission cycles) between two interleaved transmissions from the same station, which is equal in this case to the delay between frame transmissions from the source. This delay is created by the TCP congestion control mechanisms.

Consecutive backoff The performance of this test differs from that of the previous one for the reasons mentioned in the description of the test (Section 9.2.2) and confirmed by simulations.

In the UDP traffic case, the results of the test are of no use (the curves are superimposed with the x-axis and therefore are omitted) because, in this case, the measured average consecutive backoff rapidly decreases with the number of stations. The comparison of small values becomes inaccurate, thus seriously affecting the test significance.

In the TCP traffic case, the test yields good results as Figure 9.10b shows. This is due to the presence of other sources that do not allow the source with the inter-frame delay (induced by congestion control) to transmit two frames consecutively without having queued the second one, i.e., the delay does not affect the idle time between two consecutive non-interleaved transmissions from the source. Otherwise, if there is no frame ready in the queue, another source takes control over the channel and transmits at least one frame between two successive frames of the first source.

Complete mechanism The descriptions of the *actual backoff test* and the *consecutive backoff test* in Section 9.2.2, as well as the simulation graphs presented so far, have shown that each test performs well in specific traffic scenarios. The complete mechanism is thus a combination of both.

It is worth noting that, as long as there is enough traffic on the channel to satisfy the assumption in Test 5, only the type of the sender traffic determines which test works and thus misbehavior in mixed-traffic scenarios (TCP/high rate UDP) can also be accurately detected. If the traffic on the channel is low, misbehavior does not yield substantial throughput benefits, hence its detection is not necessary.

Downlink traffic

In this section the only simulations we report are related to the second of the two attacks on the downlink described in Section 9.2.1, i.e., referring to Figure 9.4, Mc jams the TCP packets from server S to M, transmitted by the AP, and sends MAC-ACKs on behalf of M.

In order to save energy and to make detection harder for the AP, Mc can jam only a proportion X of the downlink traffic (e.g., when $X = 1$, Mc jams all frames from the AP to M). This proportion of jammed packets can be either uniformly distributed in time, or applied in bursts. In the latter case, Mc jams the channel during D seconds, for each period T with $D < T$ (therefore $X = D/T$). We refer to this method as "bursty jamming."

Let us consider the throughputs of the cheater (Mc) and the well-behaved node (M). Sources S and Sc start transmitting at the same time, using TCP-Reno with 1000-byte packets. To make sure that the throughput reaches a steady state, it is assumed that the cheater begins jamming 60 s after Mc and M begin transmitting. The results are averaged over 35 simulations. Three factors help Mc increasing its throughput:

- reducing the number of competing flows entering the queue at the AP;
- reducing the collision rate on the wireless channel (TCP-DATA transmitted by the AP with TCP-ACKs transmitted by M and Mc);

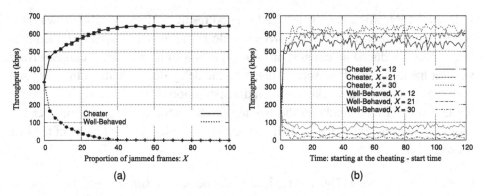

Figure 9.11. Cheating performance using regular jamming on downlink traffic. Used with permission, from [322], © IEEE, 2006.

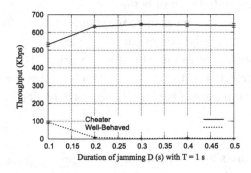

Figure 9.12. Cheating performance using bursty jamming on downlink traffic. Used with permission, from [322], © IEEE, 2006.

- reducing the queueing delays at the AP (the jammed packets are not retransmitted).

Figure 9.11a shows the throughput of Mc and the throughput received by M as a function of the proportion X of jammed frames. We can see that jamming 30% of the frames is enough to reduce M's received throughput to zero, and increase Mc's received throughput to the maximum available data rate. The evolution of Mc's throughput and M's received throughput in time is shown in Figure 9.11b. Note the low throughput of Mc when it first starts jamming M's frames and forging MAC-ACKs. Later on, this overhead is reduced, since M receives decreased throughput, therefore Mc jams less frames and forges less MAC-ACKs, increasing its efficiency. This transient period lasts less than 10 s.

Figure 9.12 shows the same metrics when the cheater applies bursty jamming. Using for example $T = 1$ s, inspired by [243, 7], the same jamming $D/T = X$ proportion leads to a better throughput for Mc and lower consumption for M, thus making the attack even more devastating.

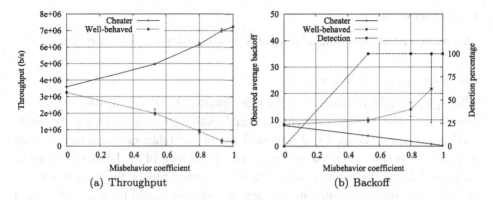

(a) Throughput (b) Backoff

Figure 9.13. Experimental results for (a) the throughput and (b) the backoff of the cheating and the well-behaved stations obtained by using a prototype implementation of DOMINO. On the x-axis, the *misbehavior coefficient* takes values 0, 0.53, 0.8, 0.93, and 1. Used with permission, from [322], © IEEE, 2006.

9.2.4 Implementation

We briefly describe here an implementation of one of the cheating techniques based on backoff manipulation and of a prototype of DOMINO. We consider a simple scenario where two stations are sending UDP traffic to the AP.

The performed experiments correspond to several values of the cheater contention window, which should be of the form $2^n - 1$, where n is an appropriate integer. Specifically, both CW_{min} and CW_{max} are set to 0, 1, 3, 7, 15 (the default value of CW_{min} is 15 and that of CW_{max} is 1023 on the wireless cards that have been used), which correspond to *misbehavior coefficients* of 1, 0.93, 0.8, 0.53, and 0, respectively.[11] The resulting throughput (Figure 9.13a) and backoff (Figure 9.13b) of the cheating and well-behaved stations are observed.

In Figure 9.13a we can see that the cheater obtains higher throughput, at the expense of the well-behaved station, by increasing her misbehavior. The corresponding observed backoff values are shown in Figure 9.13b along with the detection curve. When the misbehavior percentage increases, the cheater's average backoff decreases (thus increasing her chances to grab the channel first and boosting its throughput); this can be easily detected by DOMINO, as the detection curve shows. In the meantime, the average backoff of the well-behaved station increases with the misbehavior percentage (owing to collisions and the subsequent increase of the contention window); this explains her decreasing throughput.

---

[11] In conventional IEEE 802.11 WLANs, setting these values requires one to overwrite the corresponding bytes in the firmware of the adapter. But if the Access Point is QoS-enabled (hence compliant with IEEE 802.11e), some adapters allow to set these values by a simple command.

9.2.5 Discussion

This section addresses some additional issues related to the detection system.

Throughput as a detection metric

Although throughput seems to be the most intuitive metric for distinguishing stations using higher shares of the channel bandwidth than other stations, it cannot be used as a metric. Indeed, if two stations have different data rates and delays, such as VoIP versus streaming video sources, the throughput of the latter will be naturally much larger than that of VoIP. Hence, we cannot rely on throughput without knowing the application running on each station (this would require each station to declare its currently communicating applications to the AP, in violation with the principle of layering in protocols).

Experimental studies (e.g., in [29] and [383]) have shown that the throughput of a UDP source in a wireless network is affected by many factors, such as packet overhead, Signal-to-Noise Ratio (SNR), network and host hardware, device drivers, and network protocol implementations in the operating system. The authors of [169] prove that the decrease of the bit rate of a single station (due to a bad channel) decreases the bit rates of all the other stations to values close to that of the disadvantaged station. The negative effect of SNR on channel capture is explored in [374] (according to the authors, the results obtained in the infrastructure mode are identical to those observed in the ad hoc mode). All these factors lead to high differences in throughput even among stations sending at equal rates.

The performance of TCP over wireless networks is studied experimentally in [383]. The authors explain that TCP coupled with the IEEE 802.11 MAC protocol results in performance degradation. Among the factors that contribute to the degradation are the congestion window, recovery mechanism, packet size, and timeout values of TCP, as well as the acknowledgements, retransmission retry limit, and backoff mechanism of IEEE 802.11 MAC.

Hence, although the fairness of wireless networks has been evaluated [41, 226, 284] typically using Jain's fairness index [199] (which in turn uses channel bandwidth shares), throughput is far from being the optimal misbehavior metric in our case.

Hidden terminals

Hidden terminals can have a negative effect on DOMINO. For example, if two stations A and B are seen by the AP but hidden from each other, A can sense the medium idle while the AP senses it busy because B transmits. As a result, A will keep decrementing its backoff counter and then transmit a frame whose backoff measured at the AP will appear smaller than the actual value. After several repetitions of this scenario, the detection mechanism will output a wrong suspicion of A. Both simulation scenarios for uplink traffic (UDP and TCP) have been rerun with hidden terminals (by changing the reception range); by choosing appropriate values for detection thresholds (specifically, α_{ac} and α_{co} defined in Section 9.2.2), i.e., by tolerating some misbehavior, the results indicate that it is possible to reduce false positives in the presence of hidden terminals.

Security

It should be noted that DOMINO can be exploited to create *hybrid* attacks, taking advantage of both security flaws and MAC vulnerability. For example, a cheater can impersonate a well-behaved station to provoke its punishment and, possibly, its disconnection from the network by the operator. But a de-authentication attack [124], which is easier to perpetrate, would yield a similar effect without relying on the punishment policy. In addition, the adoption of security mechanisms, such as WPA (W-Fi Protected Access) and IEEE 802.11i (see Chapter 1), would limit the efficiency of these hybrid attacks. In fact, the cheater cannot transfer useful data in the faked frames because it does not know the encryption key of the impersonated host. In addition, such an attack would incur on the cheater an overhead owing to the dummy frames it sends.

The solution to these attacks lies in the use of enhanced security mechanisms jointly with DOMINO. This issue further illustrates the fact, already mentioned at the beginning of this book, that greediness and security must be jointly addressed.

Adaptive cheating

We call *adaptive cheating* the set of misbehavior techniques that exploit some knowledge about the way DOMINO works. For example, a cheater can switch frequently enough between several techniques described in Section 9.2.1, in such a way that DOMINO fails to collect enough data to detect misbehavior. But as the cheater does not know the detection parameters, such as the monitoring period and the thresholds, it will be hard to adapt to the detection system in order to avoid being caught.

It is possible that if system administrators set the default detection parameter values during installation, cheaters could use these to adapt their techniques. But as explained earlier, different environments result in different parameter values, even considering default values. Hence, it would be hard for the cheater to adapt to different AP and their parameters.

Another way of tricking DOMINO would consist in employing techniques to disable some tests. For example, a cheater might intentionally create collision-like signals (e.g., by emitting scrambled frames at high power) to fail the *actual backoff* test and never transmit two consecutive non-interleaved frames to fail the *consecutive backoff* test. But such techniques obviously increase the cheater's overhead (e.g., in terms of inter-frame delay) that might not be compensated by a significant throughput advantage over other stations.

Choice of the detection parameters

The choice of the right parameters affects the performance of DOMINO. As this choice depends on the environment in which the system operates, the parameters should be set during the installation of the AP. In practice, system administrators have to run a series of tests anyway. The values we have provided are a good trade-off between high detection and low misdetection ratios in the simulated environment. Hence using other values would degrade either one or both of the ratios; the use of a larger monitoring period is not necessary and will only contribute to a slower response of the detection mechanisms. In a real setting, the administrators could start with default values and then tune them appropriately. The default values can be obtained by techniques similar to site surveys run by cellular and WLAN operators.

Monitoring period

To avoid overloading the AP with per-frame computations, the data required for detection are collected during configurable intervals of time; at the end of each interval, the detection mechanism is run. Another advantage of this method over a per-frame detection approach is the ability to collect more statistical data and hence increase the accuracy. In addition, it is shown in [41, 226, 284] that the binary exponential backoff algorithm of IEEE 802.11 is unfair in the short term. This would result in false positives if stations were monitored over short term periods (even in the absence of misbehavior). Therefore the monitoring period has to be large enough to rely on long term backoff fairness.

Taking into account the typical bit rates, monitoring periods can be short enough (as was shown in the simulations) to prevent the cheater from gaining large benefits before being detected. For example, assuming 500-byte packets and 7Mbps data rate (this is the maximum effective IEEE 802.11b rate) equally divided among 50 stations, the AP can collect in 10 s 350 backoff values per station.

Practical relevance

As mentioned, it is in principle relatively easy for a cheater to increase her amount of traffic on the uplink, typically by altering the way the backoff is computed. However, this alteration has to be done on proprietary pieces of software, meaning that it is not trivial to implement, especially on non-Linux machines. Attacks against downlink traffic are even more complicated to mount.

This means that the described threat is not as severe as it looks at first sight.[12] As we pointed out, the interesting dimension of this section is not the specific case addressed here, but rather the description of how a real system can be abused by greedy participants. The more software-based the radio systems will become, the easier it will be to perpetrate the kind of misdeeds described in this section.

It is also important to note that if the Access Point supports Quality of Service (in compliance with the IEEE 802.11e standard), then the greedy behavior by means of an IEEE 802.11e enabled adapter is straightforward.

9.3 Selfish behavior in pure ad hoc networks

In the previous section, we have seen that a selfish user connected to an IEEE 802.11 access point can easily increase the amount of her bandwidth, at the expense of the other users of this access point. We have also explained how such misbehavior can be detected by the access point.

In this section, we will consider a set of mobile stations communicating with each other in *ad hoc mode* (hence we assume that there is no access point). More specifically, we consider that a selfish user (*cheater*) makes use of the easiest (and yet highly rewarding) cheating technique among those that we have just described: she deliberately does not respect the random deferment of the transmission of her packets.

Although this cheating technique is straightforward, we will see that studying its implications is far from trivial. In the following analysis, each node is a player, the throughput it enjoys is its payoff, and the size of its contention window represents its move. We study several scenarios. First, we consider the simple case of a network with a single cheater. We then assume the presence of several cheaters and identify two families of *Nash equilibria* in a *single stage* (i.e., *static*) game: one family always results in a network collapse, whereas in the other family a single selfish user receives nonzero throughput. We also explain that the equilibria of the former family (*the tragedy of the commons*-equilibria) are not *robust* (with respect to arbitrarily small perturbations of the users' payoff functions).

As the Nash equilibria of the static game are either highly inefficient or highly unfair, we look for an alternative solution. In this regard, we compute the fair Pareto-optimal point of operation of such a system. We then show how to make

---

[12] A parallel can be establish with TCP (in the wireline Internet): in principle, some users could increase their share of the available bandwidth by departing from the protocol (not respecting the "multiplicative decrease algorithm". In practice, this misdeed, although observed from time to time, has not significantly affected the overall behavior of the Internet. Yet the problem addressed in this section is somewhat different, because here the shared resource is the wireless link, which is – and is likely to remain – a precious asset.

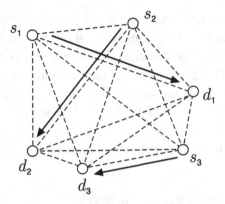

Figure 9.14. An example of a single-collision domain network with $N = 3$ communicating pairs. Dashed lines represent connectivity, while the flows between the
pairs are represented by solid arrows. Used with permission, from [80], © IEEE,
2005.

the Pareto-optimal point a Nash equilibrium point by using the theory of *repeated
games*. We introduce the notion of *cooperative players*, namely cheaters who try
to continue operating at the fair Pareto-optimal point of operation. We also propose a detection and a punishment technique for those players who exhibit a non-
cooperative behavior. Finally, we explain how the players can collectively search
for the Pareto-optimal point of operation, even if they are unaware of the number
of nodes present in the network.

   As mentioned, we heavily rely on game theory (see Appendix B).

9.3.1 System model and assumptions

Let us consider a finite number N of wireless nodes that are willing to transmit data
to N designated receivers. All the nodes use the same radio channel and share the
same collision domain, that is any node can hear any other node (see Figure 9.14);
this is to avoid complications introduced by the *hidden terminal problem*. Nodes
use a CSMA/CA based protocol to resolve contention at the MAC layer. In this
section (as in the previous one), we will be dealing exclusively with IEEE 802.11 (in
DCF mode) [104]; the analysis carried out in this setting can be extended to other
CSMA/CA based protocols. We further assume that each node has an authentic
MAC layer identifier (the MAC address). This can be achieved by means of MAC
layer authentication. Finally, we assume that the nodes are static and that they
always have packets (of the same size) to send.

   We consider a scenario where out of the N senders, a subset P of cardinality $|P|$
sending nodes deliberately deviate from the IEEE 802.11 protocol (the reason why
we call this subset "P" is that it corresponds to the set of players). Without any loss

of generality, we assume $P = \{1, \ldots, |P|\}$. We designate the nodes of subset P as *cheaters*. There can be a number of ways in which a node can cheat. For example, in violation of the standard protocol, a cheater $i \in P$ initializes her contention window size to a lower value in order to obtain a higher throughput, as we have described in Section 9.2. We will call this lower value W_i.[13] Moreover, a cheater does not respect the *binary exponential backoff* [104] principle and keeps her contention window size fixed after a collision, i.e. equal to W_i. This mode of cheating is the easiest for potential cheaters, because it does not require changes to be made in the operation of the IEEE 802.11 protocol. We stress that the main conclusions of this section are applicable to most of the cheating techniques discussed in the previous section. As mentioned, the relevance of these misbehaving techniques becomes even higher with the emerging standards that address the Quality of Service support, such as IEEE 802.11e [191]. The latter gives the users total control of the MAC parameters, therefore enabling them to cheat even more easily.

As in the previous section, we assume each cheater to be *rational*, meaning that she wants to maximize her own benefit. In this particular context, every cheater $i \in P$ seeks to maximize the average throughput τ_i she enjoys. The cheater nodes define the set P of players in this game. We define the pure-strategy set S_i of a given player i as follows

$$S_i = \{1, 2, \ldots, W_{\max}, W_\infty\}, \tag{9.1}$$

where $W_{\max} < \infty$ is a positive integer representing the highest finite value that the player considers allocating for her contention window and the symbol W_∞ means that the player i does not transmit, which is equivalent to $W_i = \infty$; note that the set S_i is finite. The strategy of each player i consists in setting her contention window $W_i \in S_i$ to a specific value. As we assume that each player i tries to maximize her own throughput, we define a player i's payoff's function[14] u_i to be equal to the enjoyed throughput $\tau_i^{(c)}$, that is

$$u_i(W) = \tau_i^{(c)}(W), \tag{9.2}$$

where $W = (W_1, \ldots, W_{|P|}, W_{|P|+1}, \ldots, W_N)$, and $W_j \in S_j$, $(j \leq |P|)$, are strategies chosen by players $j \in P$, whereas contention windows W_k, $(|P| + 1 \leq k \leq N)$, belong to the *well-behaved* nodes. Here the superscript "(c)" denotes a cheater. In the game theoretic analysis, we will often neglect the well-behaved nodes, so we will often use W to denote $W = (W_1, \ldots, W_{|P|})$. Finally, we denote the game as defined above by $G_{\text{CSMA/CA}} = \left\langle N, P, (S_i)_{i \in P}, (\tau_i^{(c)})_{i \in P} \right\rangle$ and call it the *CSMA/CA game*.

---

[13] In order to avoid confusion, we will continue to call "*CW*" the contention window when not subject to cheating.

[14] The reason why we refrain from calling this a "utility function" is explained in Appendix B.

9.3.2 Single stage game

We first analyze the problem of misbehaving from the perspective of a single cheater and then consider more complex scenarios with multiple cheaters in the system. In this subsection, we consider the interaction of multiple cheaters in a *single stage* of the CSMA/CA game $G_{\text{CSMA/CA}}$.

Characterization of the payoff $u_i(W)$

In order to characterize the payoff functions $u_i(W) = \tau_i^{(c)}(W)$, $i \in P$, we first have to understand the relationship between the contention window profiles $W = (W_1, \cdots, W_{|P|}, \cdots, W_N)$ and the resulting payoffs $\tau_i^{(c)}(W)$. As we assume that a cheater's objective is to maximize her throughput (and we assume she always has a packet to send), she will tend to use the full channel capacity (i.e., the system will operate at the saturation point).

There are different analytical models and simulation studies of the 802.11 MAC layer in saturated condition. One of the most well-known models is the so-called Bianchi model [53], published in the year 2000.[15]

The main contribution of Bianchi's model is the analytical calculation of saturation throughput in a closed-form expression. The model also calculates the probability of a packet transmission failure due to collision. It assumes that the channel is in ideal conditions, i.e., there is no hidden terminal and capture effect. We make use of this model for the saturation throughput of the IEEE 802.11 protocol.

Bianchi uses a two-dimensional Markov chain of m backoff stages in which each stage represents the backoff time counter of a node, see Figure 9.15. A transition takes place upon collision and successful transmission, to a "higher" stage (e.g., from stage $i - 1$ to stage i in Figure 9.15) and to the lowest stage (i.e., stage 0) respectively.[16]

This model adopts a discrete and integer time scale. In this time scale, t and $t + 1$ correspond to the beginning of two consecutive slot times. Each station decrements its backoff time counter at the beginning of each time slot. Note that as the backoff time decrement is stopped when the channel is busy, the time interval between t and $t + 1$ may be much longer than the defined slot time for 802.11, as it may include a packet transmission or a collision.

Each state of this bidimensional Markov process is represented by $\{s(t), b(t)\}$, where $b(t)$ is the stochastic process representing the backoff time counter for a given station and $s(t)$ is the stochastic process representing the backoff stage $(0, 1, \ldots, m)$ of the station at time t. This model assumes that in each transmission attempt,

---

[15] In 2002, Wu *et al.* [378] proposed a refinement of Bianchi's model by considering finite packet retry limits as defined in the IEEE 802.11 standard. Yet, in order to keep the presentation as simple as possible, we will stick to Bianchi's model.

[16] Actually it appears lower in the figure.

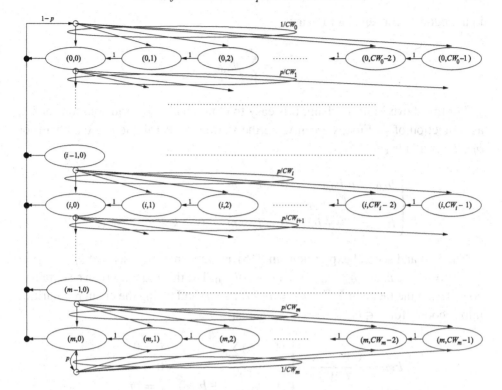

Figure 9.15. Markov chain model of backoff window size in CSMA/CA. In each stage, CW_i is the maximum value for the contention window and is equal to $2^i W_{\min}$ (Note that we define for convenience $W_{\min} = CW_{\min} + 1$ and that consequently $CW_{\max}$ is equal to $2^m W_{\min} - 1$). If a correct transmission takes place in any $(i, 0)$ state, a random backoff will be chosen between 0 and $CW_0 - 1$ with probability of $(1 - p)/CW_0$. This case is represented by states $(0, 0)$ to $(0, CW_0 - 1)$ in the Markov chain. In the case of collision (e.g., in state $(i - 1, 0)$), a random backoff will be chosen (between 0 and $CW_i - 1$, each with probability p/CW_i). This case is represented by states $(i, 0)$ to $(i, CW_i - 1)$ in the Markov chain. Used with permission, from [53], © IEEE, 2000.

regardless of the number of retransmissions suffered, each packet collides with constant and independent probability p. In other words, p is the probability that, in a time slot, at least one of the $N - 1$ remaining stations transmits as well. If at steady state each remaining station transmits a packet with probability π, p can be written as:

$$p = 1 - (1 - \pi)^{N-1} \tag{9.3}$$

Let $b_{i,k} = \lim_{t \to \infty} P\{s(t) = i, b(t) = k\}$, $i \in (0, m)$, $k \in (0, CW_i - 1)$ be the stationary distribution of the chain. A transmission occurs when the backoff time counter is equal to zero. Thus, we can write the probability that a station transmits in a randomly chosen slot time as:

$$\pi = \sum_{i=0}^{m} b_{i,0}. \tag{9.4}$$

For the above Markov chain, it is easy to obtain a closed-form solution for $b_{i,0}$ as a function of p. First, we can write the stationary distribution of the chain for $b_{i,0}$, $b_{m,0}$, and $b_{i,k}$:

$$\begin{cases} b_{i,0} = p^i b_{0,0} & 0 < i < m \\ b_{m,0} = \frac{p^m}{1-p} b_{0,0} \\ b_{i,k} = \frac{CW_i - k}{CW_i} b_{i,0} & 0 < i < m, \quad 0 < k < CW_i - 1. \end{cases} \tag{9.5}$$

The first and second expressions in (9.5) account from the fact that $b_{i-1,0} \cdot p = b_{i,0}$ for $0 < i < m$ and $b_{m-1,0} \cdot p = (1-p)b_{m,0}$. The third equation can be obtained considering the fact that $\sum_{i=0}^{m} b_{i,0} = b_{0,0}/(1-p)$ and taking the chain regularities into account (for $k \in (1, CW_i - 1))$, that is:

$$b_{i,k} = \frac{CW_i - k}{CW_i} \cdot \begin{cases} (1-p)\sum_{j=0}^{m} b_{j,0} & i = 0 \\ p \cdot b_{i-1,0} & 0 < i < m \\ p \cdot (b_{m-1,0} + b_{m,0}) & i = m. \end{cases} \tag{9.6}$$

By imposing the normalization condition and considering Equation (9.5), we can obtain $b_{0,0}$ as a function of p:

$$1 = \sum_{i=0}^{m} \sum_{k=0}^{CW_i - 1} b_{i,k}$$

$$= \sum_{i=0}^{m} b_{i,0} \sum_{k=0}^{CW_i - 1} \frac{CW_i - k}{CW_i}$$

$$= \sum_{i=0}^{m} b_{i,0} \frac{CW_i + 1}{2} = \sum_{i=0}^{m} b_{i,0} \frac{2^i W_{min} + 1}{2}$$

$$= b_{0,0} \frac{W_{min} + 1}{2} + \sum_{i=1}^{m-1} \left(b_{0,0} p^i \left(\frac{2^i W_{min} + 1}{2} \right) \right) + \left(\frac{b_{0,0} p^m}{1-p} \right) \left(\frac{2^m W_{min} + 1}{2} \right)$$

$$= \frac{b_{0,0}}{2} \left[W_{min} + 1 + \sum_{i=1}^{m-1} \left((2p)^i W_{min} + p^i \right) + \frac{p^m}{1-p} (2^m W_{min} + 1) \right]$$

$$= \frac{b_{0,0}}{2} \left[W_{min} \left(\sum_{i=0}^{m-1} (2p)^i + \frac{(2p)^m}{1-p} \right) + \frac{1}{1-p} \right]. \tag{9.7}$$

Thus $b_{0,0}$ can be written as

$$b_{0,0} = \frac{2(1-2p)(1-p)}{(1-2p)(W_{\min}+1) + pW_{\min}(1-(2p)^m)}. \tag{9.8}$$

Finally, considering equations (9.4), (9.5), and (9.8), the channel access probability π of a node is derived as a function of the number of backoff stage levels m, the minimum contention window value $W_{\min}$, and the collision probability p:

$$\pi = \sum_{i=0}^{m} b_{i,0} = \frac{b_{0,0}}{1-p} = \frac{2(1-2p)}{(1-2p)(W_{\min}+1) + pW_{\min}(1-(2p)^m)}$$

$$= \frac{2}{1 + W_{\min} + pW_{\min}\sum_{k=0}^{m-1}(2p)^k}. \tag{9.9}$$

Equations (9.3) and (9.9) form a system of two nonlinear equations that has a unique solution and can be solved numerically for the values of p and π (e.g., one can use the *solve* function in MATLAB to obtain the values for p and π). Once these probabilities are obtained, the throughput enjoyed by a given node i, which is the average information payload transmitted in a slot time over the average duration of a slot time, can be computed as follows:

$$\tau_i = \frac{E[\text{Payload information transmitted by user } i \text{ in a slot time}]}{E[\text{Duration of slot time}]}$$

$$= \frac{P_s^i L}{P^s T^s + P^c T^c + P^{id} T^{id}}, \tag{9.10}$$

where $P_i^s = \pi_i \prod_{j \neq i}(1-\pi_j)$ is the probability that the station i successfully transmits during a random time slot ($j \neq i$ is a shorthand notation for $j \in \{1, \ldots, N\}\setminus\{i\}$); L is the average packet payload size; $P^s = \sum_{j=1}^{N} P_j^s$; T^s is the average time needed to transmit a packet of size L (including the inter-frame spacing periods [53]); $P^{id} = \prod_{j=1}^{N}(1-\pi_j)$ is the probability of the channel being idle; T^{id} is the duration of the idle period (a single slot); $P^c = 1 - P^{id} - \sum_{j=1}^{N} P_j^s$ is the probability of collision; and T^c is the average time spent in the collision. Note that we must have the following satisfied $P^s + P^c + P^{id} = 1$.

T^c and T^s can be calculated for the basic transmission mode (i.e., no RTS and CTS packets) with:

$$\begin{cases} T^s = H + L + SIFS + \sigma + ACK + DIFS + \sigma \\ T^c = H + L + DIFS + \sigma, \end{cases} \tag{9.11}$$

where H, L, and ACK are the transmission times needed to send the packet header, the payload, and the acknowledgement, respectively, and σ is the propagation delay.

To describe a network with cheating nodes we use two separate Markov chains. The first, with $m = 0$ (no exponential backoff, because cheaters are assumed to fix their contention windows (Section 9.3.1)), is used to derive the channel access probabilities $\pi_i^{(c)}$ of cheaters $i \in P$. The second chain, with $m > 0$, is used to derive the access probabilities $\pi_j^{(w)}$ of well-behaved (non-cheating) nodes. The conditional collision probabilities are derived considering both well-behaved and cheating nodes access probabilities.

Because a cheater i does not respect the backoff procedure of IEEE 802.11 (i.e., $m = 0$), her channel access probability degenerates to

$$\pi_i^{(c)} = \frac{2}{W_i + 1},$$ (9.12)

where W_i is the cheater i's contention window size. The channel access probability for well-behaved nodes, $\pi_j^{(w)}$, is expressed by

$$\pi_j^{(w)} = \frac{2}{1 + W_{\min} + p^{(w)} W_{\min} \sum_{k=0}^{m-1} \left(2 p^{(w)} \right)^k},$$ (9.13)

where

$$p^{(w)} = 1 - \left(1 - \pi_j^{(w)} \right)^{N - |P| - 1} \prod_{i \in P} \left(1 - \pi_i^{(c)} \right).$$ (9.14)

Note that (9.14) is the generalization of (9.3) in the presence of cheaters. Note also that $\pi_j^{(w)}$ is the same for all the well-behaved nodes and so we set $\pi_j^{(w)} = \pi^{(w)}$. After a straightforward algebraic manipulation of expression (9.10), we obtain the following expression for the throughput $\tau_i^{(c)}$ of a cheater i:

$$\tau_i^{(c)} = \frac{\pi_i^{(c)} c_i^{(1)}}{\pi_i^{(c)} c_i^{(2)} + c_i^{(3)}},$$ (9.15)

where

$$c_i^{(1)} = p_{-i} L$$ (9.16)
$$c_i^{(2)} = p_{-i}(T^s - T^{id}) - s_{-i}(T^s - T^c)$$ (9.17)
$$c_i^{(3)} = (1 - p_{-i} - s_{-i})T^c + s_{-i}T^s + p_{-i}T^{id},$$ (9.18)

where the following substitutions have been used

$$p_{-i} = \prod_{j \in P \setminus \{i\}} \left(1 - \pi_j^{(c)} \right) \left(1 - \pi^{(w)} \right)^{N - |P|}$$

$$s_{-i} = \sum_{j \in P \setminus \{i\}} \pi_j^{(c)} \prod_{k \in P \setminus \{i,j\}} \left(1 - \pi_k^{(c)} \right) \left(1 - \pi^{(w)} \right)^{N - |P|}.$$ (9.19)

Note here, that the only parameter that a cheating node i has a control over is its own W_i. By varying W_i, a node changes its own access probability $\pi_i^{(c)} = f(W_i)$,

Table 9.2. *Simulation parameters*

Parameter	Value
Topology	100 m × 100 m, random
Receive range	240 m
Propagation	Free space
MAC	802.11b
Scheme	Basic (No RTS/CTS)
Channel capacity	2 Mbits/s
Traffic sources	CBR/UDP, 1050-byte frames every 5 ms (1.68 Mb/s)

as well as the access probability $\pi^{(w)} = f(W)$, $(W = (W_1, \ldots, W_i, \ldots, W_N)$, $W_i = W_{\min}$, $i = \{I + 1, \ldots, N\})$, of the well-behaved nodes; this follows from expressions (9.12), (9.13) and (9.14).

As we have seen at the beginning of this chapter, the contention window can take only a few integer values. However, for mathematical convenience, let us assume for the moment that W_i for every cheater $i \in P$ is a continuous variable. Although the access probabilities of the well-behaved nodes (and thus the expressions $c_i^{(1)}$, $c_i^{(2)}$ and $c_i^{(3)}$) depend on $\pi_i^{(c)}$, we neglect this dependence for a first-degree analysis. This approximation allows us to elaborate a closed-form expression of the first derivative of Equation (9.15):

$$\frac{\partial \tau_i^{(c)}}{\partial W_i} = \frac{\partial \tau_i^{(c)}}{\partial \pi_i^{(c)}} \frac{\partial \pi_i^{(c)}}{\partial W_i} = \frac{c_i^{(1)} c_i^{(3)}}{\left(\pi_i^{(c)} c_i^{(2)} + c_i^{(3)}\right)^2} \frac{-2}{(W_i + 1)^2} \leq 0. \qquad (9.20)$$

If $\pi_j^{(c)} < 1$ for all $j \in P\backslash\{i\}$, then we have a strict inequality in (9.20). Therefore, as expected, the received throughput $\tau_i^{(c)}$ is a strictly decreasing function of W_i (for $\pi_j^{(c)} < 1, \forall j \in P\backslash\{i\})$. Thus, by unilaterally decreasing its own W_i, a selfish node can increase its received throughput (except if $\pi_j^{(c)} = 1$, for some cheater $j \neq i$ – as we will see in the following subsection, this case has important implications for the set of Nash equilibria of the CSMA/CA game). We stress here that this conclusion would remain the same even if we considered the dependence of $c_i^{(1)}$, $c_i^{(2)}$ and $c_i^{(3)}$ on $\pi_i^{(c)}$. In fact, by using this approximation, we actually underestimate the benefits of the cheater (the cheater gets more throughput in reality).

This claim (and the modified Bianchi's model) will now be verified by simulations performed in *ns*-2 [131]. The simulation setup,[17] summarized in Table 9.2, consists of $N = 20$ sender nodes. A single node X deliberately departs from the

---

[17] In the rest of the section, we will mention only the changes that are made from this reference simulation setup.

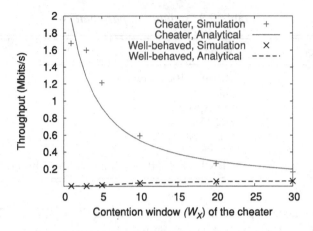

Figure 9.16. Throughputs for $N = 20$ nodes, out of which one is a cheater. Used with permission, from [80], © IEEE, 2005.

protocol and tries to misbehave following the cheating model presented in Subsection 9.3.1. The parameter values for the IEEE 802.11 protocol are chosen according to the IEEE 802.11b standard [104]. The duration for each simulation run is 50 s and the results are averaged over five simulation runs.

Figure 9.16 plots the throughput obtained by cheater X, as well as by each well-behaved node for different values of W_X. Simulation results show a good match with the analytical results. As can be observed from Figure 9.16, the throughput obtained by the cheater increases monotonically with the decrease in W_X.

Now that we have characterized the cheaters' payoff functions $u_i(W) = \tau_i^{(c)}(W)$, we next study Nash equilibria of the single stage game $G_{CSMA/CA}$.

Nash equilibria of the CSMA/CA game

In this subsection we do not consider well-behaved nodes (i.e., we assume $N = |P|$). We will focus only on pure-strategy Nash equilibria, because, as we will soon show, they exist in $G_{CSMA/CA}$ and we know by game theory that no player can do better than playing her best-response pure-strategies.

We will study the existence of Nash equilibria by making use of the concept of a player's best-response function. Let us introduce the following notations:

$$W_{-i} = \left(W_1, \ldots, W_{i-1}, W_{i+1}, \ldots, W_{|P|} \right)$$
$$S_{-i} = \{ S_1, \ldots, S_{i-1}, S_{i+1}, \ldots, S_{|P|} \},$$

where S_i are the pure-strategy sets of the players (cf. expression (9.1)). We define a player i's best-response function $br_i(W_{-i})$ as follows

$$br_i(W_{-i}) = \left\{ W_i \in S_i : \tau_i^{(c)}(W_i, W_{-i}) \geq \tau_i^{(c)}\left(W_i', W_{-i} \right) \text{ for all } W_i' \in S_i \right\}.$$

From game theory, we know that a pure-strategy profile $W^* = (W_1^*, \ldots, W_{|P|}^*)$ is a Nash equilibrium if and only if $W_i^* \in br_i(W_{-i}^*)$ for every player $i \in P$.

Lemma 9.1 *For any strategy profile W that constitutes a Nash equilibrium in $G_{\text{CSMA/CA}}$, $\exists i \in P$ such that $W_i = 1$.*

Proof Assume by contradiction that $W = (W_1, \ldots, W_{|P|})$ is a Nash equilibrium such that $W_k > 1$, $\forall k \in P$. Now, take one player, say i, and consider her best-response function $br_i(W_{-i})$. As $\tau_i^{(c)}$ is a strictly decreasing function of W_i (equation (9.20) and $W_k > 1 \Rightarrow \pi_k^{(c)} < 1$, $\forall k \in P$ (equation (9.12)), it follows readily that the only value of W_i that satisfies

$$\tau_i^{(c)}(W_i, W_{-i}) \geq \tau_i^{(c)}\left(W_i', W_{-i}\right) \text{ for all } W_i' \in S_i ,$$

is unity, that is $br_i(W_{-i}) = \{1\}$. By definition, at any Nash equilibrium $W_i \in br_i(W_{-i})$, thus we have $W_i = 1$. However, this contradicts our initial assumption that $W_i > 1$, which concludes the proof. $\square$

Theorem 9.1 *The game $G_{\text{CSMA/CA}}$ admits exactly $(W_{\max} + 1)^{|P|} - W_{\max}^{|P|}$ Nash equilibria.*

Proof Assume that for some player $i \in P$ we have $W_i = 1$. Then her access probability $\pi_i^{(c)} = 1$ and consequently for all players $k \in P \setminus \{i\}$ it follows that $\tau_k^{(c)} = 0$ for any value of $W_k \in S_k$ (Equation (9.15)). Therefore, for any value of $W_k \in S_k$ we have $W_k \in br_k(W_{-k})$, where $k \in P \setminus \{i\}$. This clearly holds for any number of players who have their contention window set to unity. Combining this with Lemma 9.1, we obtain the following characterization of Nash equilibria:

(**Nash equilibria**) At any Nash equilibrium of $G_{\text{CSMA/CA}}$ there is at least one cheater who sets her contention window to unity and all the other cheaters play any strategy from $\{1, \ldots, W_{\max}, W_\infty\}$.

Finally, the theorem follows by observing that out of the total of $(W_{\max} + 1)^{|P|}$ different strategy profiles $W = (W_1, \ldots, W_{|P|})$, $(W_j \in \{1, \ldots, W_{\max}, W_\infty\})$, exactly $W_{\max}^{|P|}$ do not contain any unity element. $\square$

It is interesting to observe that the equilibria can be classified in two families. To describe these, we define the set $\mathcal{D} = \{i : W_i = 1, i \in P\}$.

1st family: $|\mathcal{D}| = 1$, that is there is only one player $i \in P$ who plays $W_i = 1$ and receives a non-null throughput $\tau_i^{(c)} > 0$, and $\tau_k = 0$ for all players $k \in P \setminus \{i\}$.
2nd family: $|\mathcal{D}| > 1$, that is there is more than one player $i \in P$ who play strategy $W_i = 1$, in which case $\tau_k^{(c)} = 0$ for all players $k \in P$.

Note that some Nash equilibria from the first family are also Pareto-optimal. For example, a strategy profile $W = (1, W_2 = W_\infty, \ldots, W_{|P|} = W_\infty)$ is a Pareto-optimal Nash equilibrium, because players in $P\backslash\{1\}$ do not actually transmit (i.e., $W_i = W_\infty = \infty \Rightarrow \pi_i^{(c)} = 0$) and player 1 gets all the system capacity for herself. The equilibria from the second family are known as *the tragedy of the commons* [162] in economics: the selfish behavior of each player leads to a tremendous misuse of the public good.

Once an equilibrium has been identified, it is good practice to study its properties. In this specific case, we would like to know whether the equilibrium would still exist in case a very small modification is brought to the definition of the game. More precisely, we would like to check whether the game is *essential* (see the definition in Appendix B) or not, and, for this purpose, we will study the robustness of the equilibrium $W = (W_i = 1)_{i \in P}$. Let us define an approximate game $\hat{G}_{\text{CSMA/CA}}$ to the original game $G_{\text{CSMA/CA}}$ as follows

$$\hat{G}_{\text{CSMA/CA}} = \langle P, (S_i)_{i \in P}, (\hat{u}_i)_{i \in P} \rangle ,$$

$$\text{with } \hat{u}_i(W) = \tau_i^{(c)}(W) - \begin{cases} \epsilon_i, & \text{if } W_i < W_\infty; \\ 0, & \text{if } W_i = W_\infty \end{cases} , \forall i \in P ,$$

where ϵ_i is an infinitesimally small but positive constant (i.e., $0 < \epsilon_i \ll 1$) that satisfies the following: $\tau_i^{(c)}(W) > \epsilon_i$, $\forall W$ such that $\tau_i^{(c)}(W) > 0$. The existence of such a constant follows from the fact that the number of nodes in the system is finite $(N < \infty)$.

Intuitively, the *cost* term ϵ_i says that a player prefers not to transmit at all than to transmit unsuccessfully. Being infinitesimally small, the cost term ϵ_i does not significantly change the player i's payoff function u_i. Let us now look at the equilibria of the game $\hat{G}_{\text{CSMA/CA}}$.

Theorem 9.2 *A strategy profile W is a Nash equilibrium of the game $\hat{G}_{\text{CSMA/CA}}$ if and only if*

$$\exists! \, i \in P \text{ such that } W_i = 1 \text{ and } W_j = W_\infty, \, \forall j \in P\backslash\{i\} .$$

Proof It is easily seen that Lemma 9.1 applies to game $\hat{G}_{\text{CSMA/CA}}$ too. Now, consider again the case where for some player $i \in P$ we have $W_i = 1$. Then her access probability $\pi_i^{(c)} = 1$ and consequently for all players $k \in P\backslash\{i\}$ it follows that $\tau_k^{(c)} = 0$ for any value of $W_k \in S_k$ (Equation (9.15)). This further implies $u_k(W) = -\epsilon_k \leq 0$. The best-response function for player k in game $\hat{G}_{\text{CSMA/CA}}$ is

$$br_k(W_{-k}) = \left\{ W_k \in S_k : \hat{u}_k (W_k, W_{-k}) \geq \hat{u}_k \left(W_k', W_{-k} \right) \text{ for all } W_k' \in S_k \right\} .$$

Then, $br_k(W_{-k}) = \{W_\infty\}$, $\forall k \in P\backslash\{i\}$, as $u_k(W_k = W_\infty, W_{-k}) = 0 \geq -\epsilon_k$. Also, $br_i((W_k = W_\infty)_{k \in P\backslash\{i\}}) = \{1\}$ (Lemma 9.1). Therefore, a strategy profile

$$W = \left(W_i = 1, (W_k = W_\infty)_{k \in P \setminus \{i\}} \right)$$

is a Nash equilibrium. We conclude the proof by observing that this is valid for an arbitrary player $i \in P$. □

Therefore, by an infinitesimally small change in the original game's payoff functions, it is possible to create a game with a significantly different set of Nash equilibria: the set of Nash equilibria of $\hat{G}_{\text{CSMA/CA}}$ is a small subset of those of $G_{\text{CSMA/CA}}$. Actually, all the Nash equilibria in $\hat{G}_{\text{CSMA/CA}}$ are Pareto-optimal; moreover, the strategy profile $(W_i = 1)_{i \in P}$ is not even an equilibrium point in $\hat{G}_{\text{CSMA/CA}}$. We conclude the study of robustness of the Nash equilibria of the original game $G_{\text{CSMA/CA}}$ with the following theorem.

Theorem 9.3 *The Nash equilibrium $W = (W_i = 1)_{i \in P}$ of the CSMA/CA game $G_{\text{CSMA/CA}}$ is nonessential (it is not robust), and therefore the CSMA/CA game $G_{\text{CSMA/CA}}$ is nonessential.*

Proof Observe first that the Nash equilibrium $W = (W_i = 1)_{i \in P}$ of $G_{\text{CSMA/CA}}$ implies $u_i(W) = 0$, $\sigma_i(W_i = 1) = 1$ and $\sigma_i(W_i \in S_i \setminus \{1\}) = 0$, $\forall i \in P$; $\sigma_i(W_i)$ designates the probability that player i assigns to strategy $W_i \in S_i$. Let us consider the following equilibrium of game $\hat{G}_{\text{CSMA/CA}}$

$$\hat{W} = \left(\hat{W}_1 = W_\infty, \hat{W}_2 = 1, \hat{W}_3 = W_\infty, \ldots, \hat{W}_{|P|} = W_\infty \right).$$

Note that this implies $\hat{\sigma}_i(\hat{W}_i = 1) = 0$, $\forall i \in P \setminus \{2\}$.

To prove this theorem, we next calculate the distances $D(\cdot)$ and $d(\cdot)$ between the payoff vectors u and $\hat{u}$, and between the strategy vectors σ and $\hat{\sigma}$ of the games $G_{\text{CSMA/CA}}$ and $\hat{G}_{\text{CSMA/CA}}$, respectively. Using the definitions introduced in Appendix B, we have

$$D(u, \hat{u}) = \max_{i \in P, W \in \times_{i \in P} S_i} |u_i(W) - \hat{u}_i(W)|$$

$$\leq \max_{i \in P, W \in \times_{i \in P} S_i} \epsilon_i$$

$$= \eta,$$

where $\eta > 0$ is an infinitesimally small but positive value; this follows from the definition of ϵ_i. Similarly, for the distance between strategy profiles we have

$$d(\sigma, \hat{\sigma}) = \max_{i \in P, W_i \in S_i} |\sigma_i(W_i) - \hat{\sigma}_i(W_i)|$$

$$\overset{(1)}{\geq} \max_{i \in P, W = (W_i = 1)_{i \in P}} |\sigma_i(W_i) - \hat{\sigma}_i(W_i)|$$

$$= \max_{i \in P} |\sigma_i(W_i = 1) - \hat{\sigma}_i(W_i = 1)|$$

$$\overset{(2)}{=} 1,$$

where the inequality (1) follows from the fact that we reduce the maximization domain and the equality (2) follows from the two fixed Nash equilibria W and $\hat{W}$. But then it follows immediately from the definition of essential games that the Nash equilibrium W is not essential (robust) and consequently the game $G_{\text{CSMA/CA}}$ is nonessential.   □

The result of this theorem can be generalized: if at least two greedy players have set their contention window to 1, none of them benefits from the network, because all packets collide. But if there is even a minimal cost for transmitting (for example, because it consumes some of the player's resources such as her device's battery, or because each transmission attempt is charged in some way as it consumes a shared resource), each greedy player is better off by deviating from that strategy; she will simply stop attempting to transmit.

Uniqueness, fairness and Pareto-optimality

We have seen in the earlier subsection that, generally, there exist two families of Nash equilibria in the $G_{\text{CSMA/CA}}$ game. In the first family, there is great unfairness (a single player gets some positive payoff); as we have seen, some of the equilibria from the first family are Pareto-optimal. The second family contains highly inefficient equilibria resulting in a zero payoff for every player.

Clearly, none of these families is satisfactory in practice. Therefore, we look for an alternative solution to $G_{\text{CSMA/CA}}$ by allowing the players to *agree* on the strategies they will use.

A *desirable solution* of the CSMA/CA game should exhibit the following three properties.

(1) **Uniqueness** The solution should be unique. This is to avoid uncertainties with respect to what solution each player should choose.
(2) **Pareto-optimality** The solution should result in a Pareto-optimal allocation of the available bandwidth.
(3) **Fairness** The solution should result in a fair distribution of the system through-put (there exist many definitions of fairness; in our case, as we assume that all stations are willing to transmit at a maximum rate, we will consider that fairness means that all stations enjoy the same throughput).

Intuitively, we want the stations to "naturally agree" on a (common) access probability such that the radio channel is shared in a way that fulfills the three properties that we have just mentioned. The computation of such an access probability can be carried out analytically, but the derivation requires notions of game theory (cooperative games) that go beyond the relatively simple ones we restrict ourselves to in

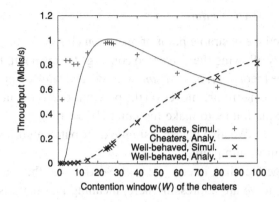

Figure 9.17. Throughput vs. contention window size of the cheaters (20 nodes, out of which 10 are cheaters). Used with permission, from [80], © IEEE, 2005.

this book (but the interested reader can refer to [80, 79]). Here we will simply rely on intuition.

Consider the case in which $N = 20$ stations are located in the same collision domain, out of which $|P| = 10$ are cheaters. Assume that all cheaters start with a contention window W of 1 (hence reaching a Family 2 equilibrium). Noticing that the network does not work, they slowly and *simultaneously* increase their contention window in order to let some traffic successfully be transmitted. Quite obviously, the throughput they enjoy will increase rapidly, while the one of the well-behaved nodes will start increasing as well, albeit at a lower pace. The channel having a finite capacity, for a given value of W the throughput enjoyed by the cheaters will reach its maximum. If they keep increasing their contention window, their throughput will begin declining, while the one of the well-behaved nodes continues increasing.

This behavior clearly appears in Figure 9.17, which exhibits the average aggregated throughput (the system throughput) obtained by 10 cheaters, all of which use the *same* contention window size; the simulation setup was described in Subsection 9.2. Note that in the simulations, the well-behaved nodes are also taken into account; they, however, do not affect the qualitative conclusions of this subsection. From this figure we can see that there exists a unique joint contention window size, which we will call W^*, maximizing the system throughput. A similar observation was already made by Bianchi in [53].

We conclude that the strategy profile $(W_i = W^*)_{i \in P}$ exhibits all the properties of a desirable point of operation in the CSMA/CA game $G_{\text{CSMA/CA}}$. In our context, this is particularly important since $(W_i = W^*)_{i \in P}$ is *not* a Nash equilibrium point (because, $W_i^* > 1, \forall i \in P$) and as such is not stable. Therefore, in the following subsection, we look at how to make the conjectured Pareto-optimal point $(W_i = W^*)_{i \in P}$ a Nash equilibrium point.

9.3.3 Repeated CSMA/CA game

Having determined the desirable point of operation $(W_i = W^*)_{i \in P}$, we now intend to devise a strategy allowing the players to converge to this point. For this purpose, we make use of the theory of *repeated games*. Repeated games capture the idea that a player can condition her future moves on the previous outcomes in the game. Using this model, we explain how to make the point $(W_i = W^*)_{i \in P}$ a Nash equilibrium of the game $G^\infty_{\text{CSMA/CA}}$. We also devise a simple distributed algorithm that leads the players to this equilibrium point.

Let us stress the fact that, in contrast with Section 9.2, there is no access point in the problem that we are addressing here, meaning that there is no authority to "catch the bad guys" (as we discussed in Chapter 3). As a consequence, the nodes must rely *exclusively on themselves* to make sure that everyone plays by the rules (whatever these rules are). As we will show, the enforcement of these rules can be realized by the threat of peer castigation.

Nash equilibria of the repeated game

Essentially, the repeated CSMA/CA game is defined as the game $G_{\text{CSMA/CA}}$ played repeatedly T times. We consider an infinitely repeated game, that is $T \to \infty$ (the game is "infinite" in the sense that none of the players knows when it will finish). We denote the repeated CSMA/CA game by $G^\infty_{\text{CSMA/CA}}$. In this new setting, the utility function of every player $i \in P$ becomes:[18]

$$u_i^\infty = \liminf_{T \to \infty} \frac{1}{T} \sum_{t=1}^{T} u_i^t \left(\pi_i^t, \pi_{-i}^t \right) \tag{9.21}$$

where $u_i^t(\pi_i^t, \pi_{-i}^t)$ denotes a stage t payoff for the player i. One of the reasons why we do not use the *discounting factor*, where "impatient" players discount future payoffs, is that, as we will show in subsection 9.3.4, the players of this game converge reasonably fast to a game equilibrium. Therefore, it is legitimate to assume that the players are "completely patient" (no discounting). The approach presented hereafter follows the principle of the Folk theorem (Appendix B).

For mathematical convenience, we assume the contention window W_i^t (and therefore also the access probability, π_i^t) to be a continuous variable for every player $i \in P$, and for all $t = \{1, \ldots, T\}$. Moreover, $\pi_i^t \in [0, 1], \forall i \in P, (t = \{1, \ldots, T\})$.

---

[18] The lim inf in this expression is in response to the fact that some infinite sequences of stage payoffs do not have well-defined average values. The reader interested in the convergence of such limits can check, for example, Fudenberg and Tirole [144], Section 5.1.1.

Let us define the following *penalty function* for every player $i \in P$

$$pf_i(\pi_i, \pi_{-i}) = \begin{cases} \varphi_i(\pi_i, \pi_{-i}), & \pi_i \in (\overline{\pi}, 1]; \\ 0, & \pi_i \in [0, \overline{\pi}], \end{cases} \quad (9.22)$$

where $\overline{\pi} \in (0, 1)$ represents the *targeted* equilibrium point and $\varphi_i(\pi_i, \pi_{-i})$ satisfies

$$\varphi_i(\pi_i, \pi_{-i}) > 0 \quad \text{and} \quad \frac{\partial}{\partial \pi_i} \varphi_i(\pi_i, \pi_{-i}) > \frac{\partial}{\partial \pi_i} \tau_i^{(c)}(\pi_i, \pi_{-i}),$$

$$(9.23)$$

for all $\pi_i \in (\overline{\pi}, 1]$ and $\pi_j < 1 \quad (j \in P \backslash \{i\})$.

Let us further define the players' per stage payoffs as

$$u_i^t(\pi_i^t, \pi_{-i}^t) = \tau^{(c)t}(\pi_i^t, \pi_{-i}^t) - pf_i^t(\pi_i^t, \pi_{-i}^t), \quad \forall i \in P. \quad (9.24)$$

We note here that any *penalizing mechanism* used to impose the penalty pf_i^t on some player i, should be designed so that it does not bring any performance degradation to the players $k \in P \backslash \{i\}$. A "nice" property of the single-channel single-collision domain CSMA/CA networks is that at any time instant only one station can successfully transmit. Therefore, in these networks, we can single out any player for punishment.[19]

Lemma 9.2 *Let* $\pi_j^t < 1, \forall j \in P \backslash \{i\}$. *Then, the stage payoff function* $u_i^t(\pi_i^t, \pi_{-i}^t)$ *has a unique maximizer* $\pi_i^t = \overline{\pi} \in (0, 1)$ *for every stage* $t = \{1, \ldots, T\}$.

Proof Since $\pi_k^t < 1, \forall k \in P \backslash \{i\}$, we have from, Equation (9.15)

$$\frac{\partial}{\partial \pi_i^t} \tau_i^{(c)t}(\pi_i^t, \pi_{-i}^t) > 0 \quad (9.25)$$

for $\pi_i^t \in [0, 1]$. Therefore, on the interval $[0, \overline{\pi}]$, $\pi_i^t = \overline{\pi}$ is the unique maximizer of the payoff $u_i^t(\pi_i^t, \pi_{-i}^t)$. For the remaining interval $(\overline{\pi}, 1]$ we have

$$\frac{\partial}{\partial \pi_i^t} u_i^t(\pi_i^t, \pi_{-i}^t) = \frac{\partial}{\partial \pi_i^t} \tau_i^{(c)t}(\pi_i^t, \pi_{-i}^t) - \frac{\partial}{\partial \pi_i} \varphi_i^t(\pi_i^t, \pi_{-i}^t) \overset{(1)}{<} 0,$$

where the inequality (1) follows from the condition (9.23). Therefore, on the interval $(\overline{\pi}, 1]$, $u_i^t(\pi_i^t, \pi_{-i}^t)$ is a strictly decreasing function in π_i^t, which concludes the proof. $\qquad \square$

Lemma 9.2 implies that the strategy profile $(\pi_i^t = \overline{\pi})_{i \in P}$ is the unique Nash equilibrium of the stage game $G_{\text{CSMA/CA}}$ played in stage t.

In order to characterize the game, we will now make use of the notion of Subgame Perfect Nash Equilibrium (SPNE), see Appendix B.

---

[19] In game theory, this property is known as *full dimensionality*.

Theorem 9.4 *The strategy profile* $(\pi_i^t = \overline{\pi})_{i \in P, t=\{1,\ldots,T\}}$ *is a subgame perfect Nash equilibrium (SPNE) of the game* $G_{\text{CSMA/CA}}^{\infty}$.

Proof For every $k \in \{1, \ldots, T\}$ and every player $i \in P$ the following holds

$$u_i^{k,\infty} = \liminf_{T \to \infty} \frac{1}{T} \sum_{t=k}^{T} u_i^t(\pi_i^t, \pi_{-i}^t)$$

$$\leq \liminf_{T \to \infty} \frac{1}{T} \sum_{t=k}^{T} \max_{\pi_i^t \in [0,1]} \left\{ u_i^t(\pi_i^t, \pi_{-i}^t) \right\}$$

$$\overset{(1)}{=} u_i \left((\pi_j = \overline{\pi})_{j \in P} \right),$$

where (1) follows from Lemma 9.2.

Therefore, by definition, $(\pi_i^t = \overline{\pi})_{i \in P, t=\{1,\ldots,T\}}$ is an SPNE of $G_{\text{CSMA/CA}}^{\infty}$.   □

Observe that $(\pi_i^t = \overline{\pi})_{i \in P, t=\{1,\ldots,T\}}$ is not the only SPNE under the averaging criterion given by (9.21). The reason is that any finite number of deviations by some player i from the equilibrium strategy $\pi_i^t = \overline{\pi}$ becomes irrelevant under the averaging criterion (9.21). Still, the best strategy for player i is $\pi_i^t = \overline{\pi}$, because otherwise her overall payoff will be strictly smaller than $u_i \left((\pi_j = \overline{\pi})_{j \in P} \right)$; in $G_{\text{CSMA/CA}}^{\infty}$, any deviation from $\overline{\pi}$ necessarily results in a smaller per stage payoff.

The following corollary is a simple implication of the penalty functions pf_i, ($i \in P$), defined by (9.22). This result is reminiscent of the *Nash Folk theorem*.

Corollary 9.1 *Any strategy profile* $(\pi_i^t = \pi)_{i \in P, t=\{1,\ldots,T\}}$, *such that* $\pi \in (0, 1)$, *can be made an SPNE.*

In our context, this result is important as we want to make the Pareto-optimal point $(W_i = W^*)_{i \in P}$, i.e., the corresponding channel access probability profile $\left(\pi_i = 2/(1 + W^*) \right)_{i \in P}$, a Nash equilibrium.

As mentioned at the beginning of this subsection, the nodes can rely only on themselves to make sure that the network operates in a desirable way; more specifically, this means that all the contention windows are (and stay) at value W^*. In the absence of any external authority, this means that they must monitor each other and react in case a node deviates in an unacceptable way; the only practical way to achieve this is to perform *selective jamming*, as will be explained in the following subsection.

Practical penalty function

Let us consider two arbitrary players k and i from set P. Let us assume that player k calculates the penalty pf_i to be inflicted on player i as follows

$$pf_i(\pi_i, \pi_{-i}) = \begin{cases} \tau_i^{(c)}(\pi_i, \pi_{-i}) - \tau_k^{(c)}(\pi_i, \pi_{-i}), & \text{if } \tau_i^{(c)}(\pi_i, \pi_{-i}) > \tau_k^{(c)}(\pi_i, \pi_{-i}); \\ 0, & \text{otherwise} . \end{cases}$$

$$(9.26)$$

It is easily seen that the penalty function (9.26) has essentially the same format as the penalty function given by (9.22) and (9.23). To see this, using the notation of the definition in (9.22), we define $\varphi_i(\pi_i, \pi_{-i}) \overset{def}{=} \tau_i^{(c)}(\pi_i, \pi_{-i}) - \tau_k^{(c)}(\pi_i, \pi_{-i})$, where $\tau_i^{(c)}(\pi_i, \pi_{-i}) > \tau_k^{(c)}(\pi_i, \pi_{-i})$. Observe that the condition $\tau_i^{(c)}(\pi_i, \pi_{-i}) > \tau_k^{(c)}(\pi_i, \pi_{-i})$ in (9.26) is equivalent to $\pi_i > \pi_k$ when $\pi_j < 1, \forall j \in P \backslash \{i\}$. Finally, for $\pi_j < 1, \forall j \in P \backslash \{i\}$, we have

$$\frac{\partial}{\partial \pi_i} \varphi_i(\pi_i, \pi_{-i}) = \frac{\partial}{\partial \pi_i} \tau_i^{(c)}(\pi_i, \pi_{-i}) + \left| \frac{\partial}{\partial \pi_i} \tau_k^{(c)}(\pi_i, \pi_{-i}) \right| \overset{(1)}{>} \frac{\partial}{\partial \pi_i} \tau_i^{(c)}(\pi_i, \pi_{-i}),$$

where (1) follows from the fact that $\pi_j < 1, \forall j \in P \backslash \{i\}$.

Therefore, we can apply Lemma 9.2 to conclude that the unique maximizer of the player i's (single stage) payoff $u_i(\pi_i, \pi_{-i})$ is $\pi_i = \pi_k$. In the context of the two players i and k, a very important property of the penalty function is that it results in the same throughputs for both player i and player k; i.e., $\pi_i = \pi_k$ implies that players i and k will receive the same throughputs.

Inspired by the penalty functions (9.26), it is possible to devise a simple penalizing scheme, in which the packets of a *deviating* player are *selectively jammed* for a short duration of time, T^{jam}, by the other players in the system. By "deviating" we mean a player that departs from the given equilibrium point. Suppose that a player $k \in P$ detects the presence of a deviating player $i \in P$. Thereafter, if the player k listens to a transmitted packet corresponding to the player k, it switches to the transmission mode and *jams* enough bits so that the packet cannot be properly recovered at the receiver.

Let the throughput obtained by the two considered players over the last *observation window*, T^{obs}, be $\tau_i^{(c)}$ and $\tau_k^{(c)}$, respectively, where $\tau_i^{(c)} > \tau_k^{(c)}$. As we have seen above, the penalty function (9.26) aims at making the throughputs received by the players i and k equal. We denote with $\tau_x^{(c)}(t)$ the instantaneous throughput of the given player x. The *average throughput* received by the players i and k should be the same over the total time duration of $T^{obs} + T^{jam}$, that is

$$\frac{1}{T^{obs} + T^{jam}} \int_t^{t + T^{obs} + T^{jam}} \tau_k^{(c)}(t) dt = \frac{1}{T^{obs} + T^{jam}} \int_t^{t + T^{obs} + T^{jam}} \tau_i^{(c)}(t) dt$$

$$\overset{(1)}{=} \frac{1}{T^{obs} + T^{jam}} \int_t^{t + T^{obs}} \tau_i^{(c)}(t) dt, \tag{9.27}$$

where (1) follows from the fact that the player k jams the player i during the period T^{jam}. Let us denote the average throughput over a time period P starting at time instant t by $\bar{r}(t, P)$, that is

$$\bar{r}(t, P) \overset{def}{=} \frac{1}{P} \int_t^{t+P} r(t) dt.$$

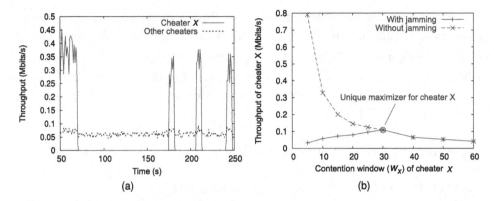

Figure 9.18. Realization of the penalty function $pf_i(\pi_i, \pi_{-i})$ by selective jamming: (a) throughputs (payoffs) obtained by the cheaters over time in the presence of the deviating cheater X and selective jamming mechanism; (b) unilateral deviation by the cheater X with and without the penalty mechanism. Used with permission, from [80], © IEEE, 2005.

Then, from the expression (9.27) we obtain

$$T^{\mathrm{jam}} = T^{\mathrm{obs}} \frac{\overline{\tau}^{(c)}_i \left(t, T^{\mathrm{obs}}\right) - \overline{\tau}^{(c)}_k \left(t, T^{\mathrm{obs}}\right)}{\overline{\tau}^{(c)}_k \left(t + T^{\mathrm{obs}}, T^{\mathrm{jam}}\right)}. \tag{9.28}$$

We note that $T^{\mathrm{jam}} < \infty$, except in the case when $\overline{\tau}^{(c)}_k \left(t + T^{\mathrm{obs}}, T^{\mathrm{jam}}\right) = 0$. But the case $\overline{\tau}^{(c)}_k \left(t + T^{\mathrm{obs}}, T^{\mathrm{jam}}\right) = 0$ never happens under the penalty functions (9.26) if all the players are rational (and the number of players is finite); by Theorem 9.4 there are strictly better outcomes than zero for every player. It is also interesting to observe that the deviating player i minimizes T^{jam} by playing $\pi_i = 0$ during the period T^{jam}. This is because $\partial \tau^{(c)}_k / \partial \pi_i < 0$, when $\pi_j < 1, \forall j \in P \setminus \{i\}$, and therefore $\overline{\tau}^{(c)}_k(t + T^{\mathrm{obs}}, T^{\mathrm{jam}})$ gets larger.

In order to illustrate the performance of the jamming mechanism, we provide hereafter the results of a simulation in *ns*-2. The simulation setup is the same as in Subsection 9.3.2 with $N = 20$ and $|P| = 10$. A cheating player, designated as cheater X, is randomly selected and her contention window size fixed to a value of 10. The contention window size for all the other cheaters in the system is fixed to the point $\overline{W} = 30$ (i.e., the corresponding $\overline{\pi}$). We use an observation window size, T^{obs}, of 20 s. Cheater X gets *detected* by the other cheaters in the network and is penalized for her deviation. We will describe the detection mechanism in Subsection 9.3.4. On Figure 9.18a we plot the throughput obtained by the cheaters in the system over time, with and without the penalizing scheme. As can be observed from Figure 9.18a, cheater X is detected and is penalized for her deviation. When penalized, cheater X's throughput drops to zero. Observe from this figure the dependency of the period

T^{jam} on the observation period T^{obs}; for better system efficiency, T^{obs} should be kept short (much shorter than 20 s as used in the simulations).

Figure 9.18b plots the average throughput obtained by cheater X, when it unilaterally deviates from the given equilibrium point $\overline{W} = 30$. The results are averaged over a duration of 1000 s. As can be observed from Figure 9.18b, after the introduction of the detection and penalizing mechanism, cheater X achieves maximum throughput by operating at the given equilibrium point $\overline{W}$, i.e., $\overline{\pi}$, which is consistent with the result of Lemma 9.2. Thus, any unilateral deviation from this point brings less payoff to the cheater X. Therefore, by definition, $\overline{W}$ is a unique Nash equilibrium of the single stage game.

9.3.4 Implementation

In this subsection, we will provide a comprehensive, distributed and efficient equilibrium coordination protocol based on the theoretical insights from Subsection 9.3.3. We have seen that the key building block for the model of repeated games is the penalization mechanism. We have already elaborated a practical penalization mechanism in Subsection 9.3.3. The penalization mechanism, however, relies on the ability of the players to estimate the difference in their payoffs. In order to empower the players with this ability, we first develop an appropriate *detection mechanism*. Then, we describe how the players should react once they are penalized. We call the scheme followed by the penalized nodes an *adaptive strategy*. Finally, we put together all the basic building blocks and simulate the behavior of such a comprehensive coordination algorithm.

Detection mechanism In this approach, each cheating node (player) measures the throughput of all the nodes, including itself. This is indeed feasible due to the broadcast nature of the wireless medium. If a cheater observes a difference in throughput with some other node, it characterizes that node as a deviating cheater. Let τ_i and τ_j be the measured throughput of nodes i and j, respectively. Because of the inherent short-time unfairness of the IEEE 802.11 MAC protocol [226], and in order to increase the efficiency of the detection mechanism, we use two parameters: the observation time-window size T^{obs} and the tolerance margin ϵ, in percentage of throughput. After measuring the throughput of each node for T^{obs} seconds, cheater i concludes that cheater j is *deviating* whenever the throughput of node j exceeds the throughput of node i, that is whenever

$$\frac{\tau_j}{\tau_i} > 1 + \epsilon.$$

This detection mechanism has been implemented in *ns*-2, with $N = |P| = 30$ nodes. The contention window size (W_j) of a single node j is varied, and the others' contention window sizes are set to 30 (i.e., $W_k = 30, \forall k \in P \backslash \{j\}$).

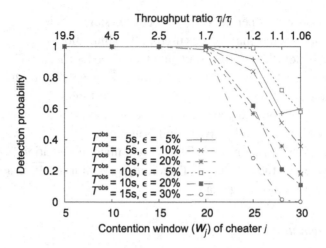

Figure 9.19. Performance of cheating detection based on throughput measurements. Used with permission, from [80], © IEEE, 2005.

Figure 9.19 shows the performance of the detection mechanism for different values of T^{obs} and ϵ. The probability of false positives corresponds to the detection probability with $W_j = 30$; at this point, cheater j uses a contention window value equal to that of node i, but still obtains a higher throughput, $\tau_j/\tau_i = 1.06$, owing to the IEEE 802.11 unfairness. Therefore, node j gets detected as deviating with positive detection probability. To reduce the false positives (at contention window size 30), we can consider large ϵ values ($> 10\%$). However, this comes at the expense of lower detection probabilities if cheater j uses contention window sizes slightly lower than 30. Similarly, large T^{obs} values ($\geq 15s$) will reduce the effect of the inherent IEEE 802.11 unfairness, and therefore the corresponding false positives. This also comes at the expense of lower detection probabilities if cheater j uses contention window sizes slightly lower than 30. Therefore, choosing appropriate values for T^{obs} and ϵ is crucial for both the described detection mechanism and the overall system performance. For very low contention window sizes of cheater j ($W_j \leq 20$), the throughput ratio τ_j/τ_i is much larger than $1 + \epsilon$, making the detection of the cheater j's deviation easy.

As we have seen in the previous section, the detection mechanism DOMINO is based on calculating the average backoff used by the nodes; it can be used in the case of heterogeneous conditions among the cheaters in the system. Although DOMINO is more appropriate for misbehaving detection at the MAC layer, we make use here of the throughput-based detection, for simplicity of implementation and presentation.

Adaptive strategy In order to reach the desired operation point, the cheaters make use of the following adaptive strategy. When cheater i observes that she is being

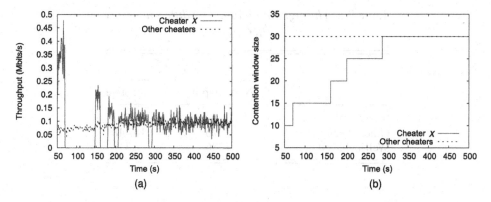

Figure 9.20. Performance of the system with the adaptive strategy: (a) throughput of the cheaters over time; (b) Contention window size of the cheaters over time. Used with permission, from [80], © IEEE, 2005.

jammed (penalized) during some period Δ, she gradually increases her contention window by steps of size γ. Note that a cheater can easily decide whether she is being jammed by observing her own throughput. The choice of Δ determines the efficiency of the system. A high value of Δ might let a deviating cheater escape from being penalized. However, choosing a small value of Δ might magnify the effect of a possible misdetection by unnecessarily causing a cheater to increase her contention window size. This will eventually lead the whole system towards an inefficient point of operation. The choice of the step size, γ, offers a trade-off between convergence time and efficiency: if we increase the contention window in large steps, although the system will stabilize in less time, the point of operation might be far away from the Pareto-optimal point (W^*), resulting in an inefficient system and vice versa.

This adaptive strategy has been implemented in *ns*-2. The simulation setup is the same as in the previous subsection ($N = 20$, $|P| = 10$, $W^{init} = 30$). A cheater is randomly selected, designated as node X, and her initial contention window size fixed to 10. The contention window size for all the other cheaters in the system is fixed to W^{init}. We fix Δ to be 5 s and γ to be 5. Figure 9.20a plots the obtained throughput by different cheaters in the system over time. Figure 9.20b plots the evolution of contention window size of node X over time. We can observe how node X adapts its contention window size by following the adaptive strategy and eventually converging to a window size of 30, equal to W^*. Thus the other cheaters in the system are successful in guiding the deviating cheater to the desired equilibrium point.

Table 9.3 summarizes the throughput averages obtained by different nodes over a time interval of 1000 s. As can be observed from Table 9.3, the jamming and

Table 9.3. *Throughput obtained by different nodes*
(bytes/s)

	Strategy	
	Non-adaptive	Adaptive
Cheater X	7650	11 577
Other cheaters	7826	11 448
Well-behaved nodes	1286	2318

Table 9.4. *Throughput obtained by different nodes*
(bytes/s) with multiple levels of misbehavior

	Strategy	
	Non-adaptive	Adaptive
Cheater X	2843	10 356
Cheater Y	2686	10 185
Cheater Z	2565	10 239
Other cheaters	2544	10 172
Well-behaved nodes	270	1981

detection mechanism combined with the adaptive strategy, besides being fair to all the cheaters in the system, is also the most efficient.

Finally, we provide an evaluation of the performance of the adaptive algorithm (in *ns*-2) for a scenario consisting of multiple levels of misbehavior in the system. The simulation setup is the same as above ($N = 20$, $|P| = 10$, $W^* = 30$). We randomly select three cheaters, designated as node X, Y and Z respectively. We fix their contention window sizes to be 5, 10 and 15, respectively. The contention window size for all the other cheaters in the system is fixed to W^*. Table 9.4 summarizes the average throughput obtained by different nodes over an interval of 1000 s. As can be observed from Table 9.4, the jamming mechanism combined with the adaptive strategy results in an optimal and fair performance, even with multiple levels of misbehavior in the system. As we predicted in Subsection 9.3.3, the deviating cheaters (players) X, Y and Z clearly have an incentive to adapt upon being penalized. In the same way, each cheater has an incentive to penalize the other cheaters.

Reaching the Pareto-optimal point Here again, we will rely primarily on intuition. The motivated reader can find a more formal presentation in [79].

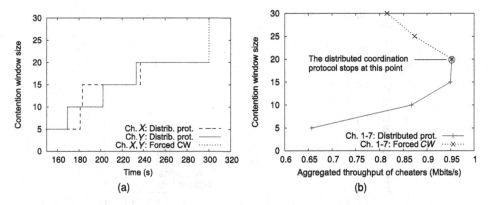

Figure 9.21. Performance of the distributed coordination protocol, with $N = 20$ and $|P| = 7$: (a) evolution of the contention windows; (b) contention window vs. average throughput (the axes in (b) are swapped for the convenience of matching them with (a)). Used with permission, from [80], © IEEE, 2005.

An accurate implementation of detection, penalizing and adaptive strategy will lead the nodes to reach a stage equilibrium point, $\left(W_i = \overline{W}\right)_{i \in P}$.

The following distributed coordination algorithm will fulfill this goal. At the onset of the system, $\left(W_i = W^{\text{init}}\right)_{i \in P}$ for all cheaters. Every cheater sets up a random timer (in the simulations this corresponds to a random value between 0 and 20 s) to increase her contention window by step size, γ. One of the cheaters, say X, will eventually increase her contention window size to $W_X^{\text{init}} + \gamma$. Based on the detection mechanism, node X will conclude that all other cheaters in the system are deviating and will begin penalizing them. If a cheater observes that she is being penalized, she will disable the timer and use the adaptive strategy described earlier. Eventually the system will stabilize, when $W_i = W_i^{\text{init}} + \gamma$ for all cheaters.

The cheaters realize that they have reached a new stable point of operation when they all begin enjoying the same throughput (in the implementation, the cheaters remain at this stable point for 20 s before continuing the search for W^*). At this point in time, every cheater $i \in P$ compares her throughput at $W_i = W_i^{\text{init}} + \gamma$ with the throughput at $W_i = W_i^{\text{init}}$; if she observes a decrease in her throughput, she will terminate the search for W^*. Otherwise she again sets up the random timer to increase her contention window size by γ. Therefore, the proposed distributed protocol simply "climbs" up the left side of the aggregate throughput curve shown on Figure 9.17, until it reaches the optimal value W^*.

This protocol has been implemented in ns-2. The simulation setup consists of 20 nodes and seven cheaters ($N = 20$, $|P| = 7$). The cheaters initialize their contention window sizes to 5 (($W_i^{\text{init}} = 5)_{i \in P}$). The cheaters continue their search for W^* only

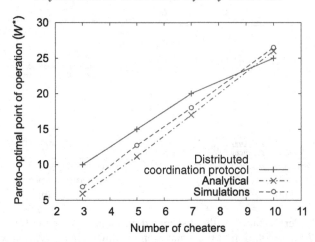

Figure 9.22. Variation of the Pareto-optimal point W^* with the number of cheaters. Used with permission, from [80], © IEEE, 2005.

if they see an increase of 10% or more in their throughput from the last stable point of operation. Figure 9.21a plots the sample evolution of the contention window for 2 cheaters, X and Y, in the system. Note that all of the cheaters follow a similar pattern and eventually converge to a window size of 20. We are unable to show their evolution in the same plot as it simply generates overlapping lines. Note also that the convergence time is relatively short, around 80 s for seven cheaters (from $t_{start} \approx 160$ to $t_{end} \approx 240$; in these simulations we used the warm-up period of around 160 s).

Figure 9.21b plots the average throughput obtained by the cheaters at different contention window sizes. As can be seen from Figure 9.21b, the throughput is maximized at $(W_i = 20)_{i \in P}$ and the cheaters will stabilize at that value. For completeness, the "dotted" curves in Figure 9.21 are obtained by deliberately forcing the cheaters to go beyond $(W_i = 20)_{i \in P}$.

We next evaluate the performance of the protocol by varying the number of cheaters in the system. The protocol is run and the window size at which all the cheaters eventually converge is measured. Thus, according to the protocol, this point of convergence is the Pareto-optimal point of operation. The actual Pareto-optimal point (W^*) has been evaluated through ns-2 simulations, under the same network settings. The Pareto-optimal point (W^*) is also evaluated analytically, using Bianchi's model. Figure 9.22 plots the obtained results. The results are averaged over five simulation runs. The results obtained by the distributed coordination protocol closely match the analytical results obtained using Bianchi's model. Note that the minimum resolution of the distributed coordination protocol is equal to the step size, $\gamma = 5$. As can be seen from Figure 9.22, the discrepancy is bounded by $\pm\gamma$, which clearly proves the efficiency of the distributed coordination protocol.

The protocol operates in a completely distributed manner, without requiring any *a priori* knowledge about the optimal point of operation or of the total number of nodes/cheaters in the system. However, we rely on the fact that the numbers of nodes and of cheaters does not change in the system. In more dynamic networks, where new nodes/cheaters can enter or existing nodes/cheaters can leave the system, a possible solution for the cheaters consists in timing out periodically and re-running the whole protocol from the beginning.

Comment on selective jamming

As we have seen, in this section we have established that selective jamming can be a useful mechanism, in the sense that it can encourage smart cheaters to demonstrate restraint in their greedy behavior. This result might sound paradoxical, because jamming is usually considered to be a plague for wireless networks; in addition, jamming is usually related to malicious, not selfish behavior.

But, as we have seen in Chapter 3, the more the networks are decentralized, the more they have to rely on rule enforcement mechanisms to ensure the appropriate behavior of all the nodes; and at the MAC layer of a self-organized network, the only possible threat is jamming.

Of course, we do not claim that jamming mechanisms ought to be implemented in upcoming wireless devices. Indeed, as already mentioned, at the time of this writing it is impossible to determine how selfishness will materialize in practice. The purpose of this section is rather to show that (unfortunately) such misdeeds are possible but also that (fortunately) it is possible to design techniques by which the nodes can monitor and police each other's activities.

An additional insight provided by this section is that nodes can *cooperatively* search for the optimal operation point of the network. We believe this to be of particular interest for the study of **cognitive radios**.

9.4 Summary

This whole chapter was devoted to selfish behavior techniques in CSMA/CA networks, and to techniques to thwart these misdeeds. We have first considered the case in which several wireless stations are attached to an Access Point, and one of them deliberately departs from the normal protocol in order to increase its share of the bandwidth (at the expense of the other stations). We have provided a detailed description of these cheating techniques; we have also described a system to be run at the Access Point, called DOMINO, which is able to detect all these misdeeds.

We have then addressed the problem of cheating in single collision domain CSMA/CA networks, in the absence of any authority. For this purpose, we have developed a game-theoretical model and corroborated the analytical conclusions by appropriate simulations. We have studied several aspects. First, we have described a formalism for the systematic study of rational cheating in CSMA/CA networks. Second, we have focused on the simple cases (i) of a single cheater and (ii) of several cheaters acting without restraint. Third, using the theory of repeated (multi-stage) games, we have shown how it is possible to transform the Pareto-optimal point into a Subgame Perfect Nash Equilibrium. Finally, we have shown that smart cheaters can collectively find this point.

9.5 To probe further

Section 9.2 is derived from papers [324, 322]; the reader interested in analytical techniques to compute the values of the nominal actual backoff can refer to them. These two papers have been inspired by the work of Kyasanur and Vaidya [298] who propose that the receiver assigns the backoff value to be used by the sender, so the former can detect any misbehavior of the latter. If the sender deviates from the assigned value, it will be assigned high backoff values on the next round to compensate its deviation. Konorski [230] proposes another misbehavior-resilient backoff algorithm. Both proposals require change to the IEEE 802.11 protocol.

Statistics of wireless use can be found in [39, 233]. We have described the subtle relationship between the MAC layer and TCP; some authors have also investigated cheating at the TCP layer, notably Akella *et al.* [15].

The detailed analytical treatment of Section 9.3 can be found in [80, 79].

Game theory has been widely applied to the study of the network layer, whereas fewer researchers have applied it to the MAC layer. MacKenzie and Wicker [263] study the problem of selfish users in Aloha from a game-theoretic point of view. They analyze the stability of the system (Nash equilibrium), and calculate the transmission probabilities that optimize each node's throughput. They assume that all nodes have the same transmission rates and costs. Moreover, every node has an *a priori* knowledge about the total number of nodes in the system. Altman *et al.* [22] reconsider the same Aloha "game" with partial information, where the transmission probability is adapted according to collision feedback only. They consider two frameworks: team work and non-cooperative game. Jin and Kesidis [207] study non-cooperative equilibria of Aloha networks for heterogeneous users. MacKenzie and Wicker have also studied the stability of multipacket slotted Aloha with selfish users and perfect information [264].

Alpcan *et al.* [20] apply game theory for uplink power control in cellular networks. In [381], Xiao, Schroff and Chong describe a utility-based power control framework for a cellular system. In [155], Goodman and Mandayam introduce the concept of network-assisted power control to equalize signal-to-interference ratio between the users.

In this chapter (and in all the other chapters of Part III), we assumed a non-cooperative behavior of the players. In some cases, however, it can be realistic to assume that players try to find an agreement by *bargaining* with each other. An example of that approach is provided by the work of Heather Zheng and her team [310, 399, 83, 398].

Finally, let us mention the work by Zander [390], published as early as 1990, describing jamming and jamming avoidance in a slotted Aloha network as a zero-sum game.

9.6 Questions

(1) In IEEE 802.11, why isn't SIFS=0?
(2) If a hot-spot contains both UDP and TCP nodes cheating with backoff, how does DOMINO detect each of them?
(3) Is IEEE 802.11 fair or unfair? How does this affect DOMINO? Hint: fairness can be consider either over a short-term or over a long-term interval.
(4) Why can't a cheater increase her bandwidth share of downlink traffic only by playing with the MAC layer, as is the case for the uplink traffic?
(5) Why is it more advantageous (for a cheater) to cheat with UDP traffic than with TCP traffic?
(6) DOMINO is an external system that makes IEEE 802.11 protocol resilient to cheating. What are possible modifications that can make IEEE 802.11 inherently resilient to cheating?
(7) Take a look at the IEEE 802.11e standard (to be retrieved from the IEEE Web site). Is it possible to detect greedy behavior in such networks? Why?
(8) What is the rationale behind the $(CW_i - k)/CW_i$ factor in Equation (9.6)?
(9) Prove that the system of two equations (9.3) and (9.9) with unknown variables π and p has a unique solution.
(10) What would be the modifications in Bianchi's model if the mobile nodes used RTS/CTS before transmitting the packets (to avoid the hidden terminal problem)?
(11) IEEE 802.11 WGs provide a multi-rate physical layer capability which is obtained by employing different sets of modulation and channel coding at PHY layer. How should Bianchi's model be modified to consider different data rates?
(12) Does Bianchi's model take into consideration the wireless channel conditions for mobile users? How can we incorporate this parameter in Bianchi's model?

(13) Equation (9.14) calculates the probability of collision for the well-behaved nodes, i.e. $\pi^{(w)}$. How is this equation obtained?

(14) Equation (9.15) expresses the throughput of one cheater node i. Explain the meaning of numerator and denominator in this equation.

(15) As is mentioned in Section 9.3.2, the Pareto optimality can be obtained directly from Bianchi's model. Considering Equation (9.10), explain how we can obtain this point directly from Bianchi model?

(16) What is the main purpose of the repeated game definition in Section 9.3.3?

(17) In the definition of approximate game (i.e., $\hat{G}_{\text{CSMA/CA}}$), why is $N < \infty$ a condition required for the existence of ϵ_i?

(18) Research question. Let us assume that a cheater adopts an on-off strategy: she cheats, then behaves, then cheats again, and so on. Using game theory, can you model the resulting game between cheaters and detector (DOMINO)? Who wins and under what conditions?

10

Selfishness in packet forwarding

In the previous chapter, we have studied selfish behavior at the MAC layer. We will now focus on the network layer. For this purpose, we will consider self-organized wireless ad hoc networks. As we have explained in Chapter 2, in such networks the networking services are provided by the nodes themselves. As a fundamental example, the nodes must make a mutual contribution to packet forwarding in order to ensure an operable network. If the network is under the control of a single authority, as is the case for military networks and rescue operations, the nodes cooperate for the critical purpose of the network. However, if each node is its own authority, cooperation between the nodes cannot be taken for granted; on the contrary, it is reasonable to assume that each node has the goal of maximizing its own benefits by enjoying network services and at the same time minimizing its contribution. This selfish behavior can significantly damage network performance.

In this chapter, we focus on the most resource demanding operation of the network layer, namely packet forwarding. We address the case of self-organized wireless ad hoc networks, in order to derive some fundamental results. In particular, we will see that a network without incentives for cooperation is very likely to collapse. In Chapter 12, we will describe incentive techniques to solve this problem.

The question underpinning this chapter is the following: when a node is requested to forward a packet by one of its neighbors, will it do so, if no mechanism is in place to enforce this cooperative behavior? We define a model in a game theoretic framework and identify the conditions under which an equilibrium based on cooperation exists. As the problem is involved, we deliberately restrict ourselves to a static configuration (this is the reason why we talk here about a "self-organized wireless ad hoc network" and not about a "self-organized mobile ad hoc network").

10.1 Game theoretic model of packet forwarding

Let N be the set of the nodes of an ad hoc network ($n = |N|$). Each node has a given power range and two nodes are said to be neighbors if they reside within the power range of each other. We represent the neighbor relationship between the nodes with an undirected graph, which we call the *connectivity graph*. Each vertex of the connectivity graph corresponds to a node in the network, and two vertices are connected with an edge if the corresponding nodes are neighbors.

Communication between two non-neighboring nodes is based on multi-hop relaying. This means that packets from the source to the destination are forwarded by intermediate nodes. For a given source and destination, the intermediate nodes are those that form the shortest path between the source and the destination in the connectivity graph.[1] We call such a chain of nodes (including the source and the destination) a *route*. We call the topology of the network with a given set of communicating nodes a *scenario*. To simplify the treatment, throughout the chapter we will assume that each node is the source of a single route.

We use a discrete model of time where time is divided into slots. We assume that both the connectivity graph and the set of existing routes remain unchanged throughout the life of the system. We assume that the duration of the time slot is much longer than the time needed to relay a packet from the source to the destination. This means that a node is able to send several packets within one time slot. This allows us to abstract away individual packets and to represent the data traffic in the network with *flows*. We assume flows at fixed rate, which means that a source node sends the same amount of traffic in each time slot. Note, however, that this amount can be different for every source node and every route.

10.1.1 Forwarding game

We model the operation of the network as a game, which we call the *forwarding game*. The players of the forwarding game are the nodes. In each time slot t, each node i chooses a *cooperation level* $m_i(t) \in [0, 1]$, where 0 and 1 represent full defection and full cooperation, respectively (we use the letter "m" to designate this cooperation level because it is actually the *move* of the node in game theoretic parlance). Here, defection means that the node does not forward traffic for the benefit of other nodes, whereas cooperation means that it does. Thus, $m_i(t)$ represents the fraction of the traffic routed through i in t that i actually forwards. Note that i has a single cooperation level $m_i(t)$, which it applies to every route in which it is

---

[1] In other words, we abstract away the details of the routing protocol, and we model it as a function that returns the shortest path between the source and the destination. If there are multiple shortest paths, then one of them is selected at random.

involved as a forwarder. We prefer to not require the nodes to be able to distinguish the flows that belong to different routes, because this would require identifying the source–destination pairs and applying a different cooperation level to each of them; this would probably increase the computation at the nodes significantly.

Let us consider a route with source node s and with ℓ forwarding nodes $f_1, f_2, \ldots, f_\ell$. Let us denote by T_s the constant amount of traffic that s wants to send on the route r in each time slot. The throughput $\tau(r, t)$ experienced by the source s in t is defined as the fraction of the traffic sent by s (on r) in t that is delivered to the destination. As we are studying cooperation in packet forwarding, we assume that the main reason for packet losses in the network is the non-cooperative behavior of the nodes. In other words, we assume that the network is not congested and that the number of packets dropped owing to the limited capacity of the nodes and the links is negligible. Hence, $\tau(r, t)$ can be computed as the product of T_s and the cooperation levels of all intermediate nodes:

$$\tau(r, t) = T_s \cdot \prod_{k=1}^{\ell} m_{f_k}(t). \tag{10.1}$$

In addition, we define the normalized throughput $\hat{\tau}(r, t)$ as follows:

$$\hat{\tau}(r, t) = \frac{\tau(r, t)}{T_s} = \prod_{k=1}^{\ell} m_{f_k}(t). \tag{10.2}$$

We will use the normalized throughput later as an input of the strategy function of s.

The **benefit** $b_s(t)$ of s in t depends on the experienced throughput $\tau(r, t)$. In general, $b_s(t) = \mathcal{F}_s(\tau(r, t))$, where $\mathcal{F}_s$ is some non-decreasing function. We further assume that $\mathcal{F}_s$ is concave, differentiable at T_s, and that $\mathcal{F}_s(0) = 0$. We place no other restrictions on $\mathcal{F}_s$. Note that the function $\mathcal{F}$ of different nodes can be different.

The **cost** $c_{f_j}(r, t)$ in t of the jth intermediate node f_j on each route r containing the forwarding node f_j is non-positive and represents the "effort" for node f_j to forward packets on route r during time slot t. It is defined as follows:

$$c_{f_j}(r, t) = -T_s \cdot C \cdot \hat{\tau}_j(r, t), \tag{10.3}$$

where C is the cost of forwarding one unit of traffic, and $\hat{\tau}_j(r, t)$ is the normalized throughput (on r) in t leaving node j. For simplicity, we assume that the nodes have the same, fixed transmission power, and therefore C is the same for every node in the network, and it is independent from r and t. $\hat{\tau}_j(r, t)$ is computed as the product of the cooperation levels of the intermediate nodes from f_1 up to and including f_j:

$$\hat{t}_j(r, t) = \prod_{k=1}^{j} m_{f_k}(t). \tag{10.4}$$

In this model, the payoff of the destination is 0. In other words, we assume that only the source benefits if the traffic reaches the destination (information push). However, this model can be applied in the reverse case: all the results also hold when only the destination benefits from receiving traffic. An example of this case is a file download (information pull).

The total payoff $u_i(t)$ of node i in time slot t is then computed as

$$u_i(t) = b_i(t) + \sum_{r \in F_i(t)} c_i(r, t), \tag{10.5}$$

where $F_i(t)$ is the set of routes in t where i is an intermediate node.

10.1.2 Strategy space

In every time slot, each node i updates its cooperation level using a strategy function σ_i. In general, i could choose a cooperation level to be used in time slot t, based on the information it obtained in *all* preceding time slots. In order to make the analysis feasible, we assume that i uses only information that it obtained in the *previous* time slot. More specifically, we assume that i chooses its cooperation level $m_i(t)$ in time slot t based on the normalized throughput it experienced in time slot $t - 1$ on the route where it is a source:

$$m_i(t) = \sigma_i(\hat{t}(r, t - 1)), \tag{10.6}$$

where $\hat{t}(r, t - 1)$ represents the normalized throughput enjoyed by node i in time slot $t - 1$ on route r.

The strategy of a node i is then defined by its strategy function σ_i and its initial cooperation level $m_i(0)$.

Note that σ_i takes as input the normalized throughput and not the total payoff received by i in the previous time slot. The rationale is that i should react to the behavior of the rest of the network, which is represented by the normalized throughput.

There is an infinite number of possible strategies as a response to the observed traffic $\hat{t}$. Here we highlight only a few of them for illustrative purposes.

- *Always Defect (AllD):* a node playing this strategy defects in the first time slot and then uses the strategy function $\sigma_i(\hat{t}) = 0$.

- *Always Cooperate (AllC):* a node playing this strategy starts with cooperation and then uses the strategy function $\sigma_i(\hat{t}) = 1$.
- *Tit-For-Tat (TFT):* a node playing this strategy starts with cooperation, and then mimics the behavior of its opponent in the previous time slot. The strategy function that corresponds to the TFT strategy is $\sigma_i(\hat{t}) = \hat{t}$.
- *Suspicious Tit-For-Tat (S-TFT):* a node playing this strategy defects in the first time slot, and then applies the strategy function $\sigma_i(\hat{t}) = \hat{t}$.
- *Anti-Tit-For-Tat (Anti-TFT):* a node playing this strategy does exactly the opposite of what its opponent does. In other words, after cooperating in the first time slot, it applies the strategy function $\sigma_i(\hat{t}) = 1 - \hat{t}$.

If the output of the strategy function is independent of its input, then the strategy is called a *non-reactive strategy* (e.g., AllD or AllC). If the output depends on the input, then the strategy is *reactive* (e.g., TFT or Anti-TFT).

Our model requires that each source be able to observe the throughput in a given time slot on each of its routes. We assume that this is made possible with high enough precision by using some higher level control protocol above the network layer.

10.2 Meta-model

In this section, we introduce a meta-model in order to formalize the properties of the packet forwarding game defined in the previous section. In the meta-model, we focus on the evolution of the cooperation levels of the nodes; all other details of the model defined earlier (e.g., amounts of traffic, forwarding costs, and payoffs) are abstracted away. As previously, we will assume that routes remain unchanged during the lifetime of the network and that each node is the source of only one route.

Let us consider a route r. The payoff received by the source on r depends on the cooperation levels of the intermediate nodes on r. We represent this dependency relationship between the nodes with a directed graph, which we call the *dependency graph*. Each vertex of the dependency graph corresponds to a network node. There is a directed edge from vertex i to vertex j, denoted by the ordered pair (i, j), if there exists a route where i is an intermediate node and j is the source. Intuitively, an edge (i, j) means that the behavior (cooperation level) of i has an effect on j. The concept of dependency graph is illustrated in Figure 10.1.

Now we define the automaton Θ that will model the unfolding of the forwarding game in the meta-model. The automaton is built on the dependency graph. We assign a machine M_i to every vertex i of the dependency graph and interpret the edges of the dependency graph as links that connect the machines assigned to the vertices. Each machine M_i thus has some input and some (possibly 0) output links.

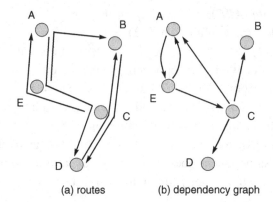

(a) routes   (b) dependency graph

Figure 10.1. Representation of a network: (a) a graph showing five routes and (b) the corresponding dependency graph. Used with permission, from [136], © IEEE, 2006.

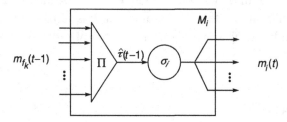

Figure 10.2. Internal structure of machine M_i. The cooperation level m_{f_k} of each relay node at time $t - 1$ on the route for which node i is the source contributes to the normalized throughput $\hat{\tau}$ of node i. The latter determines the cooperation level $m_i(t)$ of that node (on all the routes for which it is a forwarder). Used with permission, from [136], © IEEE, 2006.

The internal structure of the machine is illustrated in Figure 10.2. Please note that from now on, whenever appropriate, we will write τ only with its parameter t, because we have assumed that each node is source of a single route (hence there is no need for another parameter referring to the route). Each machine M_i consists of a multiplication gate $\prod$,[2] followed by a gate that implements the strategy function σ_i of node i. The multiplication gate $\prod$ takes the values on the input links and passes their product to the strategy function gate.[3] Finally, the output of the strategy function gate is passed to each output link of M_i.

---

[2] The multiplication comes from the fact that the experienced normalized throughput for the source (which is the input of the strategy function of the source) is the product of the cooperation levels of the forwarders on its route.

[3] Note that here σ_i takes a single real number as input, instead of a vector of real numbers as we defined earlier, because we assume that each node is source of only one route.

The automaton Θ works in discrete steps. Initially, in step 0, each machine M_i outputs some initial value $m_i(0)$. Then, in step $t > 0$, each machine computes its output $m_i(t)$ by taking the values that appear on its input links in step $t - 1$. The evolution of the values (which, in fact, represent the state of the automaton) on the output links of the machines models the evolution of the cooperation levels of the nodes in the network.

In order to study the interaction of node i with the rest of the network, we extract the gate that implements the strategy function σ_i from the automaton Θ. What remains is the automaton without σ_i, which we denote by Θ_{-i}. Θ_{-i} has an input and an output link; if we connect these to the output and the input, respectively, of σ_i (as illustrated in Figure 10.3), then we get back the original automaton Θ. In other words, the automaton in Figure 10.3 is another representation of the automaton in Figure 10.4, which captures the fact that from the viewpoint of node i, the rest of the network behaves like an automaton: The input of Θ_{-i} is the sequence $\overline{m}_i = m_i(0), m_i(1), \ldots$ of the cooperation levels of i, and its output is the sequence $\overline{\tau}_i = \hat{\tau}_i(0), \hat{\tau}_i(1), \ldots$ of the normalized throughput values for i.

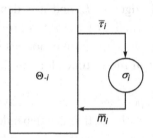

Figure 10.3. Model of interaction between node i and the rest of the network represented by the automaton Θ_{-i}. Used with permission, from [136], © IEEE, 2006.

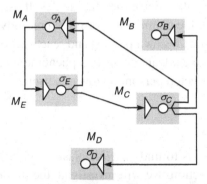

Figure 10.4. Automaton corresponding to the dependency graph of Figure 10.1. Used with permission, from [136], © IEEE, 2006.

By using the system of equations that describe the operation of Θ, we can easily express any element $\hat{t}_i(t)$ of sequence $\overline{\tau}_i$ as some function of the preceding elements $m_i(t-1), m_i(t-2), \ldots, m_i(0)$ of sequence $\overline{m}_i$ and the initial values $m_j(0)$ ($j \neq i$) of the machines within Θ_{-i}. We call such an expression of $\hat{t}_i(t)$ the tth *input/output formula* or the tth *i/o formula* of Θ_{-i}, for short. It is important to note that the i/o formulae of Θ_{-i} can involve any strategy function σ_j where $j \neq i$, but they never involve σ_i. Considering again the automaton in Figure 10.4, and extracting, for instance, σ_A, we can determine the first few i/o formulae of Θ_{-A} as follows:

$$\hat{t}_A(0) = m_C(0) \cdot m_E(0)$$
$$\hat{t}_A(1) = \sigma_C(m_E(0)) \cdot \sigma_E(m_A(0))$$
$$\hat{t}_A(2) = \sigma_C(\sigma_E(m_A(0))) \cdot \sigma_E(m_A(1))$$
$$\hat{t}_A(3) = \sigma_C(\sigma_E(m_A(1))) \cdot \sigma_E(m_A(2))$$

$$\cdots \quad \cdots \quad .$$

A *dependency loop* L of node i is a sequence $(i, v_1), (v_1, v_2), \ldots, (v_{\ell-1}, v_\ell)$, (v_ℓ, i) of edges in the dependency graph. The length of a dependency loop L is defined as the number of edges in L, and it is denoted by $|L|$. The existence of dependency loops is important: if node i has no dependency loops, then the cooperation level chosen by i in a given time slot has no effect on the normalized throughput experienced by i in future time slots. In the example, nodes B and D have no dependency loops.

There exist two types of dependency loops. These types depend on the strategies played by the other nodes in the loop. If L is a dependency loop of i, and all other nodes $j \neq i$ in L play reactive strategies, then L is said to be a *reactive dependency loop* of i. If, on the contrary, there exists at least one node $j \neq i$ in L that plays a non-reactive strategy, then L is called a *non-reactive dependency loop* of i.

Note: a very convenient property of the model captured by Figure 10.3 is that node i does not need to authenticate the other nodes. It assesses the way in which it is served by the rest of the community (or more precisely, by those of the nodes that have influence on node i's own payoff) and reacts accordingly. A drawback is that a selfish node can take advantage of this phenomenon, as it knows that there will not be retaliation directed personally towards itself.

10.3 Analytical results

Our goal, in this section, is to find *possible* Nash equilibria of packet forwarding strategies. In the next section, we will investigate the *probability of fulfillment* of the conditions for possible Nash equilibria in randomly generated scenarios. The existence of a Nash equilibrium based on cooperation would mean that there are cases in which cooperation is "naturally" encouraged, i.e. without using incentive mechanisms. In the following, we use the model and the meta-model that we introduced earlier.

The goal of the nodes is to maximize the payoff that they accumulate over time. However, the end of the game is unpredictable. Thus, we apply the standard technique used in the theory of repeated games. We model the *finite* forwarding game with an unpredictable end as an *infinite* game where future payoffs are *discounted*. The cumulative payoff $\bar{u}_i$ of a node i is computed as the weighted sum of the payoffs $u_i(t)$ that i obtains in each time slot t:

$$\bar{u}_i = \sum_{t=0}^{\infty} [u_i(t) \cdot \delta^t], \tag{10.7}$$

where $0 < \delta < 1$, hence the weights exponentially decrease with t. The *discounting factor* δ represents the degree to which the payoff of each time slot is discounted relative to the previous time slot.

We denote the route originating at node i by r_i and the amount of traffic sent by i on r_i in every time slot by T_i. Recall that F_i denotes the set of routes for which i is an intermediate node. The cardinality of F_i will be denoted by $|F_i|$. For any route $r \in F_i$, we denote the set of intermediate nodes on r upstream from node i (including node i) by $\Phi(r, i)$. Moreover, $\Phi(r)$ denotes the set of all forwarder nodes on route r, and src(r) denotes the source of route r. Finally, the set of nodes that are forwarders on at least one route is denoted by Φ (i.e., $\Phi = \{i \in N : F_i \neq \emptyset\}$).

The two following theorems can easily be derived.

Theorem 10.1 *If a node $i \in \Phi$ has no dependency loops, then its best strategy is AllD.*

Proof Node i wants to maximize its cumulative payoff $\bar{u}_i$ defined in (10.7). In our case, $u_i(t)$ can be written as:

$$u_i(t) = b_i(t) + \sum_{r \in F_i} c_i(r, t)$$

$$= b_i(T_i \cdot \hat{\tau}_i(t)) - \sum_{r \in F_i} T_{\text{src}(r)} \cdot C \cdot \prod_{k \in \Phi(r,i)} m_k(t).$$

Given that i has no dependency loops, $\hat{\tau}_i(t)$ is independent of all the previous cooperation levels $m_i(t')$ $(t' < t)$ of node i. Thus, $\bar{u}_i$ is maximized if $m_i(t') = 0$ for all $t' \geq 0$. $\square$

Theorem 10.2 *If a node $i \in \Phi$ has only non-reactive dependency loops, then its best strategy is AllD.*

Proof The proof is similar to the proof of Theorem 10.1. Since all dependency loops of i are non-reactive, its experienced normalized throughput $\hat{t}_i$ is independent of its own behavior m_i. This implies that its best strategy is full defection. □

From this second theorem, the following corollary follows immediately.

Corollary 10.1 *Every node playing AllD is a Nash equilibrium.*

In other words, if every node j $(j \neq i)$ plays AllD, then the best response of i to this is AllD.

If the conditions of Theorems 10.1 and 10.2 do not hold, then we cannot determine the best strategy of a node i in general, because it very much depends on the particular scenario (dependency graph) in question and the strategies played by the other nodes.

Now, we will show that, under certain conditions, cooperative equilibria do exist in the network. In order to do so, we first prove the following lemma.

Lemma 10.1 *Consider a node $i \in \Phi$ and a route $r \in F_i$. If there exists a dependency loop L of i that contains the edge $(i, \text{src}(r))$ and if all nodes in L (other than i) play the TFT strategy, then the following holds:*

$$\hat{t}_i(t + \lambda) \leq \prod_{k \in \Phi(r,i)} m_k(t), \tag{10.8}$$

where $\lambda = |L| - 1$.

The intuition of the proof is provided by Figure 10.5, which illustrates a dependency loop of length 5 (i.e., $\lambda = 4$). According to Lemma 10.1, if nodes

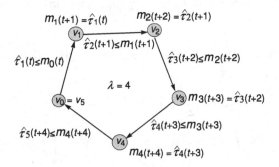

Dependency graph

Figure 10.5. Example to illustrate the propagation of behavior as expressed in Lemma 10.1. The considered node is denoted as v_0. The throughput $\tau_1(t)$ of node v_1 is upper bounded by the move $m_0(t)$ of node v_0. As node v_1 plays TFT, it will choose a cooperation level $m_1(t + 1)$ corresponding to the level of traffic that it has previously enjoyed, namely $\tau_1(t)$. Used with permission, from [136], © IEEE, 2006.

v_1, v_2, v_3, and v_4 play TFT, then the normalized throughput enjoyed by node i in time slot $t + 4$ is upper bounded by its own cooperation level in time slot t. Intuitively, this means that if node i does not cooperate, then this defection "propagates back" to it on the dependency loop. The delay of this effect is given by the length of the dependency loop.

Theorem 10.3 *The best strategy for a node $i \in \Phi$ is full cooperation in each time slot, if the following set of conditions holds:*

(1) for every $r \in F_i$, there exists a dependency loop $L_{i,\mathrm{src}(r)}$ that contains the edge (i, src(r));

(2) for every $r \in F_i$,

$$\frac{b_i'(T_i) \cdot T_i \cdot \delta^{\lambda_{i,\mathrm{src}(r)}}}{|F_i|} > T_{\mathrm{src}(r)} \cdot c \tag{10.9}$$

where $b_i'(T_i)$ is the value of the derivative of $b_i(\tau)$ at $\tau = T_i$, and $\lambda_{i,\mathrm{src}(r)} = |L_{i,\mathrm{src}(r)}| - 1$;[4] and

(3) every node in Φ (other than i) plays the TFT strategy.

The proof of Theorem 10.3 can be found in [136].

We have derived *necessary* conditions for spontaneous cooperation from Theorem 10.1 and 10.2. The fulfillment of the three conditions of Theorem 10.3 is *sufficient* for cooperation to be the best strategy for node i. We now discuss these three conditions one by one. *Condition* (1) requires that node i has a dependency loop with all of the sources for which it forwards packets. *Condition* (2) means that the maximum forwarding cost for node i on every route where i is a forwarder must be smaller than its possible future benefit averaged over the number of routes where i is a forwarder. Finally, *Condition* (3) requires that all forwarding nodes in the network (other than node i) play TFT. This implies that all the dependency loops of node i are reactive.

We note that the reactivity of the dependency loops can be based on other reactive strategies, different from TFT (for example Anti-TFT), but in that case the analysis becomes very complex. The analysis of the case in which every node plays TFT is made possible by the simplicity of the strategy function $\sigma(x) = x$, characteristic of the TFT strategy. If all three conditions of Theorem 10.3 are satisfied, then node i has an incentive to cooperate, otherwise its defective behavior will negatively affect its own payoff. However, as we will show in Section 10.4, *Condition* (1) is a very strong requirement that is virtually never satisfied in randomly generated scenarios.

Both the AllC and TFT strategies result in full cooperation if the conditions of Theorem 10.3 hold. However, node i should not choose AllC, because AllC is a

---

[4] Recall the assumption that b_i is differentiable at T_i.

non-reactive strategy, and this might cause other nodes to change their strategies to AllD, as we will show in Section 10.4. Hence, we can derive the following corollary for cooperative Nash equilibria.

Corollary 10.2 *If the first two conditions of Theorem 10.3 hold for every node in* Φ, *then all nodes playing TFT is a Nash equilibrium.*

In Section 10.4, we will study *Condition* (1) of Theorem 10.3, more specifically, the probability that it is satisfied for all nodes in randomly generated scenarios. Now, we briefly comment on *Condition 2*. As it can be seen, the following factors make *Condition 2* easier to satisfy.

- *Steep benefit functions.* The steeper the benefit function b_i expressing the benefit (and therefore the function u_i expressing the payoff) of node i as a function of the normalized throughput at $\tau = T_i$ is, the larger the value of its derivative is, which, in turn, makes the left side of (10.9) larger.
- *Short dependency loops.* In *Condition* (2), $\lambda_{i,\mathrm{src}(r)} + 1$ is the length of *any* dependency loop of node i that contains the edge $(i, \mathrm{src}(r))$. Clearly, we are interested in the shortest of such loops, because the smaller $\lambda_{i,\mathrm{src}(r)}$ is, the larger the value of $\delta^{\lambda_{i,\mathrm{src}(r)}}$ is, which, in turn, makes the left side of (10.9) larger. It is similarly advantageous if δ is close to 1, which means, in general, that the probability that the game will continue is higher and thus possible future payoffs count more.
- *Small extent of involvement in forwarding.* The left side of (10.9) is increased if the cardinality of F_i is decreased. In other words, if node i is a forwarder on a smaller number of routes, then *Condition* (2) is easier to satisfy for i.

The first two theorems state that if the behavior of node i has no effect on its experienced normalized throughput, then defection is the best choice for i. In addition, Corollary 10.1 says that if every node always defects, then this is a Nash equilibrium. Theorem 10.3 leads to Corollary 10.2, which shows the existence of a cooperative equilibrium (each node playing TFT) under certain conditions.

Figure 10.6 shows a classification of scenarios from the cooperation perspective. In the figure, set D denotes the set of all possible scenarios; indeed, we know from Corollary 10.1 that all nodes playing AllD is a Nash equilibrium in any possible scenario. Set $C2$ contains the scenarios where the conditions of Corollary 10.2 hold. Hence, all nodes playing TFT is a Nash equilibrium in every scenario in $C2$. Finally, set C contains those scenarios where the condition of Theorem 10.1 does not hold for any of the nodes in Φ, or, in other words, where every node in Φ has at least one dependency loop. Determining the Nash equilibria in the scenarios that belong to set $C \setminus C2$ is still an open research problem. In the next section, we will describe simulation results that quantify the "size" of the above sets.

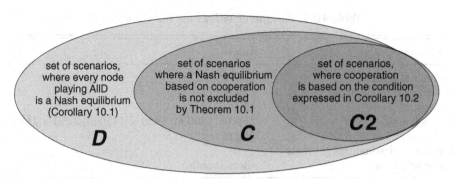

Figure 10.6. Classification of scenarios defined by our analytical results. Used with permission, from [136], © IEEE, 2006.

10.4 Simulation results

A set of simulations will now help us to determine the likelihood that the conditions of the theorems and their corollaries hold. In particular, our goal is to estimate the probability that the first condition of Theorem 10.3 holds for every node in randomly generated scenarios.[5] In addition, we also estimate the probability that the condition of Theorem 10.1 does not hold for any of the nodes in randomly generated scenarios. These probabilities quantify the size of sets $C2$ and C, respectively.

In these simulations, the nodes are placed randomly on a toroid area.[6] Then, for each node, a number of destinations is randomly chosen and a route to these destinations is selected using a shortest path algorithm. If there exist several shortest paths to a given destination, then one of them is randomly chosen. From the routes, we build up the dependency graph of the network. The simulation parameters are summarized in Table 10.1.

Note that we increase the network size and the simulation area in parallel in order to keep the node density at a constant level. All the presented results are the mean values of 1000 simulation runs.

So far, in this chapter, we have considered that a single route originates from each node. As the simulations will show, the proportion of scenarios where all nodes have a dependency loop under this assumption is extremely small. Consequently, the simulations show also the case in which multiple routes originate from such nodes. As will be shown, the likelihood for *Condition* (1) to be fulfilled increases,

---

[5] The second condition of Theorem 10.3 is a numerical one. Whether it is fulfilled or not very much depends on the actual benefit functions and parameter values (e.g., amount of traffic and discounting factor) used. As, by appropriately setting these parameters, the second condition of Theorem 10.3 can always be satisfied, we make the (optimistic) assumption that this condition holds for every node in Φ.

[6] We use this area type to avoid border effects. In a realistic scenario, the toroid area can be considered as an inner part of a large network.

Table 10.1. *Parameter values for the simulation*

Parameter	Value
Number of nodes	100, 150, 200
Distribution of the nodes	random uniform
Area type	torus
Area size	1500 × 1500 m, 1850 × 1850 m, 2150 × 2150 m
Radio range	200 m
Number of destinations per node	1–10
Route selection	shortest path

albeit very progressively. This is important, because all the results presented in this chapter can be extended to this case.

In the first set of simulations, we investigate the probability that the first condition of Theorem 10.3 holds for every node (the size of the set $C2$ in Figure 10.6). Among the 1000 scenarios that we generated randomly, we observed that there was not a single scenario in which the first condition of Theorem 10.3 was satisfied for all nodes. Thus, we conclude that the probability of a Nash equilibrium based on TFT as defined in Corollary 10.2 is very small.

In the second set of simulations, we investigate the proportion of random scenarios, where cooperation of all the nodes is not excluded by Theorem 10.1. Figure 10.7 shows the proportion of scenarios, where each node in Φ has at least one dependency loop (the scenarios in set C in Figure 10.6) as a function of the number of routes originating at each node. We can observe that for an increasing number of routes originating at each node, the proportion of scenarios, where each node has at least one dependency loop, increases as well. Intuitively, as more routes are introduced in the network, more edges are added to the dependency graph. Hence, the probability that a dependency loop exists for each node increases. Furthermore, we can observe that the proportion of scenarios in which each node has at least one dependency loop decreases, as the network size increases. This is due to the following reason: the probability that there exists at least one node for which the condition of Theorem 10.1 holds increases as the number of nodes increases.

Figure 10.7 shows that the proportion of scenarios, where cooperation of all nodes is not excluded by Theorem 10.1 (set C) becomes significant (with respect to set D) only for cases in which each node is a source of a large number of routes. This implies that the necessary condition expressed by Theorem 10.1 is a strong requirement for cooperation in realistic settings (i.e., for a reasonably low number of routes per node).

Now let us consider the case in which the nodes for which Theorem 10.1 holds begin to play AllD. This non-cooperative behavior can lead to an "avalanche effect"

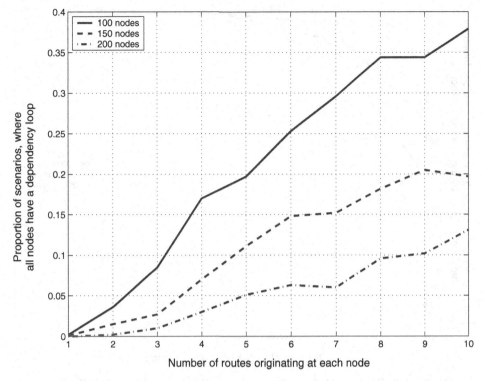

Figure 10.7. Proportion of scenarios, where each node that is a forwarder has at least one dependency loop. From [136], © IEEE, 2006.

if the nodes iteratively optimize their strategies: nodes that defect can cause the defection of other nodes. We examine this avalanche effect in a simulation setting as follows.

Let us assume that each node is a source on one route. First, we identify the nodes in the set of forwarders Φ that have AllD as the best strategy due to Theorem 10.1. We denote the set of these defectors by Z_0. Then, we search for sources that are dependent on the nodes in Z_0. We denote the set of these sources by Z_0^+. As the normalized throughput of the nodes in Z_0^+ is less than or equal to the cooperation level of any of their forwarders (including the nodes in Z_0), their best strategy becomes AllD as well, due to Theorem 10.2. Therefore, we extend the set Z_0 of defectors, and obtain $Z_1 = Z_0 \cup Z_0^+$. We extend the set Z_k of defectors iteratively in this way until no new sources are affected (i.e., $Z_k \cup Z_k^+ = Z_k$). The remaining set $\Phi \setminus Z_k$ of nodes is not affected by the behavior of the nodes in Z_k (and hence the nodes in Z_0); this means that they are potential cooperators. Similarly, we can investigate the avalanche effect when the nodes are sources of several routes. In this case, we take the pessimistic assumption that the defection of a forwarder causes

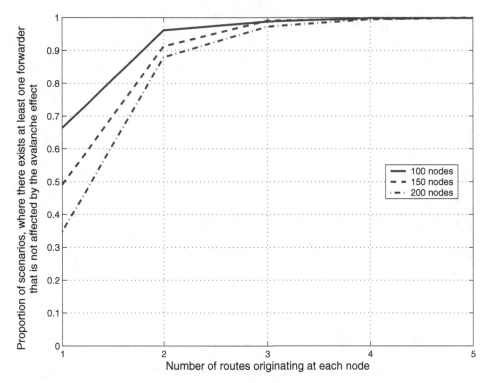

Figure 10.8. Proportion of scenarios, where at least one node is not affected by the defective behavior of the initial nodes. Used with permission, from [136], © IEEE, 2006.

the defection of its sources. Then, we can iterate the search for the nodes that are affected by defection in the same way as above.

In Figure 10.8, we present the proportion of scenarios, where there exists a subset of nodes that are not affected by the defective behavior of the initial AllD players. We can see that this proportion converges rapidly to 1 as the number of routes originating at each node increases. The intuitive explanation is that increasing the number of routes per source (i.e., adding edges to the dependency graph) decreases the probability that Theorem 10.1 holds for a given node. Thus, as the number of routes per source increases, the number of forwarders that begin to play AllD decreases, as well as the number of nodes affected by the avalanche effect.

Additionally, we present in Figure 10.9 the proportion of forwarder nodes that are not affected by the avalanche effect. The results show that if we increase the number of routes originating at each node, the average number of unaffected nodes increases rapidly. For a higher number of routes per node, this increase slows down, but we can observe that the majority of the nodes are not affected by the defective behavior of the initial AllD players.

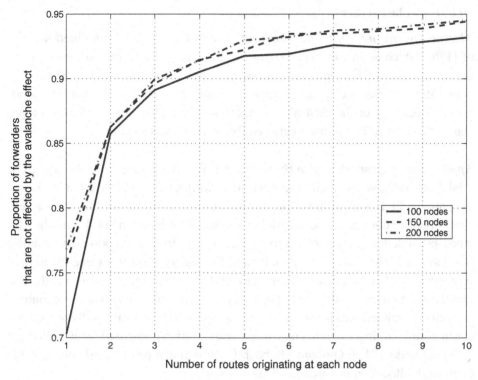

Figure 10.9. Average proportion of forwarder nodes that are not affected by the avalanche effect. Used with permission, from [136], © IEEE, 2006.

10.5 Summary

In this chapter, we presented a game theoretic model to investigate the conditions for cooperation in wireless ad hoc networks, in the absence of incentive mechanisms. Because of the complexity of the problem, we restricted ourselves to a static network scenario. We then derived conditions for cooperation from the topology of the network and the existing communication routes. We introduced the concept of dependency graph, based on which we proved several theorems. As one of the results, we proved that cooperation solely based on the self-interest of the nodes can exist *in theory*. However, the simulation results show that *in practice* the conditions of such cooperation are almost never satisfied. We conclude that there will be, with a very high probability, some nodes that have AllD as their best strategy and that therefore these nodes need an incentive to cooperate. We also showed that the behavior of these defectors affects only a fraction of the nodes in the network; hence, local subsets of cooperating nodes are not excluded.

10.6 To probe further

Abstracting the network The treatment of this chapter is a simplified version of [136], which addresses also analytically the more general case of nodes being the source of multiple routes. An alternative approach is provided by Srinivasan *et al.* [354]; but this second approach is more abstract, in the sense that it does not take the topology of the network into account. The reader interested in this topic can find a detailed comparison between the two approaches in [136].

Application of game theory to the network layer Game theory has been used to model network layer issues in fixed networks. Korilis, Lazar and Orda [231] address the problem of allocating link capacities in routing decisions; in [232], Korilis and Orda suggest a congestion-based pricing scheme. Roughgarden [332] quantifies the worst-possible loss in network performance arising from non-cooperative routing behavior. In [384], Yaiche, Mazumdar and Rosenberg present a game theoretical framework for bandwidth allocation; they study the centralized problem and show that the solution can be distributed in a way that leads to a system-wide optimum.

A closely related area is the one of pricing. For fixed networks, Kelly proposes a scheme for charging and rate control for elastic traffic, particularly well suited for ATM networks [219]. Qiu and Marbach [319] define a price-based approach for bandwidth allocation in wireless ad hoc networks.

Reputation Several authors have studied the application of reputation systems to packet forwarding, see e.g. [69] by Buchegger and Le Boudec. The idea is that nodes observe each other's behavior and retaliate against those exhibiting non-cooperative behavior, for example by dropping the packets originating from them. To be effective, these schemes need to be combined with authentication mechanisms, see Chapter 4.

10.7 Questions

(1) How many dependency loops are there in Figure 10.1?
(2) Assume that in Figure 10.1 node E plays AllD, whereas nodes B, C and D play TFT. What is the best strategy of node A? Answer the same question for node A if node C plays AllD and nodes B, D and E play TFT.
(3) Assume that node A in Figure 10.1 falsely detects that node C was dropping some of its packets. What is the result? (Establish the connection with the previous question.)
(4) Assume that in Figure 10.4 all nodes play TFT. What is the evolution of the machine outputs in the first five rounds? In the first ten rounds? Answer the

same question if node A starts with the cooperation level $m_A(t_0) = 0.5$ (and all the other nodes with cooperation level $m_i(t_0) = 1$).

(5) The defection of some nodes causes the defection of other nodes. Why do we call this the "avalanche effect"? What kind of implications does it have for the stability of cooperation?

(6) Throughout this chapter it is assumed that packet dropping is exclusively due to the selfish behavior of the nodes. Why is this assumption necessary?

(7) What type of games would you use to model the packet forwarding problem if packet dropping were possible due to link errors or congestion? (Hint: see Appendix B.)

11

Wireless operators in a shared spectrum

In the previous two chapters, we have focused on the behavior of selfish nodes and shown how this can be modeled by means of game theory. In this chapter, we will consider the co-existence of several operators in shared spectrum. We will first discuss multi-domain sensor networks and then address the more involved case of cellular operators.

It is important to stress, as we already did at the beginning of Part III, that the cases presented hereafter must be understood as *examples* that capture the interactions of operators in shared spectrum, and not as the scenarios on which the manufacturers and operators will focus in the very near future.

11.1 Multi-domain sensor networks

An important design criterion for sensor networks is the minimization of the sensors' energy consumption. The sensors are often battery powered and it is impractical (and, in some cases, even impossible) to change or recharge the batteries once the sensors have been deployed. It is known that the energy required to transmit a data packet increases (at least) as the square of the distance of the transmission. In practice, this means that, as far as energy consumption is concerned, it is often more advantageous to transmit a packet in several small hops than to transmit it in a single large hop. Hence, if there are numerous sensors near each other, then they could transmit the packets together and thus increase the lifetime of their batteries radically.

In today's research of sensor networks it is generally assumed that all the sensors and base stations belong to a single authority who can control the whole network. In this section, we depart from this common assumption, and consider sensor networks that are deployed at the same physical area, but controlled by different authorities. In such a situation, the sensors belonging to one authority can reduce their transmission energy even further if their packets are forwarded by sensors that belong to another

350

authority; an act that we call *cooperation*. There is a risk, however, that the sensors belonging to the other authority are not willing to help and they drop the foreign packets.

We study this problem in a game theoretic setting. The main question we are interested in is the following: can cooperation emerge spontaneously in multi-domain sensor networks based solely on the self-interest of the nodes (or more precisely of the authorities to which the nodes belong)? To put it in another way: is the objective of increasing the lifetime of the network enough to foster cooperation between co-located sensor networks? The analytical and simulation studies presented in this section show that, in most cases, the answer to these questions is affirmative.

We will first present a simple model, from which we will derive some encouraging analytical results. We will then show how the model can be extended and present related simulation results.

11.1.1 Simplified model

We begin to study the problem of spontaneous cooperation in a simplified model. We assume that there are only two sensor networks that co-exist at the same physical location and that each consists of a single base station and a single sensor. The placement of the base stations and the sensors is illustrated in Figure 11.1.

Now, we describe the operation of this simple system. We assume that time is divided into discrete time slots. In each time slot, each sensor wants to send a single data packet to its own base station that contains its measurement data. We also assume that data packets are equal in size.

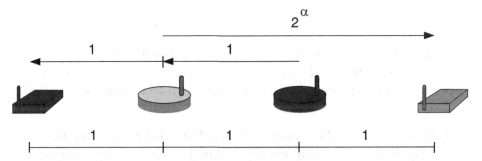

Figure 11.1. Simple sensor network: the distance between a base station and the sensor of the opponent's network is equal to 1. The distance between the two sensors is equal to 1, too. To transmit a packet to its base station, a sensor can either decide for a single hop, which will cost it an amount of energy of 2^α, or for two hops, in which case the amount of energy spent by each of the two sensors will be equal to $1^\alpha = 1$. Used with permission, from [73], with kind permission of Springer Science and Business Media.

The packet can be sent to the base station directly in a single hop, or via the other sensor in two hops. Thus, at the beginning of each time slot, every sensor has to decide the following:

- whether to request the other sensor to help in forwarding its own packet,
- whether to help in forwarding the other sensor's packet if it requests help.

The decision made by the sensor defines its *move* in the time slot. Hence, we have four possible moves, each of which is denoted by a pair of letters as follows:

CC means that the sensor tries to get help from the other sensor and helps if the other sensor requests it;

CD means that the sensor tries to get help from the other sensor, but it refuses to help if the other network requests it;

DC means that the sensor does not ask for help but rather sends its packet directly to its base station; however, it helps if the other sensor requests help;

DD means that the sensor does not ask for help from the other sensor and it refuses to help if the other network requests it.

C stands for cooperation and D stands for defection. The first letter of the move defines how the sensor behaves concerning its own packet, whereas the second letter defines how it behaves when the other sensor's packet is concerned. For instance, making the move CD means that the node tries to obtain cooperation when sending its own packet, but defects when the other sensor asks it to forward its packet.

Each pair of moves has a cost for both sensors, shown in Table 11.1. The costs are related to the energy consumption of the sensors and they are determined as follows:

- asking the other sensor to forward the packet has a unit cost, because this only requires sending the packet to a unit distance;
- forwarding the other sensor's packet also has a unit cost for similar reasons;

Table 11.1. *Costs and successes in the simple network (cost of row player, cost of column player; success of row player, success of column player)*

	CC	CD	DC	DD
CC	2, 2; 1, 1	2, 1; 0, 1	1, $1+2^\alpha$; 1, 1	1, 2^α; 0, 1
CD	1, 2; 1, 0	1, 1; 0, 0	1, $1+2^\alpha$; 1, 1	1, 2^α; 0, 1
DC	$1+2^\alpha$, 1; 1, 1	$1+2^\alpha$, 1; 1, 1	2^α, 2^α; 1, 1	2^α, 2^α; 1, 1
DD	2^α, 1; 1, 0	2^α, 1; 1, 0	2^α, 2^α; 1, 1	2^α, 2^α; 1, 1

- sending the packet directly to the base station has a cost of 2^{α}, where α is the path loss exponent (with usual values between 2 and 5), because this requires sending the packet to a distance of two units;
- dropping a packet has no cost.

In reality, the cost of communication not only depends on the distance, but there are also fixed costs associated with the reception and the transmission of packets. In this simplified model, we neglect these fixed costs.

The cells of Table 11.1 contain not only the costs for the two sensors, but also indicators of success where "1" means that the packet reached the base station (success) and "0" means that it did not (failure). As an example, let us consider the pair of moves $CC - CD$. In this case, the first sensor tries to send its packet via the other sensor that drops it, whereas the other sensor's packet will be sent via the first sensor to the base station successfully. Hence, the cost of the first sensor is 2 (1 for asking for forwarding its own packet and 1 for forwarding the other's packet), and the cost of the other sensor is 1 (the cost for asking for forwarding). Moreover, the first sensor records a failure and the other records a success.

We assume that the sensors record the results (success or failure) of the last few time slots in a buffer that we call *history*; this history is represented as a binary vector of a fixed length. We assume that each sensor's next move is a function of its history. We call this function the *strategy* of the sensor.

Here we make an important restriction on the strategy space. We assume that each sensor wants to keep the weight of its history (i.e., the number of successful slots in the recent past) above a threshold, which we call the *reception threshold*. Intuitively, this means that we do not want to allow too many unsuccessful slots in the history, because that would mean that the base station does not receive measurement data with a high enough rate (this rate being characteristic of the application). Therefore, when the weight of the history approaches the reception threshold, the sensor is not allowed to make risky $C*$ moves (i.e., CC and CD), but it is required to send its data directly to the base station (i.e., to make a $D*$ move). This situation is called the *constrained state*. Using strategies that suggest $D*$ moves in the constrained state guarantees that the reception threshold is never violated.

Note that a longer history with a lower reception threshold results in a system with more freedom, whereas a shorter history with a higher threshold results in a much stricter system.

Each sensor has some initial battery level B. In each time slot, the battery levels of the sensors decrease. The amount of this decrease depends on the pair of moves made in the time slot and their associated costs. When a sensor runs out of its battery energy, it dies. The other sensor can continue to send data to its base station if it still has some battery energy.

Note that the above mentioned concepts describe together an extensive game, where the players are the sensors, the possible moves (made simultaneously by both players) in each round (except for the constrained states) are CC, CD, DC, DD, the information sets are defined by the content of the histories, and the set of strategies are the functions that assign a move to every possible history with the restriction that only $D*$ moves are assigned to a history that represents a constrained state. The game ends when the batteries of both sensors run out. The payoff for a player is its lifetime, which is represented by the number of rounds it survived. Lifetime is a good payoff function because the authorities want to run their network as long as possible while successfully transmitting a reasonable number of packets.

Once the game is defined, we can look for Nash equilibria with the highest possible lifetime. It is quite reasonable to choose one of these Nash equilibria as an operating point in real systems. If there is more than one Nash equilibrium, the equilibrium with the highest lifetimes is chosen.

In order to make the analysis feasible, we further restrict the strategy space. Let us consider first the *two-step strategies*. These strategies suggest a fixed move if the player is not in a constrained state (independently of the actual weight of the history), and another fixed move if the player is in a constrained state. A two-step strategy is denoted by m/m', where m is the move chosen in an unconstrained state and m' is the move chosen in a constrained state. For instance, the strategy CC/DD selects CC in an unconstrained state and DD in a constrained state. Therefore, we have eight two-step strategies, because in a constrained state only $D*$ moves are possible.

By performing an exhaustive search on this strategy space (there are $8 \times 8 = 64$ pairs of strategies to consider), it is possible to identify Nash equilibria. It appears that CC/DD and CD/DD dominate the other strategies. The CC/DD strategy is a cooperative strategy, while the CD/DD is an uncooperative one. By eliminating the dominated strategies, we get a reduced game. The lifetimes of the sensors in this reduced game are shown in Table 11.2, where ρ denotes the reception threshold and B denotes the initial battery level. Values ϵ_1 and ϵ_2 come from transient states such as starting and ending the game. Both are very small positive numbers in most practical cases (that is for α between 2 and 5, and for reasonably large B).

There are two Nash equilibria: $(CC/DD, CC/DD)$ and $(CD/DD, CD/DD)$.[1] The first one results in full cooperation, whereas the second results in full defection. However, if $\rho > 1/3$ (and $\alpha \geq 2$, which is a fundamental condition in this model), then the cooperative equilibrium results in a longer lifetime for both players.

---

[1] Under very special circumstances, both ϵ_1 and ϵ_2 can be zero, in which case there may be only one Nash equilibrium (the cooperative one).

Table 11.2. *Best lifetimes with two-step strategies (lifetime for row player; lifetime for column player), B initial battery, ρ reception threshold, α path loss exponent, $\epsilon_{1,2}$ payoff from transient states*

	CC/DD	CD/DD
CC/DD	$\frac{B}{2}$; $\frac{B}{2}$	$\frac{B}{\rho 2^{\alpha}+(1-\rho)}$; $\frac{B}{\rho 2^{\alpha}+(1-\rho)}+\epsilon_1$
CD/DD	$\frac{B}{\rho 2^{\alpha}+(1-\rho)}+\epsilon_1$; $\frac{B}{\rho 2^{\alpha}+(1-\rho)}$	$\frac{B}{\rho 2^{\alpha}+(1-\rho)}+\epsilon_2$; $\frac{B}{\rho 2^{\alpha}+(1-\rho)}+\epsilon_2$

A more interesting class of strategies are the *weight aware strategies*. These strategies choose the next move as a function of the weight of the history. Thus, a weight aware strategy can be represented as $m_1/m_2/\ldots/m_k$, where m_1 is the move that is chosen when the weight of the history is maximal, and m_k is chosen when the weight of the history is just above the reception threshold. Here k is a parameter whose value depends on the history size and the value of the reception threshold. This class contains more complex and more reactive strategies.

After running 20 different exhaustive simulations, with different parameter sets, it appeared that the strategy that achieves the best Nash equilibrium is always the same: $(CD/CD/\ldots/CD/CC/DD)$. We call it the *smart strategy*. The smart strategy tries to ask for help in the first steps (the CD moves, which are inexpensive moves), but provides help (the CC move before the DD move) only in a state when the reception threshold is nearly violated in the hope that its nice behavior will be reciprocated. In other words, the smart strategy first tries to exploit the other. If this is successful, then it will never cooperate. But if the other strategy is not exploitable, then it will change to a cooperative behavior. In the long run, the strategy keeps the actual weight of the history near to the reception threshold, which means that it cooperates only as much as necessary. This turns out to be a very effective behavior to save battery energy and leads to a rational cooperation.

In summary, we can see that in the simplified model, which contains two base stations and two sensor nodes, cooperative Nash equilibria exist based on smart strategies that try to optimize the amount of cooperation. In the next section we will investigate if the same is true in a more general model.

11.1.2 Generalized model

After the encouraging results of the simplified model in Section 11.1.1, we will now examine much larger and more complex systems. We will make use of a simulator that corresponds to the model described in the first part of Section 11.1.1 with some extensions.

The generalized model considers the co-existence of two network operators, each of them possessing one base station. Each network operator also possesses the same (high) number of sensors, randomly placed on the playground (with uniform distribution). The possible moves are the same as those in the simplified model, but in the generalized model each pair of moves has a cost that depends not only on the distance of the transmissions and the path loss exponent α, but also on some fixed costs associated with the sending and receiving of packets. The fixed cost of sending and the fixed cost of receiving are constant values that represent the energy consumption necessary to connect to the communication channel and to process the packets.

The principle of routing in the model is finding the minimum energy path towards the base station [386]. This means that every node has to forward on the path which has the minimum energy cost among all the possible paths. Every node maintains three paths: one in its own network (for the defective moves) and two in the global network (i.e., where all the nodes are possible forwarders). The global network paths are maintained in order to allow cooperative moves. The two distinct cooperative paths are towards the two base stations. These three paths can be the same depending on the placement.

Both networks have a threshold value (*success threshold*) that defines the minimum number of packets that the base station has to receive in each time slot, and the time slot is considered successful only if at least that number of packets reach the base station. The lifetime of a network is the total number of time slots that elapse until the weight of the history becomes zero. The objective of the game is to reach the best possible lifetime under the constraint that the reception threshold of the history has to be respected.

Example. In an office building it is usual to deploy temperature and movement sensors. The temperature sensors measure the actual temperature and forward it to the air conditioning system. The movement sensors gather information about which zone is visited or abandoned and forward it to the security system. The two systems ask for information regularly (once every second) but it is not crucial to get the information in every time slot. The temperature can be controlled and the security can be guaranteed with enough accuracy if some of the measurements are successful (let us say three out of the last five). The systems can work properly if they get enough measurement data in a time slot. Although the sensors are usually deployed redundantly, a given proportion can execute the task (say 80% of the sensors). If the given proportion of data arrives at the control systems, then the missing information can be deduced.

Two main type of scenarios are investigated. In one of them (common base scenario), there is a single common base station that collects the information from all of the nodes (independently from the authority they belong to). In the other

Table 11.3. *Parameters for the simulations (the parameters are motivated in the* example *and in [345])*

Parameter	Value
Number of sensors per domain	10–20–40 (20)
Distribution of the sensors	uniformly random
Area size	100 × 100 m
Position of the base (common base)	[50, 50]
Position of the bases (separate bases)	[45, 50] and [55, 50]
Initial battery	10 million units
Reception fix cost	3000 units
Sending fix cost	2000 units
Success threshold	0.7–0.8–0.9 (0.8)
Reception threshold	0.6
History length	5
Path loss exponent (α)	2–3–4 (3)

(separate base scenario), both networks have their own base stations. In the common base model, the base station is placed in the middle of the playground, whereas in the separate bases model, the base stations are the same distance from the theoretical middle of the playground.

One hundred simulation runs were performed for each parameter setting with different topology. The concrete values for the simulations are shown in Table 11.3. The values in parenthesis are the defaults. For each run we made an exhaustive search in the strategy space to find the best strategy pairs (i.e., those that form a Nash equilibrium and generate the highest lifetimes).

In the extended model, it is not so easy to determine which equilibrium is a cooperative equilibrium. Two strategies can act in a cooperative way in the case of one topology and in an uncooperative way in the case of another topology. In other words, the topology and the strategies both can influence the cooperation. Therefore, we establish the following classification, based on the observed behavior at Nash equilibrium.[2]

- *Class 0:* none of the players forwards a packet for the other (no cooperation).
- *Class 1:* one of the players forwards some packets for the other (semi-cooperation).
- *Class 2:* both players forward some packets for the other (full cooperation).

The simulation results are shown in Figures 11.2, 11.3, and 11.4. In each figure, the left-hand side chart shows the results of the common base scenario, and the

---

[2] If the game has more than one Nash equilibrium, then the most cooperative of these equilibria is considered to determine the class.

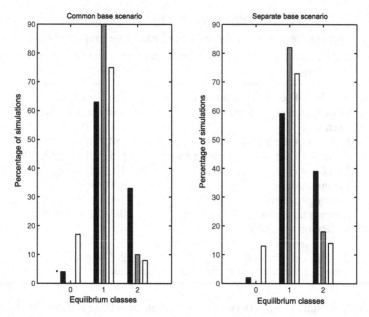

Figure 11.2. Distribution of equilibrium classes (number of nodes per domain = 10 (black), 20 (gray), 40 (white)). On the x-axis, we show the equilibrium classes (0, 1, 2) and, on the y-axis, the percentage of simulations where the best Nash equilibria fell in a given equilibrium class. The left-hand side chart shows the results of the common base scenario, and the right-hand side chart shows the results of the separate base scenario. Used with permission, from [73], with kind permission of Springer Science and Business Media.

right-hand side chart shows the results of the separate base scenario. On the x-axis, we show the equilibrium classes (0, 1, 2) and, on the y-axis, the percentage of simulations where the best Nash equilibria fell in a given equilibrium class.

Figure 11.2 shows how the distribution of the different equilibrium classes depends on the number of nodes. We can see that in most cases the best Nash equilibria result in some kind of cooperation, although semi-cooperation has a higher probability than full cooperation.

Figure 11.3 shows how the distribution of the different equilibrium classes depends on the path loss exponent α. If α is high, then full cooperation is the best choice, because it takes more battery energy to send to a far sensor. If full cooperation occurs, then the average sending distance is smaller, which is very advantageous when the path loss exponent is large.

Figure 11.4 shows how the distribution of the different equilibrium classes depends on the success threshold. We can see that the success threshold does not have much influence on the distribution. If the success threshold is higher, than a little more fully cooperative Nash equilibria occur, but the success threshold seems to be a less relevant parameter than the path loss exponent or the number of nodes.

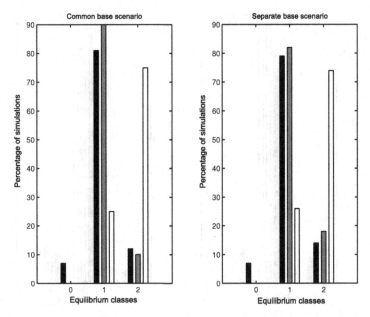

Figure 11.3. Distribution of equilibrium classes ($\alpha = 2$ (black), 3 (gray), 4 (white)). On the x-axis, we show the equilibrium classes (0, 1, 2) and, on the y-axis, the percentage of simulations where the best Nash equilibria fell in a given equilibrium class. The left-hand side chart shows the results of the common base scenario, and the right-hand side chart shows the results of the separate base scenario. Used with permission, from [73], with kind permission of Springer Science and Business Media.

As we have seen above, when co-located sensor networks are allowed to collaborate in the packet forwarding effort, some form of cooperation can emerge spontaneously, by which we mean that in the best Nash equilibria, at least one of the networks forwards some packets on behalf of the other network. It is clear that this cooperative behavior is more advantageous (meaning results in a longer lifetime) for the cooperating network than a defective behavior, given the strategy of the other network, because a Nash equilibrium consists of best response strategies. In order to quantify this advantage, the following experiment was performed. For each simulation run, the following two values were determined: (i) the networks' lifetimes when both networks ignore each other and use only their own nodes for forwarding, and (ii) the networks' lifetimes in the best Nash equilibrium when the networks are allowed to collaborate. In both (i) and (ii), the smaller lifetime value (i.e., the lifetime of the network that is shorter) was taken, and the ratio of the values obtained was computed. Finally, the ratio values over the 100 simulation runs (for each parameter setting) were averaged. We can interpret the result of this computation as the average gain in lifetime when the networks are allowed

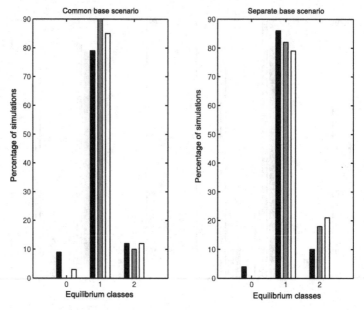

Figure 11.4. Distribution of equilibrium classes (success threshold = 0.7 (black), 0.8 (gray), 0.9 (white)). On the x-axis, we show the equilibrium classes (0, 1, 2) and, on the y-axis, the percentage of simulations where the best Nash equilibria fell in a given equilibrium class. The left-hand side chart shows the results of the common base scenario, and the right-hand side chart shows the results of the separate base scenario. Used with permission, from [73], with kind permission of Springer Science and Business Media.

Table 11.4. *Average gain in lifetime in the common base scenario and in the separate base scenario.*

Non-default parameter	Separate base scenario	Common base scenario
–	6.5%	6.1%
$n = 10$	15.5%	15.6%
$n = 40$	1.5%	0.6%
$\rho = 0.7$	4.4%	3.2%
$\rho = 0.9$	8.7%	7.8%
$\alpha = 2$	1.9%	2.2%
$\alpha = 4$	34.7%	31.0%

to collaborate compared to the case when they operate independently from each other.

The results are shown in Table 11.4. Each row of the table belongs to a particular parameter setting, where all but one of the parameters have the default values shown in Table 11.3, and the first cell of the row shows the non-default parameter value. The second and the third columns of the table contain the average gain in lifetime in the common base and in the separate base scenarios, respectively. As we can see, the average gain in lifetime can be as high as 34% in the common base scenario and 31% in the separate base scenario when $\alpha = 4$.

In the case that we have described in this section, each operator (player) could make discrete moves (to forward or not to forward; to request collaboration or not). In the next section, the space of possible moves will be continuous.

11.2 Border games in cellular operators

Today's cellular networks operate on separate frequency bands to avoid interference between them. The operators of these networks obtain an exclusive right to use a given frequency band in their respective country. However, the division based on frequency bands does not apply across national borders. The operators have to resolve their conflicts across the borders themselves. One of the issues is when mobile users of one operator attach to the network of the operator of the other country while still being in their own country. This problem is referred to as *accidental roaming* [198, 252]. There exist many examples of cities residing close to a national border such as Geneva, Basel or Aachen in Europe; San Diego and Detroit in the USA; or Hongkong and Singapore in Asia. Often, the operators make mutual agreements to resolve these problems, but these agreements are difficult to enforce, because they require the mutual cooperation of the operators.

In this section, we consider the problem of strategic behavior of operators on the border of their cellular networks. We consider 3G cellular networks, such as the *Universal Mobile Telecommunication System (UMTS)*, that are based on the *Code Division Multiple Access (CDMA)* technology [170, 321, 340].[3] Note, however, that the problem we highlight in this section applies to any CDMA network. In these networks, the base stations emit pilot signals to help users to assess the available channel quality and to attach to the base station with the best offered quality. According to the current definition in the UMTS standard, the pilot power for the base stations is determined at the network dimensioning phase and remains fixed afterward. However, as the number of users changes, the operators may adjust the network parameters. This *slow adaptation* of the pilot signal power is part of the network re-dimensioning process and hence it exists on a large time scale. On the other hand, the technology enables the base stations to quickly adapt their pilot signals to the actual usage. This *fast adaptation* technique is commonly referred to as *cell breathing* [170, 321, 340].

---

[3] In contrast with Chapter 9 in which we provided a thorough introduction to CSMA/CA and IEEE 802.11, we refrain from providing here an equivalent description of CDMA and UMTS, because this information can be found in the mentioned textbooks.

In this section, we assume that the operators want to adjust the power of the pilot signal of their base stations to attract more users. Several methods (e.g., cell breathing [321, 340]) have been proposed to implement fast adaptation in CDMA networks. In this section, however, we focus on the slow adaptation problem. We study how the network operators can fine-tune their pilot power in the presence of other operators given a certain user distribution. We investigate whether this situation leads to a game and we study the properties of the equilibria of power control strategies.

11.2.1 Model

We consider a scenario with two cellular network operators A and B. We assume that their networks are separated by a national *border*. The operators operate their network based on the principles of the CDMA method. We assume that the two operators acquired the *same* frequency band for their networks in their respective country. This means that their networks interfere along the border. We assume that each operator i controls a set of *base stations* B_i, where $i \in \{A, B\}$. We refer to the set of all base stations as $B = \bigcup_i B_i$. We also assume a set of *users* $\mathcal{M}$ equipped with *wireless devices* who access the communication network. For the sake of convenience, we assimilate the operators with their base stations and the users with their devices. In order to get an insight, we study the case in which each operator has one base station and we refer to the base stations by the letters of their operators (i.e., base station A and B). This single-cell model is often considered in the literature [200, 260]. The network scenario is shown in Figure 11.5.

We assume that the radios of the base stations and the mobile devices are compatible, meaning that any user is able to access the network via any of the base stations. We further assume that the antennas of the base stations and wireless devices are omnidirectional. Note that the results are still valid if the operators use directional antennas that point towards the national border.

We assume that the users are not associated with any of the operators (i.e., they are roaming users) and thus they attach to the base station with the best signal quality.

In CDMA networks, power control is used to mitigate the near–far effect [321], to optimize the transmission power of the devices, and to reduce interference. We focus on the *downlink (or forward link) power control of the pilot signals* emitted by the base stations. The pilot signal helps the wireless devices to perform the following tasks:

- detection of the available base stations,
- synchronization with them, and
- estimation of the channel quality and handover decision based on this estimation.

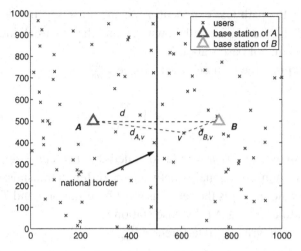

Figure 11.5. Network scenario with two base stations. Used with permission, from [137], © IEEE, 2007.

In particular, we look at how the network operators can determine the pilot signal power that will potentially attract the highest number of users.

CDMA We will now briefly present the physical model of CDMA. As mentioned earlier, the pilot signal is used to attract users. If several users attach to a given base station, their transmissions are performed on different *channels*. In CDMA-based cellular networks, unlike GSM networks, channels are not separated in different frequencies, but use different codes. Hence each transmission uses the same frequency band. In theory, the codes from one base station are orthogonal, meaning that the transmissions to different receivers do not interfere with each other. In practice, however, there exists some interference between concurrent transmissions from a given base station because of multipath propagation. This interference is called the *own-cell interference*. In addition, there is an interference caused by the transmissions of other base stations, called the *other-cell interference*.

Let us consider the scenario shown in Figure 11.5. According to the physical model of signal propagation in a CDMA system [170], we can write the *signal-to-interference-plus-noise ratio (SINR)* of the pilot signal of base station $i \in \{A, B\}$ to user $v \in \mathcal{M}$ as:

$$SINR_{iv}^{\text{pilot}} = \cdot \frac{G_{\text{p}}^{\text{pilot}} \cdot P_i \cdot d_{iv}^{-\alpha}}{N_0 \cdot W + I_{\text{own}}^{\text{pilot}} + I_{\text{other}}^{\text{pilot}}}, \tag{11.1}$$

where $G_{\text{p}}^{\text{pilot}}$ is the *processing gain* for the pilot signal, P_i is the power of the transmitted pilot signal of base station i, d_{iv} is the distance between base station i

and user v,[4] α is the path loss exponent, N_0 is the noise spectral density, W is the bandwidth, and $I_{\text{own}}^{\text{pilot}}$ and $I_{\text{other}}^{\text{pilot}}$ are the own-cell and the other-cell interferences that affect the pilot signal of base station i.

Let us first express the own-cell interference $I_{\text{own}}^{\text{pilot}}$:

$$I_{\text{own}}^{\text{pilot}} = \zeta \cdot d_{iv}^{-\alpha} \left(\sum_{w \in \mathcal{M}_i} T_{iw} \right), \tag{11.2}$$

where ζ is the *orthogonality factor* (also called the *own-cell interference factor*) that expresses the non-orthogonality between the different transmissions from base station i. Furthermore, $\mathcal{M}_i$ is the set of users at base station i and T_{iw} is the traffic power assigned to user $w \in \mathcal{M}_i$ by base station i.

Similarly, we can write the interference $I_{\text{other}}^{\text{pilot}}$:

$$I_{\text{other}}^{\text{pilot}} = \eta \cdot \sum_{j \neq i} d_{jv}^{-\alpha} \left(P_j + \sum_{w \in \mathcal{M}_j} T_{jw} \right), \tag{11.3}$$

where η is the *other-to-own-cell interference factor*, and d_{jv} is the distance between base station j and user v. Furthermore P_j is the pilot signal power of base station j, whereas $\mathcal{M}_j$ is the set of users at base station j and T_{jw} is the traffic power assigned to user $w \in \mathcal{M}_j$ by base station j.

Similarly to (11.1), we can express the SINR for the traffic signal T_{iv}:

$$SINR_{iv}^{\text{tr}} = \frac{G_{\text{p}}^{\text{tr}} \cdot T_{iv} \cdot d_{iv}^{-\alpha}}{N_0 \cdot W + I_{\text{own}}^{\text{tr}} + I_{\text{other}}^{\text{tr}}}, \tag{11.4}$$

where G_{p}^{tr} is the *processing gain* for the traffic signal.

Let us write the own-cell interference $I_{\text{own}}^{\text{tr}}$ for the traffic signal as

$$I_{\text{own}}^{\text{tr}} = \zeta \cdot d_{iv}^{-\alpha} \left(P_i + \sum_{w \neq v, w \in \mathcal{M}_i} T_{iw} \right), \tag{11.5}$$

and the interference from other base stations j as:

$$I_{\text{other}}^{\text{tr}} = I_{\text{other}}^{\text{pilot}} = \eta \cdot \sum_{j \neq i} d_{jv}^{-\alpha} \left(P_j + \sum_{w \in \mathcal{M}_j} T_{jw} \right). \tag{11.6}$$

---

[4] The rigorous reader has probably noticed that there is an abuse of notation (there seems to be an inconsistency of physical units) here. The proper interpretation of this writing is the following: d_{iv} should be understood as the ratio between the actual distance between i and v and a reference distance d_0.

Table 11.5. *UMTS parameters*

traffic type	required SINR	processing gain	required CIR
pilot	≈ -6 dB	14.3 dB	-20 dB
audio, 12.2 kbps	5 dB	25 dB	-20 dB
video, 144 kbps	1.5 dB	14.3 dB	-12.8 dB
data, 384 kbps	1 dB	10 dB	-9 dB

Furthermore, we can express the *carrier-to-interference ratio (CIR)* as a function of SINR:

$$CIR_{iu}^{\text{pilot}} = \frac{SINR_{iv}^{\text{pilot}}}{G_p^{\text{pilot}}}. \tag{11.7}$$

Similarly, we can write the CIR of the traffic signal:

$$CIR_{iv}^{\text{tr}} = \frac{SINR_{iv}^{\text{tr}}}{G_p^{\text{tr}}}, \tag{11.8}$$

where G_p^{tr} is the *processing gain* for the traffic signal from base station i to user v.

UMTS As mentioned, UMTS networks make use of CDMA. In UMTS systems, the processing gain for the pilot signal is $G_p^{\text{pilot}} = 256 \approx 14.3$ dB. The processing gain of the traffic signal G_p^{tr} depends on the bitrate of the application running on the user device. In this section, we refer to different types of communication as the *traffic type*, namely audio (12.2 kbps), video (144 kbps) and data (384 kbps) flows.[5] Accordingly, we distinguish different requirements for different traffic types typical of UMTS (see e.g. [170], Table 8.2.1). We summarize these parameters in Table 11.5. These are empirical values, primarily derived from field tests.

In wireless networks, the regulation authorities impose a transmission power limit to the devices. In UMTS networks, the base stations must emit their signal below 43 dBm $= 20$ W [170]. This limit is called the *downlink power budget*. In addition, this power budget must be split between the control channel signals, such as the pilot signal, and the traffic channel transmissions. The actual utilization of the power budget is called the *load* of the base station. As the load increases, the *bit-error-rate (BER)* at the user devices increases exponentially [170]. Hence, the base station load is typically kept such that the BER does not exceed a certain threshold, for example 10^{-3}. We assume here that the base station load is kept below 10 W.

In order to determine the average usage of the two networks, a numerical simulator was developed in MATLAB. We summarize the parameters of the simulation

---

[5] For simplicity, we consider only constant bitrate traffic.

Table 11.6. *Simulation parameters*

Parameter	Value
simulation area size	1 km^2
base station positions	(250 m, 500 m) and
	(750 m, 500 m)
default distance between base stations, d	500 m
user distribution	random uniform
number of simulations	500
default path loss exponent, α	4
base station max power	43 dBm = 20 W
base station max load	40 dBm = 10 W
base station standard power, P^s	33 dBm = 2 W
base station min power	20 dBm = 0.1 W
power control step size, P_{step}	0.1 W
orthogonality factor, ζ	0.4
other-to-own-cell interference factor, η	0.4
user traffic types	audio (12.2 kbps)
	video (144 kbps)
	data (384 kbps)
required CIR (audio, video, data)	−20 dB, −12.8 dB, −9 dB
expected incomes (θ_{audio}, θ_{video}, θ_{data})	10, 20, 50 CHF/month

in Table 11.6, which is populated notably with values typical of UMTS networks ([170]).

In each simulation run, the users are uniformly distributed. The number of users that attach to each of the base stations is estimated, based on the physical model developed in this section (i.e., using Equations (11.1)–(11.8) and the requirements shown in Table 11.5). This experiment is repeated several times for each power setting thus providing the average number of users at each base station.

We model competitive power control using game theory. We define a two-player non-cooperative *power control game G* with the operators as *players*. In this game, the *strategies* of the operators determine the pilot transmission power of their base stations. Formally, we can write the strategy of operator i as the pilot signal power value of its base station:

$$s_i = P_i, \tag{11.9}$$

where $0\,W < P_i < 10\,W$ is the pilot signal power of base station i. If they follow the UMTS standard, the base stations transmit their pilot signal with approximately $33\,dBm = 2\,W$. We denote this standard pilot power by P^s. We call the set of strategies of all players a *strategy profile* $s = \{s_1, s_2\}$.[6] In our game, the players have the same strategy set S.

The operators define their strategies in order to maximize their *expected payoff* u_i:

$$u_i = \sum_{v \in \mathcal{M}_i} \theta_v, \qquad (11.10)$$

where θ_v is the *expected income* obtained by serving user v of a certain traffic type. Suppose that each user has the same traffic type, for example audio. Then the expected payoff obtained at base station i is:

$$u_i = |\mathcal{M}_i| \cdot \theta_{\text{audio}}. \qquad (11.11)$$

We further assume that the income per user increases according to the data rate of the given service, thus $\theta_{\text{audio}} < \theta_{\text{video}} < \theta_{\text{data}}$.[7] The expected income is obtained by performing several simulation runs with various pilot power settings as described in the previous section. This results in an *expected payoff matrix* for the two players. We express the payoffs of the players in *Swiss Francs (CHF)* to emphasize the monetary advantage.[8]

The results are presented using a symmetric scenario of the base stations and assuming that the users are uniformly distributed in the simulation area. Note that the results qualitatively hold for any base station placement and any user distribution. Naturally, in these cases, the Nash equilibrium strategies and payoffs will be asymmetric.

11.2.2 Power control game

In this section, we study the behavior of the operators in a single-stage game. We first assume that one of the operators does not play and show that the other operator has an incentive to be strategic.[9] Second, we consider the case in which both operators have the possibility to adjust their pilot power and show that they are better off by doing so.

---

[6] Note that one can easily extend the definitions in the power control game to several base stations and operators.

[7] Note that the income is defined by the total amount of downloaded data, which can vary according to the length of communication sessions. If we change these income values, the results only change quantitatively, but not qualitatively.

[8] Please note that, in spite of this monetary expression, the payoff refers to a technical unit of networking (coverage) and is therefore compliant with the definition of selfish behavior that we provided in Chapter 3.

[9] Because of symmetry, we only show the results for player A.

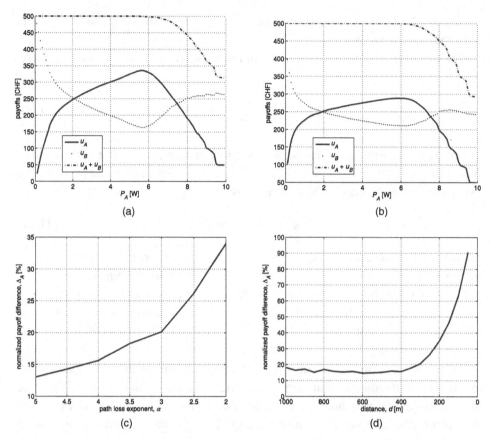

Figure 11.6. Payoffs of the players as a function of the pilot power of player A: (a) for $\alpha = 2$ and (b) for $\alpha = 4$. Normalized payoff difference Δ_A as a function of (c) the path loss exponent α and (d) the distance d between the two base stations. Used with permission, from [137], © IEEE, 2007.

Only player A is strategic

First, we consider the case where only operator A is strategic and adjusts the pilot power of its base station to attract more users, whereas operator B operates its base station according to the standard pilot power of $P^{\mathrm{s}} = 2\,\mathrm{W}$. To quantify the advantage of the strategic player, we define the concept of *normalized payoff difference* Δ_i.

Definition 11.1 The *normalized payoff difference* Δ_i is the normalized difference between the maximum payoff of player i and its payoff using the standard power P^{s} assuming that the other player j uses P^{s}:

$$\Delta_i = \frac{\max_{s_i}\left(u_i(s_i, P^{\mathrm{s}})\right) - u_i(P^{\mathrm{s}}, P^{\mathrm{s}})}{u_i(P^{\mathrm{s}}, P^{\mathrm{s}})}. \tag{11.12}$$

Suppose that there are on average 10 users of the data traffic type in the simulation area. We show the payoffs of players A and B as a function of the pilot signal power P_A as well as the sum of their payoffs in Figure 11.6. Figure 11.6a shows these payoffs for $\alpha = 2$, whereas Figure 11.6b presents the same results for $\alpha = 4$. We observe that in both cases the operators are able to serve all users in the area using certain power values. If all users are served, then the game is a zero-sum game: if player A adjusts its pilot power and obtains an increase Δ_A, the payoff of the non-strategic player B decreases by the same value Δ_A. Furthermore, the payoff function of operator A has a single maximum point. It is interesting to observe that the maximum payoff point requires a higher pilot power than $P^s = 2\,\text{W}$. Hence, we conclude that operator A has an incentive to adjust its pilot signal. Note that the same qualitative results can be obtained for different user traffic types. Figures 11.6a and b also show that, for high values of P_A, the payoff of A declines, because a high value of P_A reduces the capacity of base station A.

The significant difference between Figure 11.6a and Figure 11.6b shows that the value of the normalized payoff difference Δ_A depends on the parameter α. This dependency is illustrated in Figure 11.6c. One can observe that Δ_A increases as α decreases. The reason is that by low α values, the pilot signals propagate more easily, giving a higher gain to A if it uses higher pilot power. The value of Δ_A also depends on the distance d between the two base stations as shown in Figure 11.6d. As the distance decreases, Δ_A increases exponentially. The reason for this increase is the same as discussed before. In the remainder of the section, the conservative default values $\alpha = 4$ and $d = 500\,\text{m}$ are chosen for the simulations. We will see that even with these conservative values, the players have an incentive to fine-tune their pilot powers.

Both operators are strategic

Assume now that both operators adjust their pilot power. We still consider 10 data users in the simulation area. The payoff of player A as a function of its pilot power P_A is provided in Figure 11.7a, containing the different payoff curves as the pilot power of the other base station P_B increases. We can observe that each of the payoff functions has a unique maximum point for P_A. Moreover, this maximum point depends on the pilot power of the other base station, P_B. For low values of P_B, the maximum payoff value decreases as P_B increases. Figure 11.7b shows the payoff surface for operator A as a function of the pilot power values of the two base stations.

Using the two payoff surfaces, the best response functions (i.e., the set of maximum payoff points) for the operators are derived and shown in Figure 11.8 for two different user densities. We can identify the Nash equilibria in the power control game as shown in Figure 11.8a for 10 data users and in Figure 11.8b for 100

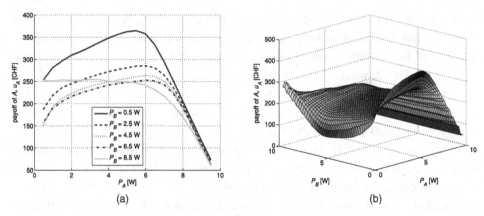

(a)   (b)

Figure 11.7. Payoff of player A as a function of its pilot power. Both operators are now strategic, hence this payoff is presented for various values of P_B in (a). The complete payoff surface is shown in (b). Used with permission, from [137], © IEEE, 2007.

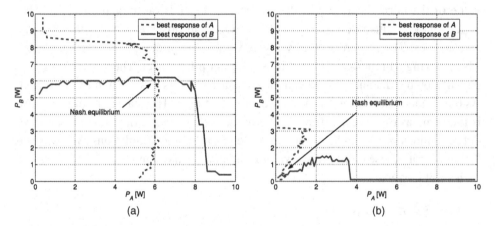

(a)   (b)

Figure 11.8. Best response functions for the two players with (a) 10 data users, (b) 100 data users. Used with permission, from [137], © IEEE, 2007.

data users. We see that there exists a unique Nash equilibrium point defined as the crossing point of the two best response functions. Note that for 10 data users, the Nash equilibrium strategy profile defines $P_A = P_B = 6\,\text{W}$, which are higher than the standard pilot powers. For 100 data users, the Nash equilibrium strategy profile defines $P_A = P_B = 0.5\,\text{W}$. The reason is that the capacities of the base stations saturate by using a relatively small power and hence there is no motivation for them to go above these pilot power values.

Next, we study the pilot power values in the Nash equilibrium as a function of the number of users. The results are shown in Figure 11.9. Because of the symmetry in the user distributions, the Nash equilibrium pilot power is the same for both players.

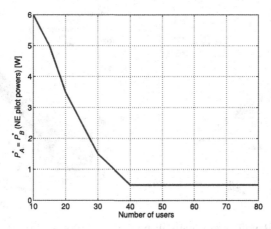

Figure 11.9. Nash equilibrium pilot power values as a function of the user density. Used with permission, from [137], © IEEE, 2007.

We observe that the Nash equilibrium pilot powers decrease as the number of users increases. For high user densities, the Nash equilibrium pilot powers stabilize at the value of 0.5 W.

We then study the *efficiency* of the system in a Nash equilibrium with respect to the case in which the players both use the standard power P^s. To this end, we investigate the *payoff region*, i.e. the payoff values for various pilot power levels. We identify the payoffs corresponding to the Nash equilibrium, the standard pilot power setting using P^s and the payoffs that correspond to Pareto-optimal strategy profiles. In particular, we can define the *Pareto boundary* as the set of Pareto-optimal payoff points. In our case, the Pareto-optimal payoff points characterize the system-efficient solutions.

Figure 11.10a shows the achieved payoffs as a function of the pilot power values P_A and P_B for 10 data users. We observe that in this case the Pareto boundary defines a straight line, because in a Pareto-optimal strategy profile each user in the system is attached to one of the base stations. Furthermore, the standard pilot powers and the Nash equilibrium strategy profile result in the same payoffs for the players and in addition they both lie on the Pareto boundary. This means that the players achieve a desirable state from the system point of view. Recall, however, that in this case the Nash equilibrium strategy profile requires higher pilot powers than the standard setting.

We present the payoffs for 100 data users in Figure 11.10b. In this case the Pareto-optimal points do not form a straight line anymore, because some users cannot be served. Another observation is that the Nash equilibrium is still close to Pareto-optimality, but the standard solution becomes very inefficient.

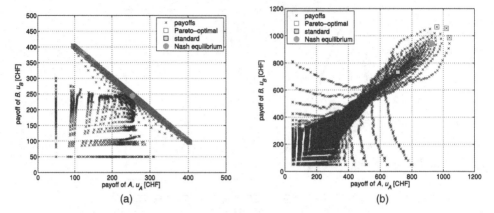

Figure 11.10. Payoff region with all possible payoffs for (a) 10 data users and (b) 100 data users. The Nash equilibrium, the payoff of the standard powers and all Pareto-optimal points are highlighted. Used with permission, from [137], © IEEE, 2007.

Following the previous experiment, we formally express the efficiency of the standard and the Nash equilibrium solutions compared to the best Pareto-optimal point (i.e., the Pareto-optimal strategy profile in which the sum of the payoffs for the two players is maximized). To this end, let us define the following two concepts.

Definition 11.2 The *price of anarchy* [234] is the ratio between the total payoff achieved by the two players in the best Pareto-optimal point and in the Nash equilibrium.

Definition 11.3 The *price of conformance* is the ratio between the total payoff achieved by the two players in the best Pareto-optimal point and when using the standard pilot powers P^s (i.e., being non-strategic).

A set of experiments was performed to measure these values for increasing user densities. Figure 11.11 presents the price of anarchy and the price of conformance as a function of the user density assuming they have data traffic. We see that both prices increase as the number of users increases. As we have seen in Figure 11.10a, both the standard payoff point and the Nash equilibrium achieve Pareto-optimality if there is a small number of users. Hence, the two prices are very close to one. As the user density increases, we observe that both prices increase and then stabilize around a constant value. Note, however, that the price of anarchy stabilizes close to one, whereas the price of conformance stabilizes around 1.4. This shows that for a high number of users, the players can achieve a higher payoff if both of them are strategic.

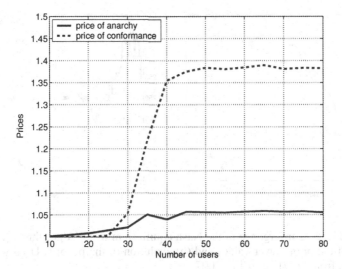

Figure 11.11. The price of anarchy and the price of conformance as a function of the user density. Used with permission, from [137], © IEEE, 2007.

11.2.3 Convergence to a Nash equilibrium

We have just seen that the expected payoff function for a certain player is continuous and has a single maximum point. We will now describe a distributed algorithm to achieve the Nash equilibrium in a given scenario. The algorithm is similar to the better-response dynamics [142], i.e., where each player tries to improve her payoff in each step. We provide the pseudo-code as shown in Algorithm 1.

Algorithm 1 Distributed convergence to the NE

1: for all player i do
2:     set pilot power $P_i = 0.1W$
3:     set the direction of optimization $dir_i = +1$
4: end for
5: set power control step size $P_{\text{step}} = 0.1W$
6: while () do
7:     for all player i do
8:         update P_i with a probability $0 < q < 1$
9:         $P_i = P_i + dir_i \cdot P_{\text{step}}$
10:        if u_i decreased then
11:            {the optimization passed the maximum payoff value}
12:            $dir_i = -dir_i$
13:        end if
14:    end for
15: end while

Figure 11.12a shows the evolution of the pilot power values applying Algorithm 1. We observe that the pilot power values follow the linear increase defined in

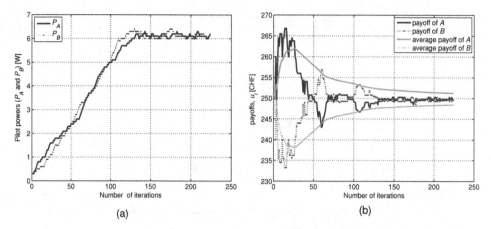

Figure 11.12. Convergence to the Nash equilibrium using Algorithm 1. (a) Evolution of the pilot power values and (b) evolution of the payoffs. Used with permission, from [137], © IEEE, 2007.

the algorithm. After reaching the Nash equilibrium pilot power values, the algorithm stabilizes after a certain number of steps.

Figure 11.12b shows the evolution of the payoffs during the convergence process. We see that the algorithm deviates from the Nash equilibrium payoffs while the pilot powers increase. As soon as the pilot powers reach the Nash equilibrium strategies, the payoffs remain close to the Nash equilibrium payoffs as well.

11.2.4 Power control game with power cost

We have seen that the operators are able to serve all users in the area if the user density is low. We observe, however, that the Nash equilibrium pilot powers are higher than the standard value. Recall that the payoff function defined in (11.10) does not include the possible cost due to the operation with high pilot power. Let us now extend the expected payoff function defined in (11.10) to capture this important aspect. We introduce two cost values for each player. The first cost denoted by C_i^{op} represents the *operating cost* of a base station i. This includes the aging of devices and hence the maintenance costs. The other cost, C_i^{subj}, expresses the *subjective cost* of player i. This covers every other aspect such as the risk of lawsuits or potential bad reputation due to high emission power. Without loss of generality, we assume that these costs are an increasing function of the downlink transmission power of the base stations.

According to the above description, we can extend the notion of expected payoff as:

$$u_i = \left(\sum_{v \in \mathcal{M}_i} \theta_v \right) - C_i^{op} - C_i^{subj}. \tag{11.13}$$

We define a non-cooperative power control game with the new expected payoff function introduced in (11.13) and denote it by $\hat{G}$. We assume that the players are able to calculate the Nash equilibrium of the original game G. We define the strategy in the extended game $\hat{G}$ as the choice between the standard power and the Nash equilibrium strategy P_i^*. Formally, we can write the strategies in $\hat{G}$ as:

$$s_i = \{P^s, P_i^*\}. \tag{11.14}$$

Let us call U the expected payoff that the players obtain by serving half of the total number of users. As we have seen in Section 11.2.2, if the players choose the Nash equilibrium strategy profile by low user densities, then it requires a higher pilot power from each operator. Without loss of generality, we denote by C^* the additional cost imposed by the Nash equilibrium compared to the standard pilot power setting P^s. The cost C^* includes both the operating and the subjective costs. Recall that we defined the normalized payoff difference Δ_A in Section 11.2.2. Because of the symmetry, $\Delta_A = \Delta_B$, and we denote both by Δ. In the extended game $\hat{G}$, we assume that the normalized payoff difference is higher than the corresponding cost of using higher pilot power, thus $\Delta > C^*$.

We present the payoff matrix of the game $\hat{G}$ in Table 11.7. In each payoff pair, the first payoff belongs to player A, whereas the second to player B.

To emphasize the structure of the payoff matrix, let us substitute the values $U = 3$, $\Delta = 2$ and $C^* = 1$. Inserting these values in Table 11.7, we obtain Table 11.8. From the payoff matrix, one can realize that the game $\hat{G}$ is equivalent to the Forwarder's Dilemma and therefore to the Prisoner's Dilemma. Strategy P^s corresponds to cooperation, whereas strategy P_i^* corresponds to defection. This means that in the Nash equilibrium, each player uses high power and the resulting payoffs are lower than if both had complied.

Table 11.7. *Payoff matrix of the game* $\hat{G}$

		Player B P^s	Player B P_B^*
Player A	P^s	U,U	$U - \Delta, U + \Delta - C^*$
	P_A^*	$U + \Delta - C^*, U - \Delta$	$U - C^*, U - C^*$

Table 11.8. *The extended power control game*
$\hat{G}$ *corresponds to the Prisoner's Dilemma.*

		Player B P^s	Player B P_B^*
	P^s	3,3	1,4
Player A	P_A^*	4,1	2,2

11.3 Summary

In this chapter, we studied the behavior of wireless operators having to co-exist.

In Section 11.1 we examined whether cooperation is possible without the usage of incentive mechanisms in multi-domain sensor networks. First, we analyzed a simple network consisting of two sensors and two base stations and showed that in this simple setting, the best Nash equilibria (where the lifetime of the sensors is the highest) consist of cooperative strategies. Then we generalized the model from two nodes to many nodes and used a two dimensional layout. We classified equilibria into non-cooperative, semi-cooperative, and fully cooperative. We showed that in most cases, the best Nash equilibria belong to the cooperative classes. In particular, when the path loss exponent is large, full cooperation is the best strategy.

In Section 11.2 we studied the problem of competitive pilot power control in two CDMA networks that reside on the two sides of a national border, assuming that each network is comprised of a single base station. We investigated whether the operators of these networks have an incentive to adjust their pilot signal powers. We first assumed that only one operator can adjust the pilot signal power of its base station. We saw that in this case it has an incentive to be strategic and we have quantified the effect of various parameters on the increase of its payoff. We have also seen that if both operators are strategic and the user density is low, then being strategic or not results in similar payoffs. But if the user density is high, then the Nash equilibrium is more efficient than using the standard pilot powers, which suggests that the operators have again an incentive to be strategic. Finally, we extended the payoff function to include the cost of using high pilot powers. We established the analogy between the power control game with power cost (in case of low user densities) and the Forwarder's Dilemma.

11.4 To probe further

Section 11.1 is derived from a paper by Buttyan, Holczer, and Schaffer [73]. Little attention has been devoted so far to the problem of cooperation between wireless sensor network operators; the interested reader can find a slightly different approach in [135].

Section 11.2 is derived from a paper by Felegyhazi *et al.* [137].

Power control has been extensively studied in the context of cellular networking. Baccelli *et al.* [37] consider downlink power allocation and admission control in CDMA networks relying on stochastic geometry. Hanly and Tse [161] as well as Catrein *et al.* [94] consider power control and capacity in CDMA networks. But, there are only a few papers about pilot power optimization [222, 364].

Game theory is used to study the power control of user devices in wireless networks, notably in cellular systems as studied in [21, 155, 183, 205, 263, 273, 253, 381] and [402]. A general framework for resource allocation in wireless networks is addressed in [121].

Recently, the coexistence of multiple Internet Service Providers (ISPs) was studied by Shakkottai and Srikant in [348]. They consider both transit and customer prices for the ISPs. They show that if the number of ISPs competing for the same customers is large, then it can lead to price wars. In another paper [347], Shakkotai *et al.* consider the problem of non-cooperative multi-homing in WLANs. Zemlianov and de Veciana [393] study a scenario in which users are able to choose between a cellular network and a WiFi network. They show that congestion sensitive strategies are better than proximity-based strategies. Felegyhazi and Hubaux [134] consider the competition between different operators in terms of pilot power control of their base stations. They show that in the pilot power control game a socially desirable Nash equilibrium exists and that it can be enforced by punishments.

Haykin provides a comprehensive overview [165] of the current tendencies and research challenges in shared spectrum communications in general. One of the challenges, namely opportunistic spectrum access, is addressed in the paper of Wang *et al.* [371].

A discussion about a possible increase of the proportion of the unlicensed spectrum is available in [50, 132].

11.5 Questions

(1) Why are the lifetimes equal to $B/2$ in the upper left cell of Table 11.2?

(2) Assume that in Figure 11.1, the distance between the base station to the left and the sensor next to it is 2 instead of 1 (the other distances are unchanged). Write the *costs* similar to Table 11.2. What can you say about cooperation in this scenario?

(3) In Section 11.1, why is the cooperation gain higher when α is higher?

(4) Why does formula (11.1) contain the processing gain G_p^{pilot}? (Note: this question and the following two assume some basic background in UMTS.)

(5) How is the processing gain G_p^{tr} calculated for data traffic?
(6) We can observe in Table 11.5 that the required SINR decreases as the data rate of the traffic increases. Why is the required CIR increasing nevertheless?
(7) Explain why the normalized payoff difference is higher for lower α.
(8) The Pareto boundary in Figure 11.10 is a straight line. Why? What type of game does this indicate?
(9) In Subsection 11.2.4, we assume that $\Delta > C^*$. Why is this assumption necessary (look at Table 11.7 for the implications of $\Delta = C^*$)?

12

Secure protocols for behavior enforcement

So far, in Part III of the book, we have shown through examples (MAC layer, packet forwarding, and co-existence of wireless operators) how to model selfish behavior. We have also explained how it is possible to enforce a desirable behavior by observing other players' behavior. For this purpose, we have made extensive use of game theory. Yet, the security techniques that we have shown in Part II can also be of help in this framework; for example, authentication of the wireless nodes is necessary in order to thwart selfish behavior at the MAC layer.

In Chapter 3, we have explained that it is difficult to provide a fully satisfactory definition of malicious and selfish behavior, because the two notions are strongly intertwined. In this chapter, we will make a fundamental additional step and show how security and game theoretic techniques can be combined to thwart misbehavior in wireless networks.

In compliance with the other chapters of Part III, we will articulate our development around an example. As we have seen in Chapter 10, cooperation does not happen "naturally" for packet forwarding in self-organized ad hoc networks. This means that cooperation must be encouraged. There are several ways to achieve this goal. One of them consists in relying on micropayments. In this chapter, we will follow the micropayment technique developed by Zhong *et al.* [400].

12.1 System model

We consider a system very similar to the one of Chapter 10: a set of N nodes is deployed in a given area. Each node is its own authority and selfishly tries to maximize its own payoff. In contrast with the previous model, however, we will now assume that there exists a micropayment system, by which any node performing packet relaying (for the benefit of an arbitrary source and destination pair) receives a certain reward. In order to obtain this reward, a relay node has to prove that it did indeed relay the packets, by providing evidence of this contribution (typically *a posteriori*) to the source node of the communication or to a network "bank" (as we will see, such a bank is needed to resolve possible disputes, meaning that the network that we consider here is not really self-organized).[1] In this way, a source node has to pay the relay nodes (either directly or indirectly through the micropayment system). It is also assumed that every node has a (public, private) key-pair and that all public keys are certified. Each node is thus able to establish a symmetric key with any other node of the network.

A further assumption is that each node i has a discrete set $\mathcal{P}_i$ of power levels at which it can transmit. For each ordered pair of nodes (i, j), there is a minimum power level P_{ij} at which a packet sent by node i can be correctly received by node j. If j cannot be reached by i, we will write $P_{ij} = \infty$. The transmission model is therefore a binary one.

Finally, we consider that the connectivity graph of the network is at least bi-connected, meaning that the removal of a single edge or vertex does not disconnect the graph.

Compared with Chapter 10, the approach here is more ambitious: not only do we consider packet forwarding, but also the process of *route selection*. Let us call "AdHocGame" the game modeling this situation: in this game, each player is a node who can participate in routing and packet forwarding. Let a_i designate the action that node i chooses. In the case of packet forwarding, for each packet the protocol requires node i to forward, a_i can indeed forward it at an appropriate power level, or withhold it, or replace it with an arbitrary message and transmit the new message at an arbitrary power level.[2] We will denote by a the actions of all nodes and by a_{-i} the actions of all nodes except node i.

We are now in a position to define the payoff of a node:

$$u_i = b_i - c_i,$$

where b_i stands for node i's *benefit* (expressed as a micropayment) for forwarding the packet and c_i for node i's *cost*. Only data packets are considered here, as they are usually larger and much more numerous than control packets. The cost c_i represents the energy consumed in forwarding data packets: when a relaying node forwards at power level l, the corresponding node cost is $l \cdot \alpha_i$, where α_i is a parameter representing the cost of energy, and is therefore influenced by the remaining battery energy.

It is important to note that both b_i and c_i depend on the actions of all players.

---

[1] The details of this micropayment system are not provided here; the interested reader can refer to Section 12.7 for pointers to possible solutions.

[2] For the sake of simplicity, the case in which i sends several arbitrary messages in lieu of one is not considered in this model.

Referring to the definition of dominant strategy, we are now in a position to define the notion of forwarding-dominant protocol.

Definition 12.1 In an AdHocGame, a *forwarding-dominant protocol* is a protocol in which (i) a subset of the nodes are chosen to form a path from the source to the destination, (ii) the protocol specifies that the chosen nodes should forward data packets, and (iii) following the protocol is a dominant action.

The non-existence of forwarding-dominant protocols for AdHocGames can be demonstrated (see the questions at the end of this chapter), and it underpins the fact that cooperation cannot be taken for granted. More precisely, it can be shown that there always exist instances of AdHocGames for which there is no forwarding-dominant protocol. This result is consistent with the conclusion of Chapter 10. Hence, a protocol that makes use of incentives for cooperation is required.

12.2 Cooperation-optimal protocol

We have seen in Chapter 10 that, in the absence of rewards, cooperation is extremely unlikely to happen. The question is now how to properly design the rewarding scheme.

Route establishment and packet forwarding occur in two subsequent stages. Accordingly, each node's action can be divided into two subactions: its participation in the route establishment and its decision in terms of packet forwarding.

This can be written as $a_i = (a_i^{(r)}, a_i^{(f)})$, where $a_i^{(r)}$ is node i's subaction in the routing stage (what it is *supposed to do* in the routing stage) and $a_i^{(f)}$ is node subaction in the forwarding stage (what each node *does* in the forwarding stage).

A routing decision $\mathcal{R}$ is determined by the routing subactions of all nodes $(a^{(r)})$. Consequently, each node's prospective payoff is determined by the routing decision $\mathcal{R}$ and by the nodes' actual subactions $a^{(f)}$:

$$u_i = u_i(\mathcal{R}, a^{(f)}).$$

Definition 12.2 Given a routing decision, the *prospective routing payoff* of a node is the payoff that it will achieve under the routing decision, assuming that all nodes are faithful in the packet forwarding subaction to the one they have declared in the routing subaction:

$$u_i^{(\mathcal{R})} = u_i(\mathcal{R}, a^{(r)}).$$

We are now in a position to introduce the important notion of dominant subaction.

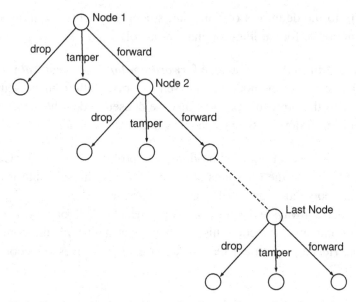

Figure 12.1. Game tree representing packet forwarding actions for a given packet along a given route. Used with permission, from [400], © ACM, 2005

Definition 12.3 In a routing stage, a ***dominant subaction*** of a potential forwarding node is one that maximizes its prospective payoff, no matter what subactions other players choose in this stage. Formally, we will say that $a_i^{(r)}$ is node i's dominant subaction in the routing stage if

$$u_i^{(\mathcal{R})}(a_i^{(r)}, a_{-i}^{(r)}) \geq u_i^{(\mathcal{R})}(\bar{a}_i^{(r)}, a_{-i}^{(r)}), \quad \forall \bar{a}_i^{(r)} \neq a_i^{(r)}, \forall a_{-i}^{(r)}.$$

The previous definition considered a single node. We now use it to introduce a definition referring to the whole route.

Definition 12.4 A routing protocol is a ***routing-dominant protocol to the routing stage*** if following the protocol is a dominant subaction of each potential forwarding node in the routing stage.

Note that it is assumed that the route computation is performed by the destination of the route, in a reliable way.

The previous definitions referred to the routing stage. Let us now consider the packet forwarding stage. Figure 12.1 represents packet forwarding as an extensive game (this game is similar to the Joint Packet Forwarding Game described in Appendix B).

Definition 12.5 A forwarding protocol is a ***forwarding-optimal protocol to the forwarding stage*** under routing decision $\mathcal{R}$ if all packets are forwarded to their destinations and following the protocol is a subgame perfect equilibrium under routing decision $\mathcal{R}$ in the forwarding stage.

As mentioned in Appendix B, a subgame perfect equilibrium is a Nash equilibrium for every subgame. In Figure 12.1, each subtree of the game tree corresponds to a subgame and each path from the root down to a leaf corresponds to a possible set of decisions by the nodes in the packet forwarding stage.

We can now combine these definitions in order to encompass both stages.

Definition 12.6 A protocol is a *cooperation-optimal protocol* to an ad hoc game if (i) its routing protocol is a routing-dominant protocol to the routing stage and (ii) for a routing decision $\mathcal{R}$ generated by the preceding subactions, its forwarding protocol is a forwarding-optimal protocol to the forwarding stage under $\mathcal{R}$.

12.3 Protocol for the routing stage

The protocol for the routing stage relies on two fundamental operations. The first is the estimate of how much should be paid for each link of the route. The second is to make sure that nodes cannot cheat about these estimates.

12.3.1 VCG payment

A crucial question in a scheme such as this one based on micropayments is to determine the appropriate reward level for each packet forwarding operation. Of course, the price should take into account the real burden supported by the forwarding node (e.g., how much energy it will have to spend to carry out this operation). But it is interesting to also include in the price a component representing what the price would be if that node were *not* included in the route.

In order to do so, we will make use of the well-known VCG mechanism, named after economists Vickrey, Clarke, and Groves. VCG is a second-best sealed auction. Its purpose is to let the bidders express the "real" value of the auctioned good. Its application to ad hoc networks is investigated by Anderegg and Eidenbenz [24]. We summarize here the principles.

Consider a node i intending to transmit a packet to a node j. The transmission range of node i depends on the transmitting power at which it transmits. As mentioned, we call P_{ij} the minimum power at which node j can receive the packet from node i. Let l be the lowest power value right above P_{ij}. As we have seen, the cost for node i is then equal to $l \cdot \alpha_i$.

Consider now a source S willing to start a session of packet sending to a destination D. Assume that the destination is able to collect the cost for each node to reach each of its neighbors. Let us denote the lowest claimed cost path from source

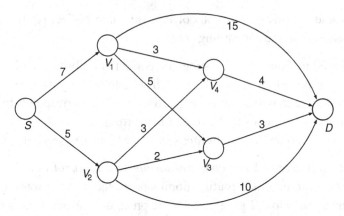

Figure 12.2. Example of ad hoc wireless network with the cost of each edge. Used with permission, from [24], © ACM, 2003

to destination by $LCP(S, D)$. Let us also denote the lowest claimed cost path from S to D that does not include i by $LCP(S, D; -i)$. Then the destination chooses $LCP(S, D)$ as the packet forwarding path from S to D, and the payment to node i is given by

$$b_i = cost(LCP(S, D; -i)) - cost(LCP(S, D) - \{i\}),$$

where the function $cost()$ sums the costs of all links on a path, and $LCP(S, D) - \{i\}$ consists of all the links on the LCP except the one starting from node i.

Figure 12.2 shows an example of how payments are calculated. The least cost path from S to D is:

$$LCP(S, D) = S, v_2, v_3, D$$

with $cost(LCP(S, D)) = 5 + 2 + 3 = 10$. The least cost path without node v_2 is

$$LCP(S, D; -v_2) = S, v_1, v_4, D$$

with $cost(LCP(S, D; -v_2)) = 7 + 3 + 4 = 14$. Likewise, the least cost path without node v_3 is

$$LCP(S, D; -v_3) = S, v_2, v_4, D$$

with $cost(LCP(S, D; -v_3)) = 5 + 3 + 4 = 12$. From this we obtain the VCG payments

$$b_2 = 14 - 10 + 2 = 6$$
$$b_3 = 12 - 10 + 3 = 5.$$

Hence these values represent the unit payment (the payment for one forwarded data packet) to nodes v_2 and v_3, respectively.

12.3.2 Prevent cheating in determining the link costs

Let us now take a close look at how the cost of transmission can be estimated, as this operation involves two nodes for each link. Consider two neighboring nodes i and j. There are two ways for node j to cheat about the power level P_{ij}.

1. Node j cheats by making P_{ij} greater.
2. Node j cheats by making P_{ij} smaller.

The first case is not beneficial for node j, because it makes node j less likely to be part of the chosen route and even if the route is chosen, node j's remuneration would be smaller (the proof of this statement is left as an exercise to the reader).

The second case can be avoided by the following cryptographic protocol. The protocol is an on-demand routing protocol (see Chapter 7 for a definition). For ease of presentation, assume that $\mathcal{P}_i = \mathcal{P}, \forall i$.

Before providing the detailed description of the protocol, we will describe it informally. The protocol makes use of two messages, TESTSIGNAL and ROUTE-INFO. The former aims at *computing* the transmission costs between neighboring nodes, and the latter is used to *transmit* information to the destination about this cost.[3] More specifically, let us describe how a given node i can become a relay node on a route between a source S and a destination D. At a given point in time, one of its neighbors (which can be the source S) begins the process of estimating the power needed to transmit to i. Let us call U this neighbor node ("U" standing for "upstream"). In order to do so, node U sends a series of TESTSIGNAL messages at different power levels (in increasing order). Of course, node i will receive only those of the messages that have been transmitted with enough power to cover the distance from U to i.

Node i then informs the rest of the network (and thus the destination) of the observed power levels by sending the message ROUTEINFO. In this way, the destination can build up a matrix of all costs of the links.[4]

In addition, if the received TESTSIGNAL mentioned previously is the first that node i receives for that specific session, node i has to perform the same operation

---

[3] This separation into two messages makes the security analysis easier. In practice, the two messages could potentially be combined, in order to reduce the overhead.

[4] This means that the network will have to convey a substantial number of ROUTEINFO message broadcasts for each of the sessions to be established. But this is typical of route establishment in reactive routing protocols, as described in Chapter 7. What makes it acceptable is the relatively small size of these messages.

that node U did previously, which is to begin estimating the power needed to reach its own neighbors. For this purpose, it will begin sending TESTSIGNAL messages in ascending order of power.

Of course, both TESTSIGNAL and ROUTEINFO need to be cryptographically protected, particularly in order to make it impossible for a forwarding node (or for anyone else) to alter the result of the power level estimation process between each pair of nodes. The details of how this is achieved are described hereafter. This security protection illustrates how protocols aimed at stimulating cooperation by means of micropayments can be secured by appropriate cryptographic protocols.

Source node's test signals

- Assume source S starts a session of M packets (all packets are assumed to be of the same size). It divides the packets into $\lceil M/\beta \rceil$ blocks, where β is the number of packets in a block.
- Source S picks a random number r_0.
- S computes $r = H^{\lceil M/\beta \rceil}(r_0)$, where H is a one-way function and $H^n(X)$ means $H(H \ldots (H(X)))$, where the number of times the H function is computed is equal to n.
- For each power level $l \in \mathcal{P}$ (in increasing order), S transmits (TESTSIG-NAL, $[S, D, r]$, $[S, h_l]$) at power level l, where h_l contains an encryption of $[S, D, r, l, \alpha_S]$ using key $k_{S,D}$ and a MAC of the encryption using the same key.

Intermediate node's operations

Upon receiving a message (TESTSIGNAL, $[S, D, r]$, $[U, h]$), from an upstream neighbor U, an intermediate node i does the following.

- It transmits (ROUTEINFO, $[S, D, r]$, $[U, i, h']$) at power level P_{ctr}, which is a power level for control messages such that if it is used by all nodes, the communication graph is connected. Here $h' = E_{k_{i,D}}(h)$. This value is very important as it contains the result of the various signal "probes" realized by TESTSIGNAL. By the way it is computed, h' is protected by two keys: $k_{U,D}$ and $k_{i,D}$, thus thwarting the possibility for node i to make P_{Ui} smaller. For integrity, this message is protected by a MAC using key $k_{i,D}$.
- If the TESTSIGNAL is the first one that node i receives for session (S, D, r), then for each $l \in \mathcal{P}$ (in increasing order), node i transmits (TESTSIGNAL, $[S, D, r]$, $[i, h'_l]$) at power level l, where h'_l contains an encryption of $[S, D, r, l, \alpha_S]$ using key $k_{i,D}$ and a MAC of the encryption using the same key.

Upon receiving a message (ROUTEINFO, $[S, D, r]$, $[U, i, h]$) for the first time, an intermediate node j retransmits it.

Destination protocol

The purpose of all these messages is to allow destination D to maintain a cost matrix for each session (S, D, r). Each entry of this matrix is an array of power levels and cost-of-energy parameters.

- Upon receiving message (TESTSIGNAL, $[S, D, r]$, $[U, h]$) from its neighbor U, D verifies the MAC by means of key $k_{U,D}$, and converts h to the corresponding power level l and cost-of-energy parameter α_U. D records (l, α_U) in the cost matrix for link (U, D).
- Upon receiving (ROUTEINFO, $[S, D, r]$, $[U, i, h]$), D decrypts h, verifies the packet's MAC by means of key $k_{i,D}$, and converts h to the corresponding power level l and cost-of-energy parameter α_U. D records (l, α_U) in the cost matrix for link (U, D).

Destination processing

Once it has collected all link cost information, D builds up the cost graph and computes the lowest cost path from S to D, using Dijkstra's algorithm. It then computes the unit payment for each intermediate node i by means of the first expression in this section. All these computations can be carried out in $O(N^3)$.

We are now in a position to formulate the following theorem.

Theorem 12.1 *If the destination is able to collect all involved link costs, then the described routing protocol is a routing-dominant protocol to the routing stage.*

The proof of this and of the other theorems of this chapter can be found in [400].

12.4 Protocol for packet forwarding

In order to be rewarded, each relay node i must be able to prove that it performed in the packet forwarding phase as it promised it would in the routing phase, in other words that $a_i^{(f)}$ is compliant with $a_i^{(r)}$.

Assume for the sake of clarity all links to be bi-directional. After it has successfully received a block, the destination gives the relay nodes a confirmation, which proves that they have successfully transmitted this block. Only after receiving this confirmation will the relay nodes accept to forward the next block.

This mechanism can be implemented in an efficient way by means of a hash chain. Let H be a one-way function. The source S picks a random number r_0 for a session it wants to start and, for each of the blocks of the session, computes $r_m = H^m(r_0)$, where m is the number of the block. There are in total $\lceil M/\beta \rceil$ blocks. Let $r = r_{\lceil M/\beta \rceil}$. The source makes r public in an authentic way and computes $r_{\lceil M/\beta \rceil - m}$ as the confirmation of the mth block and sends it with the last packet of that block. In this way, anyone (and in particular each relay node) can verify the confirmation by checking whether

$$H^m(r_{\lceil M/\beta\rceil-m}) = r.$$

It might happen that the source and an intermediate node disagree on whether or not a block was successfully transmitted. In order to solve this fair exchange problem, a possible technique is *mutual decision*. It enforces cooperation by building a cryptographic contract between the source and intermediate nodes. To eliminate the incentive to cheat, as soon as one participant disrespects the terms of the contract, the system is blocked. The interested reader can check [400] and [202].

12.4.1 Protocol description

Once it has concluded the routing discovery phase, destination D sends the following routing decision

$$([S, D, r], LCP(S, D), P_S, \{(P_i, b_i)\})$$

along the reversed path of $LCP(S, D)$, where i corresponds to all the relay nodes along that path and P_i (respectively P_S) is the power level that node i (respectively source S) should use to forward (respectively send) data packets and b_i is the unit payment that node i should receive. This message is also digitally signed by D. It is relayed backwards to the source.

When it receives this message, the source verifies the signature. If it is valid, it enters in the transmission phase. In this phase, the source and the relay nodes transmit the data packets at the power levels identified in the routing phase (P_S and P_i, respectively).

The source node sends data packets in blocks (each block containing β packets). With the last packet of the mth block, the source sends the confirmation $r_{\lceil M/\beta\rceil-m}$, encrypted with key $k_{S,D}$. Then it waits for the confirmation before starting to send the next block.

The relay nodes forward the packets along $LCP(S, D)$ to the destination. At the end of a block, they wait for the confirmation before they start forwarding the next block. When the destination has received all the packets in a block, it retrieves by decryption $r_{\lceil M/\beta\rceil-m}$ and sends it backwards in clear-text along $LCD(S, D)$. Upon receiving this confirmation, each relay node verifies that the first expression of this section holds. If this is the case, it saves this confirmation and forwards it back along $LCP(S, D)$. Once it receives the confirmation, the source starts sending the following block.

Assume that in a given session the last confirmation saved by node i is $r_{\lceil M/\beta\rceil-m}$. Then each relay node j located between S and i (including i) receives a payment of

$b_j \cdot \beta \cdot m$ from the source by submitting this confirmation (to source S directly or to the micropayment system (the "bank")). If some packets in the $(m + 1)$th block have been forwarded but the relay nodes have not received the confirmation, they submit the routing decision to the micropayment system which charges the source $\Sigma_i b_i \cdot \beta$ in addition. This amount does not go to the intermediate nodes, but to the system.

We can now formulate the following theorem.

Theorem 12.2 *Let $\mathcal{R}$ be a routing decision computed by the routing subactions triggered by the protocol described in Section 12.3.2. Assume that, for any node on the packet forwarding path, the computed payment is greater than the cost. Then the protocol presented in this section is a forwarding-optimal protocol to the packet forwarding stage under $\mathcal{R}$.*

From this theorem and Theorem 12.1, we can derive the following result (the connectivity graph is assumed to be bi-connected).

Theorem 12.3 *The complete protocol, including the routing protocol (Section 12.3.2) and the packet forwarding protocol (this section) is a cooperation-optimal protocol to AdHocGames.*

12.5 Discussion

As the careful reader has noticed, the solution described above requires a number of conditions to be met. For example, mobility should be low enough to allow all the effort carried out in the routing phase to be recouped in the packet forwarding phase. Likewise, the radio links must be stable enough so that the required minimum power estimated by the message TESTSIGNAL in the routing phase still holds in the packet forwarding phase. In addition, this model does not take possible interference of transmissions into account.

A further condition is the existence of the micropayment mechanism and of a public-key infrastructure; both are relatively heavy components, and constitute an additional burden on the system. Finally, it is important to notice that the described scheme would not necessarily resist to malicious nodes or to colluding (selfish or malicious) nodes.

We have described this example and we have mentioned these limitations because they open a more general question: to what degree should a network be protected against malice and selfishness? Should, for example, an ad hoc network be protected at all layers? Actually, too much protection can be detrimental to the protection itself (complexity being at odds with security).

There is of course no ultimate answer to this interrogation: all depends on the context of the network deployment and on the magnitude of the risk. By writing this book, it is our hope to have helped the reader to better assess what is feasible, and at what cost.

12.6 Summary

In this chapter, we have considered the problem of selfishness on both the routing and packet forwarding phases of an ad hoc network. We have shown how the problem can be studied by means of game theory. We have also explained how protocols aiming at stimulating cooperation by means of micropayments can be secured by appropriate cryptographic protocols.

12.7 To probe further

There are basically two approaches to incite nodes to participate in the network's operations: (i) by remunerating honest nodes, using for example a micropayment scheme, as we have seen in this chapter or (ii) by denying service to misbehaving nodes by means of a reputation mechanism. We provide here a brief overview of these approaches.

Several researchers propose schemes that employ micropayments to encourage cooperation. As mentioned, the approach described in this chapter is the one designed by Zhong *et al.* [400]. This work is inspired notably by a previous work by Zhong, Yang and Chen [401], where an off-line central authority collects receipts from the nodes that relay packets and remunerates them based on these receipts. They rely on public key cryptography to process each packet. A follow-up work is the one by Wang *et al.* [370], which describes optimal unicast routing systems in non-cooperative wireless networks.

Another solution based on side payments is proposed by Eidenbenz, Resta, and Santi [126]. Yet another solution, presented by Buttyan and Hubaux [75, 77], is based on a virtual currency, called the nuglet: If a node wants to send its own packets, it has to pay for it, whereas if the node forwards a packet for the benefit of another node, it is rewarded. However, some mechanisms of this solution (e.g., the generation and management of nuglets) still need to be further investigated. Finally, let us mention that the mobility of nodes can have a negative effect on an incentive mechanism, as pointed out by Figueiredo, Garetto, and Towsley [139].

The application of micropayment systems to hybrid ad hoc networks is investigated by Lamparter, Paul, and Westhoff [247]. Other solutions have been developed by Jakobsson *et al.* [203] and by Ben Salem *et al.* [46]. The former is improved by Avoine in [33]. Additional research on this topic (both from the points of view of malice and selfishness) has been carried out by Carbunar, Ioannidis, and Nita-Rotaru [91].

We already mentioned reputation systems in Chapter 7. The application of such systems has also been studied in other frameworks. Studies such as [172], [327] and [328] consider the effect of *on-line* reputation systems [110] on e-marketing and trading communities such as eBay. Reputation mechanisms are also used to foster cooperation in peer-to-peer networks [112] and in WiFi [49].

In [309], Patel and Crowcroft propose a ticket-based system that allows mobile users to connect to foreign service providers: the user contacts a *ticket server* to acquire a ticket, requests a service from a *service server* and uses the ticket to pay for that service. In [125], Efstathiou and Polyzos present a Peer-to-Peer Wireless Network Confederation (P2PWNC) where the roaming problem is considered as a peer-to-peer resource sharing problem. They propose a solution where a WISP has to allow the foreign users to access its hot-spots in order to allow its own users to connect to foreign WISPs' hot-spots.

12.8 Questions

(1) In this question, we prove by contradiction the non-existence of forwarding-dominant protocols for AdHocGames. In other words, when each node is its own authority, cooperation between the nodes cannot be taken for granted. The proof formally justifies the need of incentives for cooperation in the design of packet forwarding protocols in AdHocGames.

Assume that there exists a forwarding-dominant protocol for AdHocGames.

(a) Suppose that reliable overhearing is available and that there exists a link (i, j) such that

$$P_{i,j} < \infty$$

$$P_{i,k} = \infty, \forall k \neq j.$$

How does this assumption affect the network connectivity? Draw a simple scenario with three nodes i, j, and k.

(b) As the nodes' behavior is not coerced by a tamper-proof device, node j can have the following strategy: it forwards all packets it receives from node i except the packet with serial number 0.

What is node i's best action to maximize its payoff $u_i(a_i, a_{-i})$? Express the solution in terms of $b_i(a_i, a_{-i})$ and $c_i(a_i, a_{-i})$. Show the contradiction with the forwarding-dominant protocol definition.

(c) Now we want to assess the relevance of the assumption of the existence of such link (i, j). We replace it by assuming instead that reliable overhearing is not available anymore. Does the non-existence of forwarding-dominant protocol still stand?

(d) Finally, consider that reliable overhearing is again available and assume instead that nodes use the minimum requested transmission power to reach destination. Does the proof still stand? Conclude on the theorem's general applicability.

(2) If the system admits a forwarding-dominant protocol, then it converges to a unique solution. Explain why this property makes forwarding-dominant protocols more desirable than protocols that achieve Nash equilibrium.

(3) Section 12.3.1 introduces the VCG payment scheme. We use Figure 12.2 to illustrate the properties of the VCG protocol.

 (a) Assume that the nodes independently compute their packet forwarding cost. Explain why VCG is truthful (or strategy proof), i.e. provides an incentive to the nodes to declare their true cost of forwarding.

 (b) Assume now that the packet forwarding cost over a link is computed via a protocol between the nodes at both ends of the link (i.e. mutual computation).

 Consider the following scenario: v_2 cheats in the link cost establishment protocol and increases the cost of $v_2 \rightarrow v_3$ to 5 while the real cost is 2. How does this affect v_2's and v_3's rewards? Does the cheating influence the probability of belonging to the LCP?

 (c) What happens if v_3 reacts by decreasing the cost of $v_2 \rightarrow v_3$ to 3 ? Conclude on the truthfulness of VCG with such a link cost establishment protocol.

 (d) Explain how the secure link cost protocol depicted in Section 12.3.2 makes VCG truthful.

 (e) Finally, consider the following scenario. The cost of the link $S \rightarrow v_1$ is increased to 50.

 Compute v_2's price and explain its increase. Describe a scenario in which the cost of link $S \rightarrow v_1$ might have increased in such proportions. Conclude on the impact of the topology on the rewarding price.

(4) In the packet forwarding protocol of Section 12.4, it is possible that the source and an intermediate node disagree on whether or not the block has been successfully transmitted.

 (a) Section 12.4.1 details how block confirmation is implemented using a reverse hash chain. Why is the confirmation value encrypted and sent in the last packet of each block?

 (b) What happens if the source does not append the confirmation to the last packet of a block?

 (c) How does this mutual decision protocol eliminate the incentive to cheat?

(5) In Section 12.4 it is mentioned that r is made public in an authentic way. Please identify what misbehavior would become feasible for the source if the authenticity of r could not be verified.

Appendix A

Introduction to cryptographic algorithms and protocols

A1 Introduction

Security of information and communication systems is concerned with the prevention or the detection (if prevention is not possible or too costly) of *attacks*, where an attack is meant to be a *deliberate attempt* to compromise the system. A system is compromised if it reaches a non-desirable state or it behaves in a non-desirable way, and where the latter can be the result of the former.

A system can also reach a non-desirable state or behave in a non-desirable manner due to some random faults. Random faults, however, are usually less sophisticated than attacks; hence, coping with attacks is more difficult. For instance, a Cyclic Redundancy Code (CRC) can detect random errors in a transmitted message well enough, but it is ineffective in detecting deliberate modifications, because an attacker can compute the correct CRC value for the modified message.

Stealing a password file and cracking passwords off-line is an example of an attack. In this case, the attacker tries to put the system into a non-desirable state where some passwords are known not only to the corresponding legitimate users, but also to the attacker. This allows the attacker to login in the name of a legitimate user and to use the privileges of that user to perform some operations in the system. This then allows the attacker to further compromise the system. For instance, the attacker can modify or delete some important configuration files, which can lead to a non-desirable behavior of the system.

In general, attacks can be classified into two categories: *passive* attacks and *active* attacks. In a passive attack, the attacker does not actively interfere with the operation of the system, but she only passively monitors it. Examples of passive attacks are eavesdropping and traffic analysis. In contrast to this, in an active attack, the attacker intervenes in the operation of the system. Examples of active attacks include the modification, interception, forgery, and replay of messages, as well as jamming and tampering with devices. As a matter of fact, passive attacks are

393

difficult to detect, whereas active attacks are difficult to prevent. Hence, the usual security objective is to prevent passive attacks and detect active ones.

In information and communication systems, security objectives can be achieved by physical protection and algorithmic measures. An example of physical protection is when a server is locked in a room under continuous video surveillance. Another example is when a device is placed inside some tamper resistant packaging (e.g., smart cards). Physical protection is very effective, but it is often very expensive as well. In addition, it is not always applicable. In particular, in the case of wireless communications, the access to the wireless channel cannot be prevented by physical means. When physical protection is not feasible or very expensive, algorithmic measures can be used. Most of these algorithmic measures are based on cryptographic algorithms and on protocols that allow for secure communications over insecure channels at the cost of physically protecting a limited amount of key material only.

Usually, physical protection and algorithmic measures are combined in order to achieve the security objectives of the system. This combination depends not only on the technical requirements and constraints, but also on the available budget. In other words, we are interested in a combination that ensures the required level of security at the lowest possible cost.

In addition to the technical aspects (i.e., physical protection and algorithmic measures), the security of a system also depends on some human factors. Even the best cryptographic algorithms are ineffective if the keys are leaked. People should be educated to care about the security of the systems that they use, just like they are educated to protect their homes and other physical assets. In addition, appropriate procedures must be in place and must somehow be enforced, such as requiring users to choose hard-to-guess passwords and to change them from time to time.

In this appendix, we will focus on the algorithmic measures. We first introduce some basic cryptographic algorithms and protocols used in information and communication systems in general, then we describe some special algorithms that appear to be useful in upcoming wireless networks. Our presentation is brief; the interested reader can find more information about cryptographic algorithms and protocols by following the references given at the end of this appendix in Section A9.

We recommend that before proceeding with this appendix, the reader take a look at the common security objectives described in Section 1.2. The following algorithms and protocols are meant to achieve those objectives.

A2 Encryption

Encryption is a widely used security mechanism. Its primary purpose is to provide confidentiality services, but it can also be applied to provide authentication and integrity services.

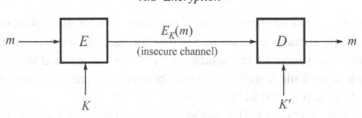

Figure A1. Classical model of encryption.

The classical model of encryption is illustrated in Figure A1. In this model, a sender wants to send a message m to a receiver over an insecure channel that can be eavesdropped by an adversary. To prevent access to the content of the message, the sender encrypts it using an encryption algorithm E and an encryption key K. The encrypted message $E_K(m)$ is called the ciphertext and the clear message m is called the plaintext. At the receiving side, the receiver decrypts the ciphertext with a decryption algorithm D and a decryption key K' and obtains the original plaintext.

The goal of the adversary is to find a way to systematically decrypt encrypted messages. One basic assumption, called Kerckhoff's principle, is that the adversary has full knowledge about the operation of the encryption and the decryption algorithms. Hence, in order to be able to systematically decrypt encrypted messages, it is sufficient for the adversary to obtain the decryption key. If the adversary finds an efficient method for determining the decryption key, then we say that the encryption scheme is broken.

The following adversary models are distinguished, depending on the amount and type of information available to the adversary.

- ***Ciphertext-only attack*** In this model, the adversary can only observe ciphertexts produced by the same encryption key. The adversary's goal is to determine the corresponding decryption key.
- ***Known plaintext attack*** In this model, the adversary can obtain corresponding plaintext–ciphertext pairs. In all these pairs, the ciphertext is produced from the plaintext with the same encryption key. The adversary's goal is to determine the corresponding decryption key.
- ***Chosen plaintext attack*** In this model, the adversary can choose plaintexts and obtain the corresponding ciphertexts. This can be modeled by assuming that the adversary has access to an encryption oracle that encrypts any set of plaintexts submitted to it with the same key. The adversary is allowed to choose a set of plaintexts, submit it to the oracle and receive the corresponding ciphertexts. The adversary's goal is to determine the decryption key that corresponds to the oracle's encryption key. In practice, any component of the system that performs

encryption and whose input can be manipulated can be used as an encryption oracle. A typical example is a smart card used for encryption based authentication. In an *adaptive* variant of this attack, the adversary is allowed to interact with the oracle in multiple rounds and choose, based on the previous responses of the oracle, the next plaintext to be submitted to the oracle.

- **Chosen ciphertext attack** This model is similar to the previous one, but instead of an encryption oracle, the adversary is assumed to have access to a decryption oracle. The adversary can submit ciphertexts to the oracle and receive the corresponding plaintexts. In this case, the adversary's goal is to find the decryption key used by the oracle. This attack has an adaptive variant, too.

- **Related key attack** In this model, the adversary can obtain ciphertexts, or plaintext–ciphertext pairs that are produced with different encryption keys, but all these keys are related in a known way to a specific encryption key. The adversary's goal is to determine the decryption key that corresponds to this specific encryption key.

We say that an encryption scheme is secure in a given adversary model if it is computationally infeasible for the adversary to determine the target decryption key under the assumptions of the given model. For many encryption schemes used in practice, no proof that they are secure exists. These schemes are used, nevertheless, because they are efficient and they resist all known attacks. Some encryption schemes are provably secure, however these schemes are often inefficient.

There are two basic types of encryption schemes: symmetric-key and asymmetric-key encryption. In the case of symmetric-key encryption, the encryption key K and the decryption key K' are the same (hence the name), or they can be computed from each other easily. In the case of asymmetric-key encryption, the encryption and the decryption keys are different, and computing the decryption key from the encryption key is difficult. In the remainder of this section, we describe the properties of both types of encryption schemes in more details.

A2.1 Symmetric-key encryption

Symmetric-key encryption itself has two types: stream encryption and block encryption. Stream ciphers operate on individual characters of the plaintext, whereas block ciphers process the plaintext in larger blocks of characters. This distinction between stream ciphers and block ciphers, however, is not very sharp: we can consider the blocks processed by a block cipher as large characters, or similarly, the characters processed by a stream cipher as tiny blocks. Nevertheless, the internal design and operation of stream ciphers and that of block ciphers are remarkably different, therefore they are usually discussed as two distinct types of symmetric-key encryption.

Stream ciphers

The heart of a stream cipher is the *key stream generator* that is used to produce a long sequence of (pseudo-)random characters that are combined with the characters of the plaintext in order to produce the characters of the ciphertext. In most cases, the characters of the key stream are simply XORed to the characters of the plaintext, one after the other. At the receiver side, the same key stream generator is used (recall that we are discussing symmetric-key encryption), thus the same key stream is produced as the one used by the sender. In order to obtain the characters of the plaintext, the characters of the key stream are XORed to the characters of the ciphertext.

A very well-known stream cipher is the *one-time pad*, where the key stream consists of truly random characters obtained, for instance, from some physical process. Perhaps, the fame of the one-time pad stems from the fact that it was proven to be unconditionally secure (against ciphertext-only attacks) by Shannon in 1949 [350]. Unconditional security means that an eavesdropper learns no information (in an information theoretical sense) about the plaintext from the observed ciphertext. In other words, when an eavesdropper observes a ciphertext, all possible plaintexts that can correspond to that ciphertext are equally likely to be the message sent by the sender. Shannon also proved that a necessary condition for a cipher to be unconditionally secure is that the length of its key is at least as large as the compressed plaintext.[1] This means that the one-time pad can require a large amount of key material to be passed securely to the receiver side, which severely limits its use in many practical applications.

Practical stream ciphers use pseudo-random key stream generators to produce the key stream from a small, random seed. In this case, only the small seed needs to be securely passed to the receiver, who can then use the same pseudo-random key stream generator as the sender to produce the same key stream as the one used for encryption. For historical reasons, many practical stream ciphers are designed for hardware implementation and for real-time applications, such as encrypted voice communications. Many of these ciphers use linear feedback shift registers (LFSR) as their basic building block. However, there exist some stream ciphers that are optimized for software implementation. An example is the RC4 stream cipher [339], which is the default cipher in the Web security protocol called Secure Socket Layer (SSL), and it is also the cipher used in the WiFi security protocol called Wired Equivalent Privacy (WEP).

A stream cipher is called *synchronous* if the key stream is generated independently from the plaintext or the ciphertext. Both the one-time pad and RC4 are synchronous stream ciphers. The advantage of synchronous stream ciphers is that

---

[1] More precisely, a necessary condition is that the entropy of the key must not be smaller than the entropy of the plaintext.

they have no error propagation, which means that if a ciphertext character is received erroneously owing to some noise in the transmission channel, then this affects only the corresponding plaintext character when the ciphertext is decrypted. More precisely, the corresponding plaintext character will have bit errors at the same positions where the errors occur in the ciphertext character; all the other characters will be decoded correctly. The disadvantage of synchronous stream ciphers is that they cannot tolerate the loss of ciphertext characters during transmission. If a ciphertext character is lost, then the sender and the receiver lose synchrony, which means that the receiver starts XORing "wrong" key stream characters to the received ciphertext characters. As a result, the receiver decodes the ciphertext characters that follow the lost character into a sequence of random characters. Hence, in the case of synchronous stream ciphers, we must ensure that de-synchronization errors are detected and, if necessary, additional mechanisms should be in place to re-synchronize the sender and the receiver.

Another class of stream ciphers is called *self-synchronizing* ciphers. In the case of self-synchronizing ciphers, the internal state of the key stream generator depends on the last few, say ℓ, ciphertext characters that have been sent by the sender (and received by the receiver). The advantage of this is that even if a ciphertext character is lost during transmission, after receiving ℓ consecutive ciphertext characters correctly, the internal state of the key stream generator at the receiver will be correct, and thus, the subsequent ciphertext characters are decoded correctly. In other words, no special re-synchronization is needed between the sender and the receiver, but the receiver re-synchronizes itself automatically. However, the disadvantage of these kinds of ciphers is that they have a larger error propagation than synchronous stream ciphers. The reason is that when a ciphertext character is received with an error, this erroneous character will affect the internal state of the key stream generator in the following ℓ rounds, thus the corresponding and the next ℓ plaintext characters are decrypted erroneously.

Both types of stream ciphers are used in practice. The choice largely depends on the application environment. For instance, self-synchronizing stream ciphers are advantageous in applications that have strict real-time constraints and that tolerate bursty errors (e.g., voice). Synchronous stream ciphers are a good choice when the channel is noisy and no error correction code is applied on the encrypted data (e.g., for efficiency reasons).

Block ciphers

Symmetric-key block ciphers are very widely used cryptographic primitives. Besides their obvious use for encryption, they are also often used as building blocks in pseudo-random number generators, hash functions, and message authentication codes. In addition, they are extensively used in entity authentication and key establishment protocols.

A block cipher is a function that takes two inputs, an n-bit plaintext block and a k-bit key, and it produces an n-bit ciphertext block. For a given key K, the block cipher can be considered to be a single variable function $E_K : \{0, 1\}^n \rightarrow \{0, 1\}^n$. For each key K, function E_K must be invertible in order to be able to reconstruct the plaintext blocks from the ciphertext blocks. This means that E_K must be a permutation over the space of the n-bit vectors. The inverse of E_K is denoted by D_K, and we have $D_K(E_K(x)) = x$ for every plaintext block x and key K.

Designing secure block ciphers is a difficult task. One common approach is to use several rounds of rather simple operations, such as small size substitutions (i.e., look-up tables) and large bit-permutations. Many well-known block ciphers are based on this approach. Examples include the Data Encryption Standard (DES) [6], and its successor, the Advanced Encryption Standard (AES)[5]. If it is done carefully, then the application of several rounds of simple operations can finally result in a complex transformation. In the case of block ciphers, the resulting transformation should satisfy the following criteria.

- **Completeness** Each bit of the output block depends on each bit of the input block and on each bit of the key.
- **Avalanche effect** If one bit is changed in the input block, then each bit in the output block changes with probability $1/2$. In other words, this means that changing one bit in the input block will change approximately half of the bits in the output block. Similarly, changing one key bit should result in the change of approximately half of the bits in the output block.
- **Statistical independence** There should not be any statistical relationship between the input and the output blocks. In other words, they should appear to be statistically independent.

The key size k of the block cipher has paramount importance with respect to the security of the cipher. If the key is too short, then the cipher is susceptible to an *exhaustive key search* attack. In this attack, the adversary first obtains a few plaintext–ciphertext block pairs $(x_1, y_1), (x_2, y_2), \ldots$, where in each pair, the ciphertext block is produced from the corresponding plaintext block using the same key K. Then, the adversary tries all possible keys until she finds K that encrypts every plaintext block x_i into the corresponding ciphertext block y_i. More precisely, for each candidate key $\tilde{K}$, the adversary does the following: She computes $y_1' = E_{\tilde{K}}(x_1)$. If $y_1' \neq y_1$, then $\tilde{K}$ is not the right key, so it can be thrown away. If $y_1' = y_1$, then the adversary computes $y_i' = E_{\tilde{K}}(x_i)$ for the remainder of the available plaintext blocks $\{x_i : i > 1\}$. If for some i, $y_i' \neq y_i$, then $\tilde{K}$ is thrown away, otherwise (i.e., if $\tilde{K}$ works for all available pairs of plaintext–ciphertext blocks) the adversary accepts $\tilde{K}$ as the right key. The accepted key is not necessarily the right one (which follows from the fact that a plaintext block can be mapped into the same ciphertext block under different keys), however, the probability of the error decreases rapidly with the number of the available plaintext–ciphertext block pairs. A similar attack works in the case when the adversary does not have plaintext–ciphertext block pairs, but she has only ciphertext blocks, and the plaintext blocks have a redundant structure (e.g., they contain parity bits) known to the adversary.

On average, in the exhaustive key search attack, the adversary finds the key after testing half of the key space. This means that the average complexity of the exhaustive key search attack is $1/2 \cdot 2^k = 2^{k-1}$. In view of this, it is clear that the key size k should be as large as possible. But, in order to reduce the amount of bits that need to be securely transmitted to the receiver to enable the decryption of encrypted messages, we would like the key to be as short as possible. The commonly accepted trade-off today, at the time of this writing, is $k = 128$.

It must be emphasized that having a large key size is only a necessary condition for the security of a block cipher, but it is not sufficient. The reason is that the cipher can be broken due to the weaknesses in its internal (algebraic) structure, even if it uses large keys. As an example, let us consider DES. The key length of DES is 56 bits, which means that the average complexity of the naïve exhaustive key search attack against DES is 2^{55}. However, this can immediately be reduced to 2^{54} due to a special algebraic property of DES, called the complementation property [272]. In addition, there are more sophisticated attacks against DES that try to exploit other weaknesses in its internal structure. Those most powerful attacks are called differential and linear cryptanalysis, the complexities of which are around 2^{47} and 2^{43}, respectively.

Block cipher modes

As we have described before, a block cipher produces an n-bit output block from an n-bit input block. However, messages to be encrypted are usually longer (e.g., files) or shorter (e.g., characters) than one block. In order to solve this problem, various operational modes have been invented for block ciphers that make it possible to encrypt plaintexts of any size efficiently.

The simplest mode is called *Electronic Code Book* (ECB) mode. In this mode, a large message is divided into blocks, and each block is encrypted with the block cipher independently from the other blocks. This mode has many disadvantages, mainly that it does not hide the statistics of the input blocks. In particular, the same input block is always encrypted into the same output block or, in other words, if two blocks are equal in the plaintext, then the corresponding blocks will be equal in the ciphertext. Another drawback is that the blocks of the ciphertext can be reordered

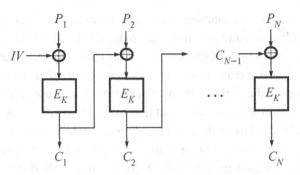

Figure A2. Using a block cipher in CBC mode.

or deleted, or new blocks can be inserted from another ciphertext that has been produced with the same key, and these modifications do not affect the decryption of the ciphertext. For this reason, the ECB mode is not recommended to be used for large messages; its common use is for the encryption of single blocks, such as passwords or session keys.

Large plaintexts are usually encrypted in *Cipher Block Chaining* (CBC) mode. In this mode, the plaintext message is divided into blocks, and the ith ciphertext block C_i is computed from the ith plaintext block P_i and the $(i-1)$th ciphertext block C_{i-1} as $C_i = E_K(P_i \oplus C_{i-1})$, where $\oplus$ denotes the bitwise XOR operation. As in the case of the first block, there is not any preceding ciphertext block yet, the first ciphertext block C_1 is computed as $C_1 = E_K(P_1 \oplus IV)$, where IV stands for an Initial Vector that must be supplied as an additional input to the encryption (and to the decryption). The operation of the CBC mode is illustrated in Figure A2.

The advantage of the CBC mode is that each ciphertext block depends on the corresponding and *all preceding* plaintext blocks. As a consequence, the reordering, deletion, and insertion of ciphertext blocks affects noticeably their decryption.[2] Moreover, XORing the plaintext blocks with the preceding ciphertext blocks before their encryption hides the statistical properties of the plaintext blocks. In particular, two equal plaintext blocks are very unlikely to be encrypted into the same ciphertext blocks, because this would require that the preceding ciphertext blocks are equal too. In addition, encrypting the same plaintext message with different IVs would result in different ciphertexts.

It can happen that the length of the plaintext message is not the multiple of the block size of the cipher. In this case, the last block of the message is shorter than the block size. In order to solve this problem, the plaintext message must be *padded* before encryption. The padding scheme should ensure that at the receiver side, the

---

[2] Nevertheless, encryption in CBC mode does not provide reliable integrity protection in general. For that purpose, standard message authentication codes or digital signatures should be used.

padding can be recognized unambiguously and removed after decryption. One commonly used padding scheme is that the last byte of the padding contains the binary representation of the length of the padding. This allows the receiver to remove the padding easily. Note that using this padding scheme means that messages are padded even if their length is the multiple of the block size (in that case, a full padding block is added to the message). The other padding bytes can be all zeros, random bytes, or they can be all equal to the last padding byte (i.e., the length of the padding).

The padding increases the length of the messages, but in some applications, this is not desirable. In sensor networks, for instance, where short messages are communicated over wireless links, the padding can be a considerable communication overhead that can shorten the lifetime of the battery powered sensor nodes. Fortunately, there exist padding schemes that do not increase the length of the messages. A simple approach is to encrypt the penultimate ciphertext block again and use the result as a key stream to encrypt the last, short-end block by XORing it with the necessary number of key stream bits.

Block ciphers can also be converted into stream ciphers. This is useful when short messages, characters, or even bits need to be encrypted as they arrive (e.g., in real-time applications). In each of the modes that convert a block cipher into a stream cipher, the block cipher is used to produce a key stream, which is then XORed to the plaintext characters. The key stream is generated by iteratively encrypting and updating some internal state. The various modes differ only in the way in which this internal state is updated.

In the *Output Feedback* (OFB) mode, the internal state is stored in a register the size of which is equal to the block length of the block cipher. In each iteration step, the content of the register is encrypted and updated with the result of the encryption. Thus, the internal state is independent from the plaintext and the ciphertext characters, which means that the block cipher in OFB mode operates as a synchronous stream cipher. The operation of the OFB mode is illustrated in Figure A3.

The *Cipher Feedback* (CFB) mode is very similar to the OFB mode, but instead of the output of the block cipher, the last ciphertext character is used to update the internal state. This is done by shifting the last ciphertext character into the register that stores the internal state. The result is a self-synchronizing stream cipher. The operation of the CFB mode is illustrated in Figure A4.

Yet another mode that converts the block cipher into a stream cipher is the *Counter* (CTR) mode (see Figure A5 for illustration). In this mode, the internal state is a counter, which is incremented in each iteration. As the state is independent of the plaintext and the ciphertext characters, this results in a synchronous stream cipher, similar to the OFB mode.

The CTR mode, however, has certain advantages compared with the OFB mode. One of these advantages is that the CTR mode supports the decryption of the

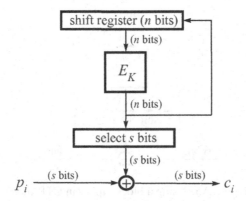

Figure A3. Using a block cipher in OFB mode.

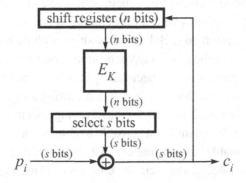

Figure A4. Using a block cipher in CFB mode.

ciphertext characters in a *random access* manner. In order to decrypt the ith character, we need to set the internal state to the ith state. In CTR mode, this can be done easily by setting the value of the counter appropriately. In contrast to this, in OFB mode, the ith state can only be reached by starting from the initial state and calling the block cipher $i - 1$ times. For similar reasons, decryption in CTR mode can be *parallelized*, whereas decryption in OFB mode can be performed only in a sequential manner.

A2.2 *Asymmetric-key encryption*

As we have mentioned earlier, in the case of asymmetric-key encryption, the encryption key K and the decryption key K' are different; moreover, computing K' from K is practically infeasible. The benefit is that the encryption key K can be made public without revealing the decryption key K'. For this reason, the encryption key is often called a *public key* and the decryption key is called a *private key* (for this reason, it is often denoted by K^{-1}). Once K is made public, anybody can encrypt messages with

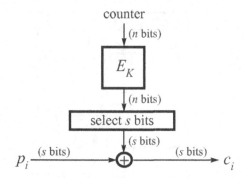

Figure A5. Using a block cipher in CTR mode.

it, whereas only the entity that possesses K' can decrypt those messages. Thus, in contrast to symmetric-key encryption schemes, where a shared secret key must be established between the parties before they can begin communicating securely, for asymmetric-key encryption no such key establishment phase is needed. This is particularly advantageous in the case of asynchronous communications, such as electronic mail, when the parties are not necessarily on-line at the same time, hence they cannot interactively set up a shared secret. Not requiring a key establishment phase before beginning to communicate can also be advantageous in wireless networks because it saves bandwidth and makes it possible to communicate even if the fixed infrastructure is not available, and on-line key distribution servers cannot be accessed.

An asymmetric-key encryption scheme consists of three algorithms: a key-pair generation algorithm, an encryption algorithm, and a decryption algorithm.[3] The key-pair generation algorithm is used to create the public key and the corresponding private key of each entity in the system. The encryption algorithm takes a message to be encrypted and the public key of the intended recipient, and it produces the encrypted messages. We usually denote the encryption of a message m with a public key K as $E_K(m)$. The decryption algorithm takes the encrypted message and the private key of the recipient, and it recovers the plaintext message.

The security of asymmetric-key encryption schemes is usually based on some well-known or widely believed hard problems, such as factoring large integers, computing discrete logarithms, decoding linear codes, or the subset sum problem. More precisely, the difficulty of breaking the encryption scheme is traced back to the difficulty of solving the underlying hard problem.

An example of a very widely used asymmetric-key encryption scheme is the RSA scheme [331]. We do not describe its operation here, because it can be found in any textbook on cryptography. We do mention, however, that the security of the RSA scheme is based on the difficulty of factoring large integers. Another well-known

---

[3] Asymmetric-key encryption schemes are also called asymmetric-key cryptosystems in the literature.

asymmetric-key encryption scheme is the ElGamal scheme [127]; its security is based on the difficulty of the discrete logarithm problem in finite fields. We must also mention the Elliptic Curve Cryptosystems (ECC), which are modifications of other schemes (e.g., the ElGamal scheme) in such a way they work in the domain of elliptic curves rather than in finite fields defined by large primes. The advantage of the ECC schemes is that they use much smaller keys than traditional asymmetric-key schemes but achieve the same level of security. Hence, their use can be beneficial in resource constrained environments, such as smart cards and sensor networks.

There is a general attack on asymmetric-key encryption schemes, which is feasible when the plaintext space (i.e., the set of possible plaintext messages) is small. Let us assume that the adversary observes a ciphertext $c = E_K(m)$, and she has previously obtained (from the context of the application or by some other means) some knowledge about the set M of the possible plaintexts. If M is small, then the adversary can try to encrypt every message in M with the publicly known key K until she finds the message m that maps into c.

The usual way to prevent this attack is to randomize the encryption. In the case of RSA, for instance, some random bytes are added to the plaintext message before encryption through the application of the PKCS #1 formatting rules [1]. The formatting rules also ensure that when the message is decrypted, the recipient can recognize and discard those random bytes.

The ElGamal encryption scheme uses another approach: it is designed to be a randomized encryption scheme in the first place. Now, we briefly describe how the ElGamal scheme works. The key generation algorithm chooses a large prime p. This determines a multiplicative group, denoted by $\mathbb{Z}_p^*$, which consists of the set $\{1, 2, \ldots, p-1\}$ as elements, and the modulo p multiplication as the operator. Then, the algorithm continues by choosing a generator element g of $\mathbb{Z}_p^*$. Being a generator element means that by iteratively powering g modulo p, one can generate the entire set of elements of $\mathbb{Z}_p^*$ (i.e., $\{g^0, g^1, \ldots, g^{p-2}\} = \{1, 2, \ldots, p-1\}$). Finally, the algorithm randomly chooses an integer a ($1 \leq a \leq p-2$), and it computes $A = g^a \mod p$. The private key is a, and the public key is (p, g, A).

A message m is encrypted as follows: The encryption algorithm randomly chooses an integer r ($1 \leq r \leq p-2$), and it computes $R = g^r \mod p$ and $C = m \cdot A^r \mod p$. The ciphertext is the pair (R, C). The encryption is randomized, because due to the random integer r, the same plaintext message is encrypted into a different ciphertext each time the encryption algorithm is called. The ciphertext is decrypted by computing $R^{p-1-a} \mod p = g^{-ar} \mod p = A^{-r} \mod p$ and $C \cdot A^{-r} \mod p = m$.

Another general attack against asymmetric-key schemes consists of substituting the public key of an honest entity with that of the adversary. As we mentioned before, encryption keys can be made publicly available, for instance, by placing them in a publicly accessible directory. However, if the integrity of this directory is not ensured, then the adversary can replace the public key of an entity with her own public key. Then, if someone wants to send an encrypted message to that entity, she will use the public key of the adversary instead of the public key of the entity, and therefore, the adversary will be able to decrypt the message. In order to prevent this and similar attacks, the general requirement is to ensure the authenticity of the public keys. The most common approach to do this is to distribute public keys in *certificates* that bind the public key to the name of its owner by the digital signature of a trusted third party called the *Certification Authority* (CA). Digital signatures are addressed in Section A5 of this appendix.

The management of certificates is a complex task that needs a considerable infrastructure, especially in large-scale applications. This infrastructure is often referred to as the *Public Key Infrastructure*, or shortly PKI. The services provided by the PKI cover the whole life cycle of the certificates, including their issuance, distribution, suspension, and revocation. A detailed description of all these services is out of the scope of this appendix; the interested reader can find more information in [13].

Asymmetric-key encryption schemes are roughly three orders of magnitude less efficient than symmetric-key encryption schemes. Therefore, they are rarely used to encrypt large messages. Instead, those messages are typically encrypted with a symmetric-key cipher and a randomly generated symmetric key that is then encrypted with the asymmetric-key cipher and the public key of the intended recipient. The encrypted message and the encrypted symmetric key are sent together to the recipient. This hybrid encryption technique is called the *digital envelope*, and it is illustrated in Figure A6.

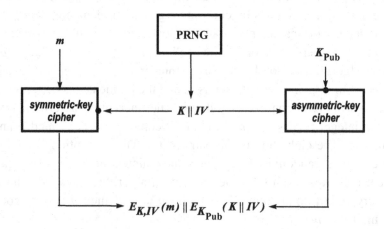

Figure A6. Digital envelope: asymmetric-key encryption is used to encrypt a randomly generated symmetric key, which, in turn, is used to encrypt the message with a symmetric-key cipher. PRNG: pseudo-random number generator.

A3 Hash functions

Hash functions take messages of arbitrary length as input and they produce outputs of a fixed length. The output of the hash function is called the *hash value* of the message or the *message digest*. By definition, there are infinitely many messages that have the same hash value (with respect to a given hash function). However, in the case of strong cryptographic hash functions, it is practically infeasible to find two messages that map into the same hash value. Therefore, the hash value of the message can represent the message for any practical purposes. The advantage of this is that hash values are shorter than messages. Thus, computationally intensive operations (e.g., a digital signature) can be performed on the hash value of the message instead of the message itself.

Hash functions are used in many applications. As mentioned above, they can be used in digital signature applications in order to reduce the size of the data to be signed. They can also be used to implement commitment schemes (see e.g., the MAD protocol in Subsection 6.2.2), one-time passwords [248], and micro-payment protocols [330]. Hash functions can be used to prove the knowledge of a value to someone who also knows that value without revealing the value itself to anybody else. This is useful, for instance, in achieving session key confirmation. Hash functions are also applicable in resource constrained wireless networks (e.g., in sensor networks) for broadcast authentication with the TESLA protocol (see Section A8 for details).

Let us now take a closer look at the properties that a cryptographic hash function should satisfy. First of all, it is desirable that the hash value can be computed efficiently (otherwise we would not gain anything by using the hash value instead of the message itself). In addition, the following properties are usually required for a hash function h.

- **Collision resistance** Collision resistance means that it is hard to find two inputs x and x' such that $h(x) = h(x')$. Such a pair of inputs is called a collision pair, or shortly a collision, hence the name of the property.
- **Weak collision resistance** Weak collision resistance means that given an input x, it is hard to find another input x' such that $h(x') = h(x)$. This property is sometimes also called *second pre-image resistance*.
- **One-way property** The hash function is said to be one-way hash function if for any hash value y (for which no pre-image is known a priori), it is hard to find an input message x such that $h(x) = y$. This property is sometimes called *pre-image resistance*.

Among the three properties, collision resistance is the strongest, by which we mean that it implies both the weak collision resistance and the one-way properties

[360]. Hence, hash functions are usually designed for collision resistance, which ensures that they satisfy the other two properties as well.

An important notion in the context of hash functions is the *birthday paradox*. The birthday paradox can be explained as follows. Let us consider a set of N elements and assume that we randomly choose elements from this set with replacement. The question we are interested in is the following: after how many trials can we expect that we choose an element that was already chosen before? According to the birthday paradox, such a repetition can be expected with a probability greater than half after around $\sqrt{N}$ trials. The paradox is that we would expect that many more trials are needed.

For a given hash function h, the set of N elements is the set of the possible hash values (i.e., $N = 2^n$, where n is the output size of h). A trial is that we randomly select an input and we compute its hash value. Due to the birthday paradox, we expect that after selecting around $\sqrt{N} = 2^{n/2}$ random inputs, we find an input that maps into a hash value that was already computed before. In other words, we can find a collision with rather high probability just by hashing $2^{n/2}$ random inputs. It follows that the output size n is an important parameter that should not be too small. At the time of this writing, $n = 160$ is considered to be acceptable, which means that the complexity of the above described birthday attack is 2^{80}.

The birthday attack on hash functions is the equivalent of the exhaustive key search attack in the case of block ciphers. We saw that choosing a large key size is a necessary condition for the security of a block cipher, otherwise the cipher is susceptible to an exhaustive key search attack. Similarly, choosing a large output size is a necessary condition for the security of a hash function, otherwise it cannot resist the birthday attack. However, we also saw that a large key size in itself cannot be sufficient to ensure that a block cipher is secure. Similarly, hash functions can be broken by exploiting their algebraic properties even if their output size is large enough.

Most hash functions used in practice are iterative hash functions. The operation of iterative hash functions is illustrated in Figure A7. The heart of an iterative hash function is its compression function, which is denoted by f in the figure. The

$M = M_1 \parallel M_2 \parallel \ldots \parallel M_k$

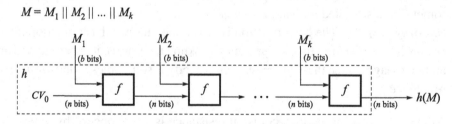

Figure A7. The operation of iterative hash functions.

compression function takes an input of length $b + n$ bits and it produces an output of length n bits. The compression function is applied iteratively to hash any input message of arbitrary length. This is done in the following way: the input is broken up into blocks of length b bits. If the length of the message is not a multiple of b, then the message is padded with zeros. Each block is input to the compression function together with the output of the previous iteration to produce the output of the current iteration. In the first iteration, the first input block is processed together with a fixed initial value CV_0. The output of the last iteration is the output of the hash function (i.e., the hash value of the input message).

Instead of padding with zeros, we can use a padding that contains the binary representation of the length of the message being hashed. This is often called Merkle-Dåmgard strengthening. If this kind of padding is used, then it can be proven formally that the collision resistance of the compression function implies the collision resistance of the hash function.

Iterative hash functions can be constructed from block ciphers by using the block cipher to implement the compression function. One problem with this approach is that the output length of the block cipher would not be sufficiently large for hash functions. For instance, if we used AES, then the output length of the hash function would only be 128 bits, which is considered to be too short to ensure collision resistance. Therefore, it is preferable to use dedicated hash functions (i.e., hash functions that use compression functions that were designed for being compression functions in the first place). A commonly used iterated hash function of this kind is the SHA-1 hash function [3].

A4 Message authentication codes

Message authentication codes (MACs) are used to provide message integrity and authentication services. The idea is to compute a MAC value from a message and then to send the message and its MAC value together to the receiver. The receiver can use the redundancy provided by the MAC value to verify if the received message is the same as the one that was sent. More precisely, the receiver computes the MAC value of the received message and compares it to the MAC value that was received together with the message. If the two MAC values are equal, then the message is considered to be intact and authentic.

These operational principles of MACs are similar to those of error detecting codes. The difference is that MACs are intended for the detection of intentional modifications of messages, whereas error detection codes are used for detecting random errors. In order to be able to detect intentional modifications, the computation of the MAC value involves some secret that is known only to the sender and the receiver. Error detection codes do not need such a secret.

$m = m_1 \parallel m_2 \parallel \dots \parallel m_N$

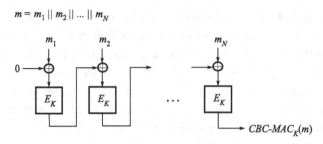

Figure A8. Computation of the CBC-MAC value.

MAC values are computed with MAC functions. A MAC function takes two inputs, a message of arbitrary length and a secret key, and it produces a fixed length MAC value. In order to be useful, MAC function should have the following security properties.

- **Key non-recovery** Key non-recovery means that it is hard to compute the secret key from observed (message, MAC value) pairs.
- **Computation resistance** Computation resistance means that even if many (message, MAC value) pairs are observed, it is hard to compute an as yet unobserved (message, MAC value) pair that verifies correctly.

Clearly, computation resistance implies key non-recovery, because if key non-recovery does not hold, then the adversary can first compute the secret key, and then use it to compute a valid MAC value for any message.

MAC functions can be built from block ciphers. A common approach is to use the block cipher in CBC mode (with a block of zeros as IV) to encrypt the message, and then to take the last encrypted block as the MAC value of the message. This scheme is called CBC-MAC [2], and it is illustrated in Figure A8. CBC-MAC has some known weaknesses. To overcome those, it is recommended to involve the length of the message in the CBC-MAC computation (e.g., by attaching a message header that contains the message length).

Another approach to constructing MAC functions is to use a hash function as the main building block. After all, hash functions are very similar to MAC functions in the sense that both types of functions produce a fixed length output from an arbitrary length input. An example of this approach is HMAC [237], which is a very widely used hash based MAC function. The operation of HMAC is so simple, that it fits in one line: $HMAC(m) = h((K^+ \oplus OP) \parallel h((K^+ \oplus IP) \parallel m))$, where h is an iterative hash function, K^+ is the MAC key padded with zeros to the length of the input block size b of the hash function, OP and IP are pre-defined constants of size b, m is the message, and $\parallel$ denotes concatenation. The size of the HMAC value is equal to the output size of the hash function h.

A5 Digital signatures

Digital signatures are very similar to MAC values: they are attached to messages to ensure their integrity and authenticity. In addition to these services, digital signatures also provide non-repudiation of message origin. Note that a valid MAC value attached to a received message does not allow the receiver to prove to a third party where the message comes from. The reason is that the receiver knows the secret key that was used to compute the MAC value of the message, hence she could have generated this message–MAC value pair herself. As the source of the message cannot be proven to a third party, the sender can deny that she sent the message. With digital signatures, this is not possible: a digital signature is generated with the help of a private key that is known to a single entity and, therefore, it cannot be repudiated.

From the above description, it follows that digital signature schemes are based on asymmetric-key cryptography. A digital signature scheme consists of a key-pair generation function, a signing function, and a verification function. The key-pair generation function is used to generate the public key and the private key of the entities. The signing function takes the message to be signed and the private key of the signer, and it produces the digital signature. The verification function takes the message, the signature, and the public key of the signer, and it outputs "accept" if the signature has been produced on the message with the private key corresponding to the given public key; otherwise it outputs "reject."

Successfully attacking a digital signature scheme means that the adversary can somehow obtain the private signing key of an entity or that she finds a method to forge signatures on messages without the signing key. For this purpose, the adversary can use the legitimate signer as an oracle to obtain valid signatures on some messages, and then, based on the obtained information, she computes the private key or generates the signature of a message that has never been submitted to the oracle. The usage of the oracle can be one-shot or interactive; the latter is called an *adaptive chosen message attack*.

As digital signature schemes use asymmetric cryptography, they are orders of magnitude less efficient than MAC functions. For this reason, we would like to limit the length of their input. The commonly used approach, called the *hash-and-sign* paradigm, is to sign the hash value of a message instead of the full message. As we have seen previously, this requires the hash function to satisfy certain properties. The hash-and-sign paradigm works as follows: The message is first hashed, and the signature is computed on the hash value using the private key of the signer and the signing function of the signature scheme. Then, the signature is attached to the original message. When the signature is verified, the verifier first computes the hash of the message and then verifies the signature using the public key of the signer and the signature verification algorithm.

It is of paramount importance regarding the security of the hash-and-sign approach that the hash function satisfies the collision resistance property (see Section A3). If h were not collision resistant, then the adversary could find two messages m and m' such that $h(m) = h(m')$. Now, the adversary can somehow obtain the signature $\sigma(h(m))$ for m from the legitimate signer, and this will be a valid signature for m' too.

Well-known signature schemes include the RSA and the ElGamal signatures. The security of the former is based on the difficulty of factoring large integers, whereas the latter is based on the difficulty of computing a discrete logarithm. The ElGamal scheme serves as the basis of the Digital Signature Algorithm (DSA) [4], a standardized signature scheme that is very widely used in practice. DSA can be implemented over elliptic curves; this is called ECDSA [380]. The advantage of ECDSA is its reduced signature length (typically 320 bits) compared to the signature length of DSA (typically 1024 bits). Thus, using ECDSA results in a considerably smaller communication overhead than using DSA, which is an important aspect to consider in wireless networks of battery-powered devices.

A6 Session key establishment protocols

Symmetric-key cryptographic primitives (symmetric-key ciphers and MAC functions) require that the communicating parties share a secret key. Shared secrets can be established between parties in various ways. For instance, the secret key can be exchanged manually between the parties, when they meet in person. This is a secure approach, but it does not scale, and it is not always practical to require a physical meeting before the parties can communicate securely over the network.

A scalable and practical approach to establish a shared secret between two (or more) remote parties is to use a session key establishment protocol. Such a protocol allows the parties to set up a shared key via the network in an on-demand manner. The typical scenario is the following: when the parties start a new communication session, they run the session key establishment protocol and set up a shared key that they use throughout the session (hence the name session key). When the session is closed, the key is deleted. When they start a new session again, they establish a new session key, and so on.

Session keys are advantageous for many reasons. First of all, using short-term session keys limits the amount of ciphertexts produced with the same key, and thus makes cryptanalytical attacks more difficult. Another benefit is that if a session key is compromised, it affects only the session in which that key was used; the other sessions are not necessarily affected. Yet another advantage is that keys are created only when they are really needed, which limits the amount of key material that needs to be kept secret at any given time.

Broadly speaking, session key establishment protocols fall into two classes: *key transport* protocols and *key agreement* protocols. In the case of key transport protocols, the session key is created by one of the protocol participants, and then it is transferred in a secure way to the parties that need it. Hence, the party that creates the session key can fully control its value. In the case of key agreement protocols, every protocol participant contributes a key share, and the parties compute the session key from these key shares. Now, no party has full control over the value of the session key, which depends on the contributions of the other parties.

Key transport protocols usually rely on some long-term key material already installed in the system for the secure transfer of the freshly generated session keys. There are two cases: either long-term symmetric-keys or long-term asymmetric-keys are used. In the former case, typically, there is an on-line server, called the *key distribution center*, in the system that already shares a long-term symmetric-key with every user of the system. The KDC is involved in every run of the key transport protocol, and the long-term symmetric keys are used to secure the transport of the session keys to the users. When long-term asymmetric-keys are used, there is no need for on-line servers. In this case, the security of the key transport protocol can be ensured with the usage of the long-term public and private keys of the users them-selves. However, off-line servers are still needed for the certification of the public keys. These are commonly called *Certification Authorities* (CAs). Key agreement protocols are typically based on asymmetric-key cryptography, hence they need the same kind of infrastructure as asymmetric-key based key transport protocols.

Besides being key transport or key agreement protocols, session key establish-ment protocols can be further classified according to the services provided to the protocol participants. There are two main services that a key establishment pro-tocol must provide: *key authentication* and *key freshness*. There are two types of key authentication: *implicit* and *explicit*. Explicit key authentication is a stronger service that provides implicit key authentication and *key confirmation*. We explain these notions in more detail.

- **Implicit key authentication** If this service is provided by the key establishment protocol to a party, then after a successful execution of the protocol, the party can be assured that the established session key *can* only be known by another well-identified party (the intended communication partner) and to some trusted third parties (e.g., to the KDC). The party cannot be sure, however, that the other party does possess the session key.
- **Key confirmation** Key confirmation is used to convince a party that the other party possesses the established session key. This can be achieved by sending the hash value of the session key to the party, or using the session key to encrypt a known message.

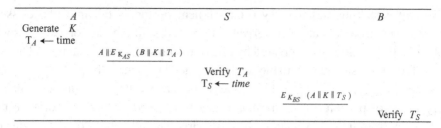

Figure A9. Operation of the Wide Mouth Frog protocol.

- **Explicit key authentication** A key establishment protocol provides explicit key authentication if it provides both implicit key authentication and key confirmation.
- **Key freshness** If this service is provided by the key establishment protocol to a party, then the party can be assured that the established session key is fresh, which means that it has never been used as a session key before.

Any of these services can be provided to one of the protocol participants or to both of them. In addition, some protocols provide a different type of key authentication to each participant.

We will now illustrate the concepts introduced through some examples. In particular, we describe and informally analyze a symmetric-key based key transport protocol, an asymmetric-key based key transport protocol, and a key agreement protocol.

Our first example is a symmetric-key based key transport protocol called the Wide Mouth Frog protocol [70]. In this protocol, there are three participants: two main parties that want to establish a session key between themselves and a KDC server. The main parties are denoted by A and B, and the server is denoted by S. It is assumed that A and B already share a long-term symmetric key with S. These keys are denoted by K_{AS} and K_{BS}, respectively. The main idea of the protocol is that A generates the session key, denoted by K, and transfers it to B via the server S. During the transfer, K is encrypted with the long-term keys K_{AS} and K_{BS}. The protocol uses timestamps to ensure key freshness.

The operation of the Wide Mouth Frog protocol is illustrated in Figure A9. A generates the session key K, and encrypts it together with the identifier of B and the current timestamp T_A. The resulting ciphertext $E_{K_{AS}}(B\|K\|T_A)$ is sent to S together with the identifier of A. Based on the clear identifier in the message, S looks up the key K_{AS} shared with A and decrypts the message. At this point, it sees that A wants to establish the session key K with B. S verifies that T_A is sufficiently close to the current time. If so, S encrypts K together with the identifier of A and a new timestamp T_S with the key K_{BS} that it shares with B. The resulting ciphertext

$E_{K_{BS}}(A\|K\|T_S)$ is sent to B. Upon reception of this message, B decrypts it and verifies if T_S is fresh enough. If so, then B accepts the key K as a fresh session key shared with A.

For A, the protocol provides implicit key authentication. This can be explained as follows: K is transferred under the protection of the keys K_{AS} and K_{BS}. Moreover, S is trusted to pass on K only to B. Hence, A can be assured that only B and S can have access to K. However, she cannot be sure that B received K, as she does not receive any acknowledgement in the protocol. In addition, the protocol ensures key freshness for A. The reason is that A generates K and therefore she knows that it is fresh.

For B, the protocol provides explicit key authentication. In order to see this, we have to show that the protocol provides both implicit key authentication and key confirmation for B. Implicit key authentication can be explained in the same way as above. Key confirmation is provided because B knows that A generated K, and therefore, she must possess it, and because B trusts S that it indeed received K from A.

Seemingly, the protocol also provides key freshness to B owing to the use of the timestamps; but in fact, it does not. An attacker can fool B into accepting an old session key as follows. The attacker quickly replays to S the second message of the protocol, extended with the identifier of B. S interprets this as the first message of a new protocol run, in which B wants to set up a key with A. If the replay is fast enough, then the timestamp T_S in the message is still acceptable for S. Hence, S responds with $E_{K_{AS}}(B\|K\|T_S')$, where $T_S' > T_S$. The attacker intercepts this message and replays it to S again, now extended with the identifier of A. For similar reasons as explained before, S responds with $E_{K_{BS}}(A\|K\|T_S'')$, where $T_S'' > T_S'$, and so on. The attacker can continue replaying messages to S for an arbitrarily long time and use S as an oracle to refresh the timestamp attached to the old key K. At some point T in time, the attacker will receive $E_{K_{BS}}(A\|K\|T)$ from S, where $T \gg T_S$, and she can send this message to B. B will accept the key K, although by this time K is quite old (and possibly compromised). It is rather easy to fix the Wide Mouth Frog protocol so that it resists this attack, and we leave this to the reader as an exercise.

Many other symmetric-key based key transport protocols have been proposed in the literature. Giving a comprehensive overview of these protocols is well beyond the scope of this appendix. However, we want to emphasize two things here. First, it is not always the case that one of the main parties generates the session key, like in the Wide Mouth Frog protocol. Indeed, in most of the protocols, the KDC generates the session key (upon request) and transfers it to both main parties. Second, key freshness is not always achieved with the help of timestamps, but rather by means of *nonces*. Nonces are "numbers that are used only once." Very often, nonces are implemented as freshly generated, unpredictable random numbers, chosen from a large space, so that it is very unlikely that they repeat. The typical usage of such

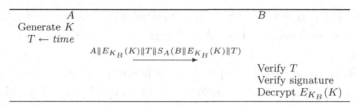

Figure A10. An asymmetric-key based key transport protocol, where the session key is encrypted with the public key of the intended receiver, and then, signed by the sender. Key freshness is achieved by use of a timestamp.

nonces is the following: one party generates a nonce and sends it to the other party. The other party includes the nonce in its response (e.g., encrypts the nonce and the session key together). When the first party receives the response, she waits for her nonce to be sent back. If so, then she knows that the response must have been generated after sending the nonce in the first message. If not too much time has elapsed between the two messages, then the first party can consider the content of the response (including the session key) to be fresh. The protocols that rely on timestamps tend to use fewer messages. However, their disadvantage is that, unlike nonce-based protocols, they require the participants' clocks to be synchronized.

Our second example is an asymmetric-key based key transport protocol, illustrated in Figure A10. The protocol has two parties, A and B. It is assumed that each party knows the public key of the other party. A generates a session key K and encrypts it with the public key K_B of B. Then, A signs the identifier of B, the encrypted session key, and a timestamp T. The encrypted session key $E_{K_B}(K)$ and the signature $S_A(B\|E_{K_B}(K)\|T)$ are sent to B. B first verifies the timestamp and then verifies the signature of A. Afterwards, she decrypts the encrypted session key and obtains K.

This protocol provides implicit key authentication for A, because she knows that only B can decrypt a message that is encrypted with the public key K_B. For B, the protocol provides explicit key authentication, because the signature of A proves that K was generated by A, and the identifier of B within the signature proves that K was indeed intended for B; moreover, B knows that A possesses K because she generated it. Key freshness for A is trivial, and key freshness for B is ensured by the timestamp, which is also covered by the signature.

Our third example is the well-known Diffie–Hellman protocol [115]. This is a key agreement protocol, the security of which is based on the difficulty of computing discrete logarithms. The protocol is illustrated in Figure A11, and its operation is explained as follows. The protocol has two parties, A and B. Both parties know the public parameters of the system: a large prime number p and a generator g ($2 \le g \le p - 2$) of the multiplicative group $\mathbb{Z}_p^*$ (see the description of the ElGamal

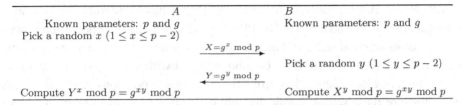

Figure A11. Operation of the Diffie–Hellman protocol.

encryption scheme in Section A2 for the definition of these terms). A generates a random number x ($1 \leq x \leq p - 2$), computes $X = g^x \bmod p$, and sends X to B. Similarly, B generates a random number y ($1 \leq y \leq p - 2$), computes $Y = g^y \bmod p$, and sends Y to A. Then, A and B can both compute the shared secret $g^{xy} \bmod p$ by computing $Y^x \bmod p$ and $X^y \bmod p$, respectively.

The Diffie–Hellman protocol is an elegantly simple protocol, but it has a serious problem: it does not provide key authentication at all. This means that the parties do not really know with whom they establish the shared secret key $g^{xy} \bmod p$. Interestingly, key freshness is still ensured for both parties, as the key depends on the parties' contributions. And if a party generates a fresh random number, then the resulting key will be fresh too.

The Diffie–Hellman protocol can be extended in various ways to provide key authentication services. One approach is to use digital signatures; an authenticated variant of the Diffie–Hellman protocol that follows this approach is called the Station-to-Station protocol. Another approach is based on the comparision of short strings as described in Section 5.4.

A7 Pseudo-random number generators

As we have seen, many cryptographic algorithms and protocols require the generation of random values, such as keys and nonces. However, the default random number generators that are provided as part of the various programming languages, such as C and Java, are not appropriate for cryptographic purposes. Although these random number generators have good statistical properties, and the sequence of values that they produce indeed appears to be random, their output is often predictable; whereas keys and nonces need to be unpredictable.

An ideal cryptographic random number generator produces truly unpredictable values. In practice, we are often satisfied with a random number generator that is not truly unpredictable, but it is practically infeasible to distinguish it from a truly unpredictable random number generator. Such a practical random number generator is called *pseudo-random* number generator, or PRNG for short.

A PRNG works, typically, as follows. The PRNG has an internal state assumed to be unknown to the adversary. The PRNG produces its next output as a one-way function of its internal state, and then it updates its internal state in a deterministic manner. In addition, the PRNG is continuously fed with samples of physical processes, such as clock values, keystroke timings, mouse positions, and disk access times. Each of these samples typically contains only a few bits of randomness, so many of them need to be collected in a so-called entropy pool. When the PRNG estimates that a sufficient amount of random samples have been collected in the pool, it hashes the content of the pool and uses the result to update its internal state. This operation is often referred to as *re-keying*.

Good PRNGs are designed in such a way that they satisfy the following properties.

- The adversary cannot compute the internal state of the PRNG, even if she has observed many outputs of the PRNG.
- The adversary cannot compute the next output of the PRNG, even if she has observed many previous outputs of the PRNG.
- If the adversary can observe or even manipulate the input samples that are fed in the PRNG, but she does not know the internal state of the PRNG, then the adversary cannot compute the next output and the next internal state of the PRNG.
- If the adversary has somehow learned the internal state of the PRNG, but she cannot observe the input samples that are fed in the PRNG, then the adversary cannot figure out the internal state of the PRNG after the re-keying operation.

An example of a widely used PRNG is the ANSI X9.17 [379] algorithm.

A8 Advanced authentication techniques

In this section, we describe some authentication techniques that can be useful in upcoming wireless networks, in particular, in resource constrained applications. More precisely, we describe the concepts of hash chains, Merkle-trees, and the TESLA protocol. Hash chains and Merkle-trees can be used to amortize the cost of a digital signature over several transactions, whereas the TESLA protocol can be used for authenticating broadcast messages efficiently using simple symmetric-key MAC functions.

A8.1 Hash chains

A hash chain is a sequence of hash values that are computed by iteratively calling a one-way hash function on an initial value. Let us denote the initial value by v_0 and the hash function by h. Then, the ith element v_i of the hash chain is computed as $v_i = h(v_{i-1}) = h^{(i)}(v_0)$.

An important property of the hash chain is that its elements can be easily computed in one direction, but not in the reverse direction. In other words, if someone knows v_i, then she can compute any $v_j = h^{(j-i)}(v_i)$ for any $j > i$, but she cannot compute any v_k for $k < i$. This property stems from the one-way property of the hash function.

A hash chain can be used for repeated authentications at the cost of a single digital signature (and at the cost of the computation and storage of the hash chain, of course). For this purpose, the entity that wants to authenticate itself first computes a hash chain $v_0, v_1, \ldots, v_n$ of length n, and digitally signs the last element v_n. By doing this, the entity commits to the hash chain. The digital signature can be verified by anyone using the public signature verification key of the entity. Later on, the entity can authenticate itself repeatedly (at most n times) by revealing the elements of the hash chain in reverse order. More precisely, at the ith authentication, the entity reveals v_{n-i}. The verifier can hash this value i times and check if the result matches v_n that has been signed by the entity. Alternatively, the verifier can remember the last used hash chain element v_{n-i+1}, and she can verify v_{n-i} with a single hash computation.

In the above described repeated authentication scheme, hash chain elements are used and accepted only once. This ensures that when v_{n-i} is accepted, the elements $v_{n-i+1}, \ldots, v_{n-1}$ can no longer be used. In addition, due to the one-way property of the hash chain, the elements $v_{n-i-1}, \ldots, v_0$ that can still be used for authentication cannot be computed by anybody else but the entity that knows v_0. This assures the verifier that if she sees any of the elements $v_{n-i-1}, \ldots, v_0$, then it must have been revealed by the entity that committed to the hash chain with its signature.

Finally, we note that hash chains can be stored efficiently with a storage complexity that is logarithmic in the length n of the hash chain. The reader is referred to [105] for the details.

A8.2 Merkle-trees

A property that limits the application of hash chains in some applications is that the elements can only be revealed sequentially. Merkle-trees overcome this problem by allowing for the pre-authentication of a set of values with a single digital signature (like in the case of hash chains) and for the revelation of those values in *any* order (unlike in the case of hash chains).

The operation of Merkle-trees can be summarized as follows: let the set of values that we want to authenticate be $v_1, v_2, \ldots, v_{2^\ell}$. First of all, we hash each value v_i into v_i' with a one-way hash function. Then, we assign the hashed values to the leaves of a binary tree. Moreover, to each internal vertex u of this tree, we assign a value that is computed as the hash of the values assigned to the two children of u.

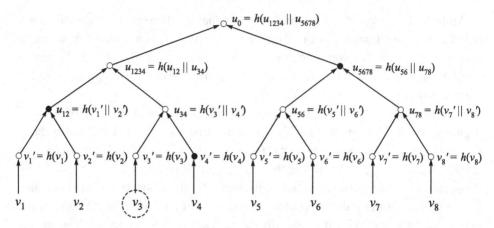

Figure A12. Illustration of the operating principles of Merkle-trees. The hash values of the values we want to authenticate are assigned to the leaves of the tree. The value assigned to an internal vertex of the tree is computed as the hash value of the values assigned to the children of that vertex. The value assigned to the root is signed and distributed. When authenticating a value, we reveal the value itself, and the values assigned to the siblings of the vertices on the path from the revealed leaf to the root. For instance, the figure shows that value v_3 is authenticated by revealing v_3, v_4', u_{12}, and u_{5678} (the black vertices). The verifier can compute $h(h(u_{12}\|h(h(v_3)\|v_4'))\|u_{5678})$ and compare the result to the value u_0 assigned to the root. If there is a match, then v_3 is authenticated.

Finally, we digitally sign the value assigned to the root of the tree. Later, when we want to authenticate any of the values v_i, we reveal v_i and all the values assigned to the siblings of the vertices on the path from v_i' to the root. The verifier can hash these values in the appropriate order and compare the result to the value assigned to the root (and signed digitally previously). If the two values match, then the value v_i that has just been revealed from the tree is accepted by the verifier as authentic (i.e., coming from the same entity who computed the tree and digitally signed the root). These operating principles are illustrated in Figure A12.

Note that, owing to the one-way property of the hash function, we cannot use the revealed value v_i and the values assigned to the siblings to compute an as yet unrevealed value v_j. This ensures that the values can be revealed in any order.

A8.3 Broadcast authentication with TESLA

Wireless communication has a broadcast nature: a message transmitted by a wireless device is received by many (potentially all) other wireless devices in the power range of the first device. Many protocols take advantage of this broadcast property. A frequently recurring problem in this context is that of authenticating broadcast messages. The requirement of broadcast authentication is that messages can be authenticated by their senders in such a way that *all* receivers can verify them.

The "standard" solution to the broadcast authentication problem is to use digital signatures. By definition, the digital signature of the sender can be verified by all other devices, with the help of the public signature verification key of the sender. However, in some resource constrained environments (e.g., in RFID systems or in sensor networks), it could be infeasible to implement a digital signature scheme. Hence, we need an alternative solution to the broadcast authentication problem that preferably uses only symmetric-key cryptographic primitives. A promising candidate is the TESLA protocol [314] that provides services that are similar to those provided by digital signatures, but it uses only hash functions and symmetric-key MAC functions. The trade-off is that the services provided by TESLA are *delayed*, meaning that the authenticity of messages cannot be verified immediately upon their reception.

The main idea of the TESLA protocol is simple, yet powerful. Each sender has a one-way key chain (i.e., a hash chain where the elements are used as cryptographic keys) and a key disclosure schedule according to which the sender releases the elements of this key chain. The elements of the key chain are released in reverse order, as with normal hash chains. When authenticating a broadcast message, the sender computes a MAC value for the message using a key K from her key chain that is not expected to be disclosed by the time the message is received by the receivers. Hence, when the message is received, the receivers cannot verify its MAC value, because they do not know K yet. Therefore, the receivers cash the message, and wait until the sender releases K. When this happens, every receiver can verify the MAC value. As the receivers know that no one but the sender knew K when the message was received, they are assured that the message was indeed sent by the sender.

The receivers must be able to verify that the key disclosed by the sender belongs to her key chain. This can be ensured in a similar way with hash chains. In other words, the last element of the key chain can be authenticated and distributed to every receiver. The authentication of the last element of the key chain can be based on a digital signature, or on a separate MAC value computed for each receiver (assuming that the sender shares a symmetric key with every receiver).

In addition, when receiving a message, the receivers must be able to verify that the key that was used to compute the MAC value has not been disclosed yet. For this reason, the receivers must know the key disclosure schedule of the sender, and their clocks must be loosely synchronized with the clock of the sender.

Due to the way the key chain is constructed, when a key K_i is disclosed, anybody can compute all the previously disclosed keys K_j ($j > i$). Thus, if a receiver missed the reception of a key K_j (e.g., owing to some interference), then she can still compute K_j later, when any key K_i ($i < j$) is released. This ensures the robustness of the protocol, even in a lossy environment.

A9 To probe further

The ultimate source of information about the operation of cryptographic algorithms and protocols is the *Handbook of Applied Cryptography* [272] written by Menezes, van Oorschot, and Vanstone. The reader can find there the descriptions of most of the cryptographic schemes that we have mentioned as examples in this appendix. Another popular source of information on cryptographic algorithms and protocols is the book of Schneier [339], which includes the source code of many cryptographic algorithms in C. There are also many good textbooks on cryptography where the main concepts are explained in a scholarly manner. An example is the textbook by Stinson [360], which is used as the basis for undergraduate Cryptography courses at many universities around the world.

A comprehensive treatment of authentication and session key establishment protocols can be found in the book of Boyd and Mathuria [65].

The TESLA protocol has been published by Perrig, Canetti, Tygar, and Song in [314]. A similar idea was described earlier by Cheung in [100]. More details on the application of hash chains and Merkle-trees in wireless networks can be found in [179].

Finally, we must note that at the time of this writing, SHA-1 is about to be broken. This means that apparently there is a way to generate collisions against SHA-1 with much less effort than that of the birthday attack. The interested reader is referred to [373, 372] for the details.

A10 Questions

(1) How does the exhaustive key search attack work when the adversary can obtain only ciphertexts, but she knows that the plaintext messages have some redundancy (e.g., contain parity bits)?

(2) In the case of the ElGamal encryption, why is it important that each message is encrypted with a different random number r?

(3) Why does collision resistance imply weak collision resistance? Try to prove it.

(4) Why is it important that the length of the message is included in the CBC-MAC computation? How would MAC forgery be possible if the length was not included?

(5) Try to correct the Wide Mouth Frog protocol so that it resists the attack described in Section A6.

(6) What are the advantages and disadvantages of nonces compared to timestamps?

Appendix B

A tutorial on game theory for wireless networks

B1 Introduction

As we have mentioned in Chapter 3, the way of the future for the proper operation of wireless networks consists in the deployment of appropriate rule enforcement mechanisms. These mechanisms should prevent or discourage malicious and selfish behavior. The design of the latter can tremendously benefit from game theoretic modeling.

Game theory [144, 150, 296] is a discipline aimed at modeling situations in which decision-makers have to choose specific actions that have mutual, possibly conflicting, consequences. It has been used primarily in economics, in order to model competition between companies: for example, should a given company enter a new market, considering that its competitors could make similar (or different) moves? Game theory has also been applied to other areas, including politics and biology.[1]

In wireless networks, the players can be either wireless stations striving to obtain as much possible bandwidth from the (shared) medium; or they can be wireless operators aimed at increasing their market share or their revenue. It is clear that in both cases, the actions of a given player can affect other players, sometimes in a negative way.

In this tutorial, we carefully explain how situations of this kind can be modeled by making use of game theory; for the sake of simplicity, we restrict ourselves in this appendix to the case in which the players are wireless stations (the reader interested in the interactions between operators should refer to Chapter 11). By leveraging on four simple running examples, we introduce the most fundamental

---

[1] The name of "game theory" itself can be slightly misleading, as it can be associated with parlor games such as chess and checkers. Yet, this connection is not completely erroneous, as parlor games do have the notion of players, payoffs, and strategies – concepts that we will introduce shortly.

concepts of non-cooperative game theory.[2] This approach should help students and scholars to quickly master this fascinating analytical tool without having to read the existing lengthy, economics-oriented books; it should also assist them in better understanding Part III of this book and in modeling problems of their own. As game theory is still rarely taught in engineering and computer science curricula, we assume the reader to have no (or very little) background in this field; therefore, we take a basic and intuitive approach.

In the examples that we will develop, the players of the game are devices willing to transmit or receive data (e.g., packets). They have to cope with a limited transmission resource (i.e., the radio spectrum), meaning that they have conflicting interests. In an attempt to resolve this conflict, they can make certain moves such as transmitting now or later; changing their transmission channel; or adapting their transmission rate.

Of course, each device is used by a (human) user, who could give it to another user, but for the sake of simplicity we will consider that each device is bound to a given user. It makes more sense to consider that the device (and not the human user) is the player, because the decisions that the device makes are conditioned by the way the device is programmed. We will thus use the two terms "device" and "player" interchangeably.

In compliance with the practice of game theory, we assume that the players are *rational*, which means that they try to maximize their *payoff* or alternatively to minimize their *costs*.[3] This assumption of rationality is often questionable, given for example the altruistic behavior of some animals, but we believe that most of the interactions (even those that seem to be irrational) can be captured using the concept of rationality, with the appropriate adjustment of the payoff function. In order to maximize their payoff, the players act according to their *strategies*. The strategy of a player can be a single *move* (as we will see in Section B2) or a set of moves during the game (as we present in Section B4).

In this tutorial, we devote particular attention to the selection of the examples so that they match our focus on wireless networks. For the sake of clarity (and in compliance with traditional examples), we define these examples for two decision-makers, hence the corresponding games are two-player games. Note that the applications of game theory extend beyond two-player games. In most networking problems, there are several participants.

---

[2] Another branch of game theory focuses on "cooperative games"; these games require additional signalization or agreements between the decision-makers and hence a solution based on them might be more difficult to realize.

[3] In game theory, one usually uses the concept of payoff maximization, whereas the cost minimization comes from control theory. As it is more appropriate for this tutorial, we use the payoff maximization objective.

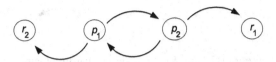

Figure B1. The Forwarder's Dilemma game.

We take an intuitive top-down approach in the protocol stack to select the examples in wireless networking as follows. Let us first assume that time is divided into *time slots* and each device can make one move in each time slot.

(1) In the first game called the *Forwarder's Dilemma*,[4] we assume that there exist two devices as players, p_1 and p_2. Each of them wants to send a packet to her receiver, r_1 and r_2 respectively, in each time slot using the other player as a forwarder. We assume that the communication between a player and her receiver is possible only if the other player forwards the packet. We show the Forwarder's Dilemma scenario in Figure B1. If player p_1 forwards the packet of p_2, it costs player p_1 a fixed *cost* $0 < c \ll 1$, which represents the energy and computation spent for the forwarding action. By doing so, she enables the communication between p_2 and r_2, which gives p_2 a *benefit* 1.[5] The payoff is the difference between the benefit and the cost. We assume that the game is symmetric and the same reasoning applies to the forwarding move of player p_2. The dilemma is the following: each player is tempted to drop the packet she should forward, as this would save some of her resources; but if the other player reasons in the same way, then the packet that the first player wanted to be relayed will be dropped. However, they could do better by mutually relaying each other's packet. Hence the dilemma.

(2) In the second example, we present a scenario, in which a sender *se* wants to send a packet to her receiver *r* in each time slot. To this end, she needs *both devices* p_1 and p_2 to forward for her. Thus, we call this game the *Joint Packet Forwarding Game*. Similarly to the previous example, there is a forwarding cost $0 < c \ll 1$ if a player forwards the packet of the sender. If both players forward, then they each receive a benefit of 1 (e.g., from the sender or the receiver). We show this packet forwarding scenario in Figure B2.

(3) The third example, called *Multiple Access Game*,[6] introduces the problem of medium access. Assume that there are two players p_1 and p_2 who want to send

---

[4] We have chosen this name as a tribute to the famous Prisoner's Dilemma game in the classic game theory literature [36, 150, 144, 296].

[5] Many authors, especially in computer science, tend to call this value the "utility." We refrain from doing so, because in game theory the function expressing the way the payoff is computed is called the "utility function," leading to a very unfortunate confusion.

[6] In game theory textbooks, this type of game is referred to as the "Hawk–Dove" game, or sometimes the "Chicken" game.

Figure B2. The Joint Packet Forwarding Game.

some packets to their receivers r_1 and r_2 using a shared medium. We assume that the players have a packet to send in each time slot and they can decide to transmit it or not. Suppose furthermore that p_1, p_2, r_1, and r_2 are in the power range of each other, hence their transmissions mutually interfere. If player p_1 transmits her packet, she incurs a transmission cost of $0 < c \ll 1$, similarly to the previous examples. The packet transmission is successful if p_2 does not transmit (stays quiet) in that given time slot, otherwise there is a collision. If there is no collision, player p_1 gets a benefit of 1 for the successful packet transmission.

(4) In the last example, we assume that player p_1 wants to transmit a packet in each time slot to a receiver r_1. In this example, we assume that the wireless medium is split into two channels ch_1 and ch_2 according to the Frequency Division Multiple Access (FDMA) principle [321, 340]. The objective of the *malicious* player p_2 is to prevent player p_1 from a successful transmission by transmitting on the same channel in the given time slot. In wireless communication, this is called *jamming*, hence we refer to this game as the *Jamming Game.*[7] Clearly, the objective of p_1 is to succeed in spite of the presence of p_2. Accordingly, she receives a payoff 1 if the attacker cannot jam her transmission and she receives a payoff of -1 if the attacker jams her packet. The payoffs for the attacker p_2 are the opposite of those of player p_1. We assume that p_1 and r_1 are synchronized, which means that r_1 can always receive the packet, unless it is destroyed by the malicious player p_2. Note that we neglect the transmission cost c in this case, because it applies to each payoff (i.e., the payoffs would be $1 - c$ and $-1 - c$) and does not change the conclusions drawn from this game.

We deliberately chose these examples to represent a wide range of problems over different protocol layers (as shown in Figure B3). There are indeed fundamental differences between these games. The Forwarder's Dilemma is a symmetric *nonzero-sum game*, because the players can mutually increase their payoffs by cooperating (i.e., from zero to $1 - c$). The conflict of interest is that they have to provide the packet forwarding service for each other. Similarly, the players have to establish the packet forwarding service in the Joint Packet Forwarding Game, but they are not in a symmetric situation anymore. The Multiple Access Game is also a nonzero-sum game, but the players have to share a common resource, the wireless

---

[7] In the classic game theory literature, this game corresponds to the game of "Matching Pennies."

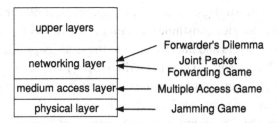

Figure B3. Classification of the examples according to protocol layers.

medium, instead of providing it. Finally, the Jamming Game is a *zero-sum game* because the benefit of one player represents the loss of the other player, meaning that $\sum_{i \in \mathcal{P}} (benefit_i - cost_i) = 0$, where $\mathcal{P}$ is the set of players. These properties lead to different games and hence to different strategic behaviors, as we will explain in the next section.

B2 Static games

In this section, we assume that each player makes a single move and that all moves are made simultaneously. In game-theoretic terms, this is called a *static game*. We will demonstrate how game theory can be used to analyze the games introduced before and to identify the possible outcomes of the strategic interactions of the players.

B2.1 Static games in strategic form

We define a game $G = (P, S, U)$ in *strategic form* (or normal form) by the following three elements. P is the set of *players*. As mentioned, we restrict ourselves to two players $p_1, p_2 \in P$ in our examples, but we present each definition such that it holds for any number of players. For convenience, we will designate by subscript $-i$ all the players belonging to P except i herself. These players are often designated as being the *opponents* of i. In our examples, player i has one opponent referred to as j.[8] S_i corresponds to the *pure-strategy* of player i. This means that the strategy assigns zero probability to all possible moves, except one (i.e., it clearly determines the move to make).

We will see in Section B2.4, that the players can also use *mixed strategies*, meaning that they choose different moves with different probabilities. We designate the joint set of the strategy spaces of all players as follows $S = S_1 \times \cdots \times S_{|P|}$. We

---

[8] The name "opponent" does not necessarily mean that the players try to decrease the payoff of the other player, but rather that each of them tries to maximize her own payoff without caring about the payoff of the other player.

will represent the pure-strategy space of the opponents of player i by $S_{-i} = S \backslash S_i$. The set of chosen strategies constitutes a *strategy profile* $s = (s_1, s_2)$. Note that our examples have two players and thus we refer to the strategy profile of the opponents as $s_{-i} = s_j \in S$. The *payoff* $u_i(s)$ (also written $u_i(s_1, s_2)$) expresses the benefit of player i given the strategy profile s minus the cost it has to incur: $u = b - c$.[9] In the examples of this appendix, we have $U = \{u_1(s), u_2(s)\}$.

At this point of the discussion, it is very important to explicitly state that we consider the game to be with *complete information*.

Definition B1 A game with ***complete information*** is a game in which each player has full knowledge of all aspects of the game.

In particular, complete information means that the players know each element in the game definition: (i) who the other players are, (ii) what their possible strategies are and (iii) what payoff will result for each player for any combination of moves. One should be careful not to confuse the concept of complete information with the concept of *perfect information* that we will present in detail in Section B3.3.

Let us first study the Forwarder's Dilemma in a static game. As mentioned before, in a static game there is only one time slot. The players can decide to forward (F) the packet of the other player or to drop it (D); this decision represents the strategy of the player. As mentioned earlier, this is a nonzero-sum game, because by helping each other to forward, they can achieve an outcome that is better for both players than mutual dropping.

Matrices provide a convenient representation of strategic-form games with two players. We can represent the Forwarder's Dilemma game as shown in Table B1. In this table, p_1 is the row player and p_2 is the column player. Each cell of the matrix corresponds to a possible combination of the strategies of the players and contains a pair of values representing the payoffs of players p_1 and p_2, respectively.

B2.2 Iterated dominance

Once the game is expressed in strategic form, it is usually interesting to *solve* it. Solving a game means predicting the strategy of each player, considering the information the game offers and assuming that the players are rational. There are several possible ways to solve a game; the simplest one consists in relying on *strict dominance*.

---

[9] As we already mentioned, the payoff is the result of the computation of a "utility function", which is the reason why it is usually designated by letter u. Note that the utility functions might be different for the two players, as for example in the Jamming Game.

Table B1. *The Forwarder's Dilemma game in strategic form, where p_1 is the row player and p_2 is the column player*

		p_2	
		F	D
p_1	F	$(1 - c, 1 - c)$	$(-c, 1)$
	D	$(1, -c)$	$(0, 0)$

Each of the players has two strategies: to forward (F) or to drop (D) the packet of the other player. In each cell, the first value is the payoff of player p_1, whereas the second is the payoff of player p_2.

Definition B2 Strategy s_i' of player i is said to be **strictly dominated** by her strategy s_i if

$$u_i(s_i', s_{-i}) < u_i(s_i, s_{-i}), \forall s_{-i} \in S_{-i}. \tag{B1}$$

Coming back to the example of Table B1, we solve the game by *iterated strict dominance* (i.e., by iteratively eliminating strictly dominated strategies). If we consider the situation from the point of view of player p_1, then it appears that for her the F strategy is strictly dominated by the D strategy (indeed, $1 - c < 1$; $-c < 0$). This means that we can eliminate the first row of the matrix, because a rational player p_1 will never choose this strategy. A similar reasoning, now from the point of view of player p_2, leads to the elimination of the first column of the matrix. As a result, the solution of the game is (D, D) and the payoff is $(0, 0)$. This can seem quite paradoxical, as the pair (F, F) would have led to a better payoff for each of the players. It is the lack of trust between the players that leads to this suboptimal solution.[10]

Not all games exhibit strictly dominated strategies, as we will see now with the Joint Packet Forwarding Game. The two devices have to *simultaneously* decide whether to forward the packet, before the source actually sends it.[11] Table B2 shows the strategic form.

In the Joint Packet Forwarding Game, none of the strategies of any player strictly dominates the other. If player p_1 drops the packet, then the move of player p_2 is indifferent and thus we cannot eliminate her strategy D based on strict dominance.

---

[10] Unfortunately, we can find many examples of this situation in the history of mankind, such as the arms race between countries.

[11] In Section B3, we will show that the game-theoretic model and its solution change if we consider a sequential move of the players (i.e., if player p_2 knows the move of player p_1 at the moment she makes her move).

Table B2. *The Joint Packet Forwarding Game in strategic form*

		p_2	
		F	D
p_1	F	$(1-c, 1-c)$	$(-c, 0)$
	D	$(0,0)$	$(0,0)$

The players have two strategies: to forward (F) or to drop (D) the packet sent by the sender. Both players p_1 and p_2 get a benefit, but only if each of them forwards the packet.

To overcome the requirements defined by strict dominance, we define the concept of *weak dominance*.

Definition B3 Strategy s_i' of player i is said to be **weakly dominated** by her strategy s_i if

$$u_i(s_i', s_{-i}) \leq u_i(s_i, s_{-i}), \forall s_{-i} \in S_{-i}, \tag{B2}$$

with strict inequality for at least one $s_{-i} \in S_{-i}$.

Using the concept of weak dominance, one can notice that the strategy D of player p_2 is weakly dominated by the strategy F. One can perform an elimination based on *iterated weak dominance*, which results in the strategy profile (F, F). Note, however, that the solution of the iterated strict dominance technique is unique, whereas the solution of the iterated weak dominance technique might depend on the sequence of eliminating weakly dominated strategies, as explained at the end of Section B2.3.

It is also important to emphasize that the iterated elimination techniques are very useful, even if they do not result in a single strategy profile. These techniques can be used to reduce the size of the strategy space (i.e., the size of the strategic-form matrix) and thus to ease the solution process.

B2.3 Nash equilibrium

The majority of the games cannot be solved by the iterated dominance techniques. As an example, let us consider the Multiple Access Game introduced at the beginning of this appendix. Each of the players has two possible strategies: either transmit (T) or not transmit (and thus stay quiet) (Q). As the channel is shared, a simultaneous transmission of both players leads to a collision. The game is represented in strategic form in Table B3.

Table B3. *The Multiple Access Game in strategic form*

p_2

	Q	T
Q	(0,0)	$(0,1-c)$
T	$(1-c,0)$	$(-c,-c)$

p_1

The two moves for each player are: transmit (T) or be quiet (Q).

It can immediately be seen that no strategy is dominated (even weakly) in this game. To solve the game, let us introduce the concept of *best response*. If player p_1 transmits, then the best response of player p_2 is to be quiet. Conversely, if player p_2 is quiet, then p_1 is better off transmitting a packet. We can express the best response of player i to an opponent's strategy vector s_{-i} as follows.

Definition B4 The *best response* $br_i(s_{-i})$ of player i to the profile of strategies s_{-i} is a strategy s_i such that:

$$br_i(s_{-i}) = \arg\max_{s_i \in S_i} u_i(s_i, s_{-i}). \qquad (B3)$$

One can see that if two strategies are mutual best responses to each other, then no player has incentive to deviate from the given strategy profile. In the Multiple Access Game, two strategy profiles exist with the above property: (Q, T) and (T, Q). To identify such strategy profiles in general, John Nash introduced the concept of Nash equilibrium in his seminal paper [285]. We can formally define the concept of *Nash equilibrium (NE)* as follows.

Definition B5 The pure-strategy profile s^* constitutes a *Nash equilibrium* if, for each player i,

$$u_i(s_i^*, s_{-i}^*) \geq u_i(s_i, s_{-i}^*), \forall s_i \in S_i. \qquad (B4)$$

This means that in a Nash equilibrium, none of the users can unilaterally change her strategy to increase her payoff. Alternatively, a Nash equilibrium is a strategy profile comprised of mutual best responses of the players.

A Nash equilibrium is *strict* [164] if we have

$$u_i(s_i^*, s_{-i}^*) > u_i(s_i, s_{-i}^*), \forall s_i \in S_i. \qquad (B5)$$

It is easy to check that (D, D) is a Nash equilibrium in the Forwarder's Dilemma game represented in Table B1. This corresponds to the solution obtained by iterated strict dominance. This result is true in general: any solution derived by iterated strict dominance is a Nash equilibrium. The proof of this statement is presented notably in [144].

In the Multiple Access Game, however, the iterated dominance techniques do not help us derive the solutions. Fortunately, using the concept of Nash equilibrium, we can identify the two pure-strategy Nash equilibria: (Q, T) and (T, Q).

Note that there can be more than one best response $br_i(s_{-i})$. For example in the Joint Packet Forwarding Game presented in Table B2, player p_2 has two best responses (D or F) to the move D of player p_1. Multiple best responses are the reason why the solutions of the iterated weak dominance technique in a given game might depend on the order of elimination.

B2.4 Mixed strategies

In the examples so far, we have considered only pure strategies, meaning that the players clearly decide on one behavior or another. But in general, a player can decide to play each of these pure strategies with some probabilities. Referring to our context, this means for example that a node decides to transmit sometimes, but not always. In game-theoretic terms such a behavior is called a *mixed strategy*.

Definition B6 The *mixed strategy* $\sigma_i(s_i)$, or for short σ_i, of player i is a probability distribution over her pure strategies $s_i \in S_i$.

Accordingly, we will denote the mixed strategy space of player i by Σ_i, where $\sigma_i \in \Sigma_i$. Hence, the notion of profile, which we defined earlier for pure strategies, is now characterized by the probability distribution assigned by each player to her pure strategies: $\sigma = \sigma_1, \ldots, \sigma_{|P|}$, where $|P|$ is the cardinality of P. As in the case of pure strategies, we denote the strategy profile of the opponents by σ_{-i}. For a finite strategy space, i.e. for so called *finite games* [144],[12] the payoff of each player i to profile σ is then given by

$$u_i(\sigma) = \sum_{s_i \in S_i} \sigma_i(s_i) u_i(s_i, \sigma_{-i}). \qquad (B6)$$

Each of the concepts that we have considered so far for pure strategies can also be defined for mixed strategies. As there is no significant difference in these definitions, we refrain from repeating them for mixed strategies.

Let us first study the Multiple Access Game. We call x the probability with which player p_1 decides to transmit, and y the equivalent probability for p_2 (this means that p_1 and p_2 stay quiet with probability $1 - x$ and $1 - y$, respectively).

---

[12] The general formula for infinite strategy space is slightly more complicated. The reader can find it in [144] or [296].

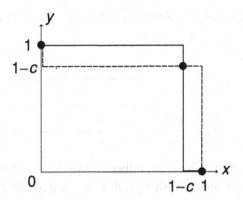

Figure B4. Best response functions in the Multiple Access Game. The best response function of player p_1 (x as a function of y) is represented by the dashed line; that of player p_2 (y as a function of x) is represented by the solid one. The two dots at the edges indicate the two pure strategy Nash equilibria and the one in the middle shows the mixed strategy equilibrium.

The payoff of player p_1 is:

$$u_1 = x(1-y)(1-c) - xyc = x(1-c-y). \tag{B7}$$

Likewise, we have:

$$u_2 = y(1-c-x). \tag{B8}$$

As usual, the players want to maximize their payoffs. Let us first derive the best response of p_2 for each strategy of p_1. In (B8), if $x < 1-c$, then $(1-c-x)$ is positive, and u_2 is maximized by setting y to the highest possible value, namely $y = 1$. Conversely, if $x > 1-c$, u_2 is maximized by setting $y = 0$ (these two cases bring us back to the two pure strategy equilibria that we have already identified). More interesting is the last case, namely $x = 1-c$, because here u_2 does not depend on y anymore (and is always equal to 0); hence, *any* strategy of p_2 (meaning any value of y) is a best response. The game being symmetric, reversing the roles of the two players leads of course to the same result. This means that $(x = 1-c, y = 1-c)$ is a mixed strategy Nash equilibrium for the Multiple Access Game.

We can graphically represent the best responses of the two players (Figure B4). In the graphical representation, we refer to the set of best response values as the *best response function*.[13] Relying on the concept of mutual best responses, one can identify the Nash equilibria as the crossing points of these best response "functions."

---

[13] From the calculus point of view, the set of best response values is not necessarily a function, because there might be several best responses to a given opponent strategy profile; yet the expression "best response function" is widely used in game theory.

Table B4. *The Jamming Game in strategic form*

		p_2 (jammer)	
		ch_1	ch_2
p_1 (sender)	ch_1	(−1,1)	(1,−1)
	ch_2	(1,−1)	(−1,1)

Note that the number of Nash equilibria varies from game to game. There are games with no pure strategy Nash equilibrium, such as the Jamming Game. We show the strategic form of this game in Table B4.

The reader can easily verify that the Jamming Game cannot be solved by iterated dominance. Moreover, this game does not even admit a pure-strategy Nash equilibrium. In fact, there exists only a mixed-strategy Nash equilibrium in this game that dictates each player to play a uniformly random distribution strategy (i.e., select one of the channels with probability 0.5).

The importance of mixed strategies is further reinforced by the following theorem of Nash [285], which represents a crucial existence result in game theory. The proof uses the Brouwer-Kakutani fixed-point theorem and is provided in [150], for example.

Theorem B1 (Nash, 1950) *Every finite strategic-form game has a mixed-strategy Nash equilibrium.*

B2.5 Equilibrium selection

As we have seen so far, the first step in solving a game is to investigate the *existence* of Nash equilibria. Theorem B1 states that in a broad class of games there always exists at least one mixed-strategy Nash equilibrium. Once we have found a Nash equilibrium, we have to determine whether it is *unique*.

Then we have to study an important property of a Nash equilibrium, namely its *efficiency*. Efficiency can also be used to select the most appropriate solutions from several Nash equilibria. *Equilibrium selection* means that the users have identified the desired Nash equilibrium profiles, but they also have to coordinate which one to choose. For example in the Multiple Access Game, both players are aware that there exist three Nash equilibria with different payoffs, but both of them try to be "the winner" by deciding to transmit (in the expectation that the other player will be quiet). Hence, their actions result in a profile which is not a Nash equilibrium. The topic of equilibrium selection is one of the hot research fields in game theory [143, 336].

B2.6 Essential games and robust equilibria

In practice it is unlikely that the game modeler will have specified payoff functions that are perfectly correct. Hence a crucial question is whether equilibrium predictions of the modeled game with payoffs u are approximate equilibrium predictions of the "real" game with nearby payoffs $\hat{u}$. We now define the notion of *proximity* in finite games ([144] Section 12.1.2). Let

$$u = (u_i(s))_{i \in \mathcal{I}, s \in S}$$

and

$$\hat{u} = (\hat{u}_i(s))_{i \in \mathcal{I}, s \in S}$$

denote two payoff profiles, and let

$$\sigma = (\sigma_i(s_i))_{i \in \mathcal{I}, s_i \in S_i}$$

and

$$\hat{\sigma} = (\hat{\sigma}_i(s_i))_{i \in \mathcal{I}, s_i \in S_i}$$

denote two mixed strategy profiles. Let

$$D(u, \hat{u}) = \max_{i \in \mathcal{I}, s \in S} |u_i(s) - \hat{u}_i(s)| \tag{B9}$$

and

$$d(\sigma, \hat{\sigma}) = \max_{i \in \mathcal{I}, s_i \in S_i} |\sigma_i(s_i) - \hat{\sigma}_i(s_i)| . \tag{B10}$$

Definition B7 A Nash equilibrium σ of game u is **essential** or **robust** if for any $\varepsilon > 0$ there exists $\eta > 0$, such that for any $\hat{u}$ such that $D(u, \hat{u}) < \eta$ there exists a Nash equilibrium $\hat{\sigma}$ of game $\hat{u}$ such that $d(\sigma, \hat{\sigma}) < \varepsilon$. A game u is **essential** if all its equilibrium points are essential.

B2.7 Pareto-optimality

So far, we have seen how to identify Nash equilibria. We have also seen that there might be several Nash equilibria, as for example in the Joint Packet Forwarding Game. One method to identify the desired equilibrium point in a game is to compare strategy profiles using the concept of Pareto-optimality. To introduce this concept, let us first define Pareto-superiority.

Definition B8 The strategy profile s is **Pareto-superior** to the strategy profile s' if for any player $i \in N$:

$$u_i(s_i, s_{-i}) \geq u_i(s'_i, s'_{-i}) \tag{B11}$$

with strict inequality for at least one player.

In other words, the strategy profile s is *Pareto-superior* to the strategy profile s', if there exists at least one player j whose payoff is higher with s than with s' while the payoff of each other player is at least as high with s as with s'. The strategy profile s' is then defined as *Pareto-inferior* to the strategy profile s. Note that from an arbitrary strategy, it is usually necessary to change the strategy of several players in order to reach a Pareto-superior profile.

Based on the concept of Pareto-superiority, we can identify the most efficient strategy profile or profiles.

Definition B9 The strategy profile s^{po} is **Pareto-optimal** if there exists no other strategy profile that is Pareto-superior to s^{po}.

Using the concept of Pareto-optimality, we can distinguish the Nash equilibria, where any improvement on a player's payoff hurts at least one other player. Hence, these Nash equilibria are more system-efficient than the others. Note that we cannot define s^{po} as the strategy profile that is Pareto-superior to all other strategy profiles, because a game can have several Pareto-optimal strategy profiles. It is important to stress that a Pareto-optimal strategy profile is not necessarily a Nash equilibrium.

We can now use the concept of Pareto-optimality to study the efficiency of pure-strategy Nash equilibria in our running examples.

- In the Forwarder's Dilemma game, the Nash equilibrium (D, D) is not Pareto-optimal. The strategy profiles (F, F), (F, D), and (D, F) are Pareto-optimal, but not Nash equilibria.
- In the Joint Packet Forwarding game, both strategy profiles (F, F) and (D, D) are Nash equilibria, out of them only (F, F) is Pareto-optimal.
- In the Multiple Access Game, both pure-strategy profiles (T, Q) and (Q, T) are Nash equilibria and Pareto-optimal.
- In the Jamming game, there exists no pure strategy Nash equilibrium, and all pure strategy profiles are Pareto-optimal.

We have seen that the Multiple Access Game (with mixed strategies) has three Nash equilibria. It is worth mentioning that the mixed strategy Nash equilibrium $\sigma = (p = 1 - c, q = 1 - c)$ results in the payoffs $(0, 0)$. Hence, this mixed strategy Nash equilibrium is Pareto-inferior to the two pure strategy Nash equilibria. In fact, it can be shown in general that there does not exist a mixed strategy profile that is Pareto-superior to all pure strategy profiles, because any mixed strategy of a player i is a linear combination of her pure strategies with positive coefficients that sum up to one.

B3 Dynamic games

In the strategic-form representation it is usually assumed that the players make their moves *simultaneously*, without knowing what the other player does. This might be a reasonable assumption in some problems, for example in the Multiple Access Game. In most of the games, however, the players might have a *sequential* interaction, meaning that the move of one player is conditioned on the move of the other player (i.e., the second mover knows the move of the first mover before making her decision). These games are called *dynamic games* [43] and we can represent them in an *extensive form*. We say that a game is with *perfect information* if the players have a perfect knowledge of all previous moves in the game at any moment they have to make a new move.

B3.1 Extensive form with perfect information

In the *extensive form*, the game is represented as a tree, where the root of the tree is the start of the game and shown with an empty circle. We refer to one level of the tree as a *stage*. The nodes of a tree, denoted by a filled circle, show the possible unfolding of the game, meaning that they represent the sequence relation of the moves of the players. This sequence of moves defines a *path* in the tree and is referred to as the *history h* of the game. It is generally assumed that a single player can move when the game is at a given node.[14] This player is represented as a label on the node. Note that this is indeed a tree, thus each node is a complete description of the path preceding it (i.e., each node has a unique history). The moves that lead to a given node are represented on each branch of the tree. Each terminal node (i.e., leaf) of the tree defines a potential end of the game called *outcome* and it is assigned the corresponding payoffs. In addition, we consider *finite-horizon games*, which means that there exists a finite number of stages.

Note that the extensive form is a more convenient representation, but basically every extensive form can be transformed to a strategic form and vice versa. However, unlike strategic-form games, extensive-form games can be used to describe *sequential* interactions more easily. In extensive form, the *strategy* of player i assigns a move $m_i(h)$ to every non-terminal node in the game tree with the history h where player i has to move. For simplicity, we use pure strategies in this section. The definition of Nash equilibrium is basically the same as the one provided in Definition B5.

To illustrate these concepts, let us consider the *Sequential Multiple Access Game*. This is a modified version of the Multiple Access Game supposing that the two

---

[14] Osborne and Rubinstein [296] define a game, where a set of players can move in one node. Also, there exist specific examples in [144], in which different players move in the same stage. For the clarity of presentation, we do not discuss these specific examples in this tutorial.

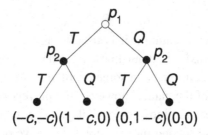

Figure B5. The Sequential Multiple Access Game in extensive form.

transmitters p_1 and p_2 are not perfectly synchronized, which means that p_1 always moves first (i.e., transmits or not) and p_2 observes the move of p_1 before making her own move.[15] We show this extensive-form game with perfect information in Figure B5. In this game, the strategies of player p_1 are to transmit (T) or be quiet (Q). But the strategy of player p_2 has to define a move given the previous move for player p_1. Thus, the possible strategies of p_2 are TT, TQ, QT and QQ, where for example TQ means that player p_2 transmits if p_1 transmits and she remains quiet if p_1 remains quiet. Thus, we can identify the pure-strategy Nash equilibria in the Sequential Multiple Access Game. It appears that there exist three pure-strategy Nash equilibria: (T, QT), (T, QQ), and (Q, TT). Please note that (Q, TQ) is not a Nash equilibrium: p_2 would be better off by playing TT in that case.

At this stage, it is useful to introduce an important existence theorem [242]. The intuition of the proof is provided in [144].

Theorem B2 (Kuhn, 1953) *Every finite extensive-form game of perfect information has a pure-strategy Nash equilibrium.*

The proof relies on the concept of backward induction, which we introduce in the following.

B3.2 Backward induction and Stackelberg equilibrium

We have seen that there exist three Nash equilibria in the Sequential Multiple Access Game. For example, if player p_2 plays the strategy TT, then the best response of player p_1 is to play Q. We notice, however, that the claim of player p_2 to play TT is an *incredible* (or *empty*) *threat*. Indeed, TT is not the best strategy of player p_2 if player p_1 chooses T in the first round.

We can eliminate equilibria based on such incredible threats using the technique of *backward induction*. Let us first solve the Sequential Multiple Access Game presented in Figure B5 with the backward induction method as shown in Figure B6.

---

[15] In fact, this is called the *carrier sense* and it is the basic technique to resolve contention in the CSMA/CA protocols [321, 340].

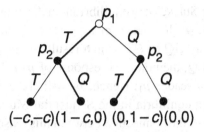

Figure B6. The backward induction solution of the Sequential Multiple Access Game in extensive form.

The Sequential Multiple Access Game is a finite game with complete information. Hence, player p_2 knows that she is the player that has the last move. For each possible history, she predicts her best move. For example, if the history is $h = T$ in the game, then player p_2 concludes that the move Q results in the best payoff for her in the last stage. Similarly, player p_2 defines T as her best move following the move Q of player p_1. In Figure B6, we represent these best choices with thick solid lines in the last game row. Given all the best moves of player p_2 in the last stage, player p_1 calculates her best moves as well. In fact, each reasoning step reduces the extensive form game by one stage. Following this backward reasoning, we arrive at the beginning of the game (the root of the extensive-form tree). The continuous thick line from the root to one of the leaves in the tree gives us the backward induction solution. In the Sequential Multiple Access Game, we can identify the backward induction solution as $h = \{T, Q\}$. Backward induction can be applied to any finite game of perfect information. This technique assumes that the players can reliably forecast the behavior of other players and that they believe that the other can do the same. Note, however, that this argument might be less compelling for longer extensive-form games owing to the complexity of prediction.

Note that the technique of backward induction is analogous to the technique of iterated strict dominance in strategic-form games. It is an elimination method to reduce the game. Furthermore, the backward induction procedure is a technique to identify *Stackelberg equilibria* in the extensive-form game. Let us call the first mover the *leader* and the second mover the *follower*.[16] Then, we can define a Stackelberg equilibrium as follows.

Definition B10 The strategy profile s is a *Stackelberg equilibrium* with player p_1 as the leader and player p_2 as the follower if player p_1 maximizes her payoff subject to the constraint that player p_2 chooses according to her best response function.

---

[16] Note that in the general description of the Stackelberg game, there might be several followers, but there is always a single leader.

Let us now derive the Stackelberg equilibrium in the Sequential Multiple Access Game by considering how the leader p_1 argues. If p_1 chooses T, then the best response for p_2 is to play QQ or QT, which results in the payoff of $1 - c$ for p_1. However, if p_1 chooses Q, then the best response of p_2 is TQ or TT, which results in the payoff of zero for leader p_1. Hence, p_1 will choose T and (T, QT) or (T, QQ) are the Stackelberg equilibria in the Sequential Multiple Access Game. We can immediately establish the connection between this reasoning and the backward induction procedure.

We have seen in the above example that the leader can exploit her advantage if the two players have conflicting goals: in this game, the leader can enforce the equilibrium beneficial to herself.

Let us now briefly discuss the extensive form of the other three wireless networking examples with sequential moves. In the extensive-form version of the Forwarder's Dilemma, the conclusions do not change. Both players will drop each other's packets. In the extensive form of the Joint Packet Forwarding Game, if player p_1 chooses D, then the move of player p_2 is irrelevant. Hence by induction, we can deduce that the Stackelberg equilibrium is (F, F). Finally, in the Jamming Game, let us assume that p_1 is the leader and the jammer p_2 is the follower. In this case, the jammer can easily observe the move of p_1 and jam. Hence, being the leader does not necessarily result in an advantage.

B3.3 Imperfect information and subgame perfect equilibria

In this section, we will extend the notions of history and information. As we have seen, in the game with perfect information, the players always know the moves of all other players when they have to make their moves. However, in the examples with simultaneous moves (e.g., the static games in Section B2), the players have *imperfect information* about the unfolding of the game. To define perfect information more precisely, let us first introduce the notion of *information set $h(n)$*, i.e. the amount of information the players have at the moment they choose their moves in a given node n. The information set $h(n)$ is a partition of the nodes in the game tree. The intuition of the information set is that a player actually in node n of the tree does not know whether she is really in node n or in some other node $n' \in h(n)$: The information set has to fulfill the following additional properties: (i) if $n, n' \in h(n)$, then the same player i has to move in both n and n', and (ii) player i must have the same moves available in n and n'. We can now formally define the concept of *perfect information*.[17]

---

[17] Note that two well-established textbooks on game theory, [144] and [296], have different definitions of perfect information. We use the interpretation of [144], which we believe is more intuitive. The authors

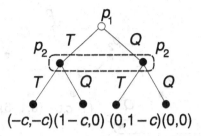

Figure B7. The original Multiple Access Game in extensive form. It is a game with imperfect information.

Definition B11 The players have *perfect information* in the game if every information set is a singleton (meaning that each player always knows the previous moves of all players when she has to make her move).

It is not a coincidence that we use the same notation for the information set as for the history. In fact, the concept of information set is a generalized version of the concept of history.

To illustrate these concepts, let us first consider the extensive form of the original Multiple Access Game shown in Figure B7. Recall that this is a game with imperfect information. The dashed line represents the information set of player p_2 at the time she has to make her move. The set of nodes in the game tree circumscribed by the dashed line means that player p_2 does not know whether player p_1 is going to transmit or not at the time she makes her own move, i.e. that they make simultaneous moves.

The *strategy* of player i assigns a move $m_i(h(n))$ to every non-terminal node n in the game tree with the information set $h(n)$. Again, we deliberately restrict the strategy space of the players to pure strategies, but the reasoning holds for mixed strategies as well [144, 296]. The possible strategies of each player in the Multiple Access Game are to transmit (T) or be quiet (Q). As we have seen before, both (T, Q) and (Q, T) are pure strategy Nash equilibria. Note that in this game, player p_2 cannot condition her move on the move of player p_1.

As we have seen in Section B3.2, backward induction and the concept of Stackelberg equilibrium can be used to eliminate incredible threats. Unfortunately, the elimination technique based on backward induction cannot always be used. To illustrate this, let us construct the game called *Multiple Access Game with Retransmissions* and solve it in the pure strategy space. In this game, the players play the Sequential Multiple Access Game, and they play the Multiple Access Game if there

---

of [296] define, in Chapter 6 of their book, a game with simultaneous moves also as a game with perfect information, where the players are substituted with a set of players, who make their moves.

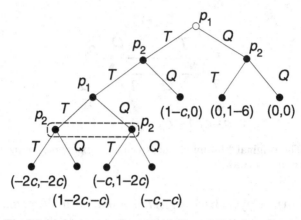

Figure B8. The Multiple Access Game with Retransmissions in extensive form. It is also a game with imperfect information.

was a collision (i.e., they both tried to transmit). We show the extensive form in Figure B8.

Note that the players have many more strategies than before. Player p_1 has four strategies, because there exist two information sets, where she has to move; and she has two possible moves at each of these information sets. For example, the strategy $s_1 = TQ$ means that player p_1 transmits at the beginning and she does not in the second Multiple Access Game. Similarly, player p_2 has $2^3 = 8$ strategies, but they are less trivial to identify. For example, each move in the strategy $s_2 = QTT$ means the following: (i) the first move means that player p_2 stays quiet if player p_1 transmitted, or (ii) p_2 transmits if p_1 was quiet and (iii) p_2 transmits in the last stage if they both transmitted in the first two stages. This example highlights an important point. *The strategy defines the moves for player i for every information set in the game where player i moves, even for those information sets that are not reached if the considered strategy is played.* The common interpretation of this property is that the players may not be able to perfectly observe the moves of each other and thus the game may evolve along a path that was not expected. Alternatively, the players may have incomplete information, meaning that they have certain beliefs about the payoffs of other players and hence, they may try to solve the game on this basis. These beliefs may not be precise and so the unfolding of the game may be different from the predicted unfolding. Game theory covers these concepts in the notion of Bayesian games [144], but we do not present this topic here.

It is easy to see that the Multiple Access Game with Retransmissions cannot be analyzed using backward induction, because the Multiple Access Game in the second stage is of imperfect information. To overcome this problem, Selten introduced a concept called *subgame perfection* in [341, 164]. In Figure B8, the Multiple

Access Game in the second stage is a *proper subgame* of the Multiple Access Game with Retransmissions. Let us now give the formal definition of a proper subgame.

Definition B12 The game G' is a *proper subgame* of an extensive-form game G if it consists of a single node in the extensive-form tree and all of its successors until the leaves. Formally, if a node $n \in G'$ and $n' \in h(n)$, then $n' \in G'$. The information sets and payoffs of the subgame G' are inherited from the original game G. This means that n and n' are in the same information set in G' if they are in the same information set in G; and the payoff function of G' is the restriction of the original payoff function to G'.

Now let us formally define the concept of subgame perfection. This definition reduces to backward induction in finite games with perfect information.

Definition B13 The strategy profile s is a *Subgame Perfect Nash Equilibrium* of a finite extensive-form game G if it is a Nash equilibrium of any proper subgame G' of the original game G.

In other words, in a subgame perfect equilibrium, there is no information set in which a player i can gain by deviating from her subgame perfect equilibrium strategy. This is called the *one-deviation property* in game-theoretic terms. A reader familiar with dynamic programming may wonder about the analogy between the optimization in game theory and in dynamic programming [45]. Indeed, the one-deviation property corresponds to the *principle of optimality* in dynamic programming, which is based on backward induction. Hence, we can give an alternative definition of subgame perfect equilibria using the one-deviation property.

Definition B14 The strategy profile s is a *Subgame Perfect Nash Equilibrium* of a finite extensive-form game G if no player i can gain by deviating from her subgame-perfect strategy s_i^* in a single stage and conform to it otherwise in any proper subgame.

Subgame perfection provides a method to solve the Multiple Access Game with Retransmissions. We can simply replace the Multiple Access Game subgame (the second one with simultaneous moves) with one of its pure-strategy Nash equilibria. Hence, we can obtain one of the game trees presented in Figure B9. Solving the reduced games with backward induction, we can derive the following solutions. In the game shown in Figure B9a, we have the subgame perfect equilibrium (QQ, TTT). In Figure B9b we obtain the subgame perfect equilibria (TT, $Q * Q$), where $*$ means any move from $\{T, Q\}$. It has to be noted that the first of the two Subgame Perfect Nash Equilibria, namely (QQ, TTT), is another example of incredible threat.

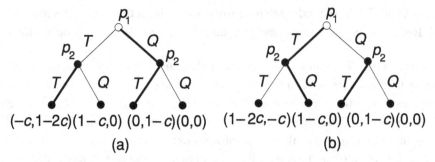

$$(-c,1-2c)(1-c,0) \quad (0,1-c)(0,0) \quad (1-2c,-c)(1-c,0) \quad (0,1-c)(0,0)$$

(a)   (b)

Figure B9. Application of subgame perfection to the Multiple Access Game with Retransmissions. In (a) the proper subgame is replaced by one of the Nash equilibria of that game, namely (Q, T). Solution (b) represents the case, where the subgame is replaced by the other Nash equilibrium (T, Q). The thick lines show the result of the backward induction procedure on the reduced game trees.

Because any game is a proper subgame of itself, a subgame perfect equilibrium is necessarily a Nash equilibrium, but there might be Nash equilibria in G that are not subgame perfect. In fact, the concept of Nash equilibrium does not require that the one-deviation property holds.

B4 Repeated games

So far, we have assumed that the players interact only once and we have modeled this interaction in a static game in strategic form in Section B2 and in part of Section B3. In addition, we have introduced the Multiple Access Game with Retransmissions, which was a first example to illustrate repeated games, although the number of stages was quite limited. As we have seen in Section B3, the extensive form provides a more convenient representation for sequential interactions. In this section, we assume that the players interact several times and hence we model their interaction using a *repeated game*. The analysis of repeated games in extensive form is basically the same as presented in Section B3, hence we focus on the strategic form in this section. To be more precise, we consider repeated games with *observable actions* and *perfect recall*: This means that each player knows all moves of the others, and that each player knows her own previous moves at each stage in the repeated game.

B4.1 Basic concepts

In repeated games, the players interact several times. Each interaction is called a *stage*. Note that the concept of stage is similar to the one in extensive form, but here we assume that the players make their moves simultaneously in each stage. The set of players is defined similarly to the static game presented in Section B2.1.

As a running example, let us consider the *Repeated Forwarder's Dilemma* that consists of the repetition of the Forwarder's Dilemma stage game. In such a repeated game, all past moves are common knowledge at each stage t. The set of the past moves at stage t is commonly referred to as the history $h(t)$ of the game. We call it a history (and not an information set), because it is uniquely defined at the beginning of each stage. Let us denote the move of player i in stage t by $m_i(t)$. We can formally write the history $h(t)$ as follows:

$$h(t) = \{(m_1(t), \ldots, m_{|N|}(t)), \ldots, (m_1(0), \ldots, m_{|N|}(0))\}. \tag{B12}$$

For example, at the beginning of the third stage of the Repeated Forwarder's Dilemma, if both players have always cooperated so far, the history is $h(2) = \{(F, F), (F, F)\}$.

The *strategy* s_i defines a move for player i in the next stage $t + 1$ for each history $h(t)$ of the game.[18] In other words, the move of player i at stage $t + 1$ is defined as:

$$m_i(t + 1) = s_i(h(t)). \tag{B13}$$

Note that the initial history $h(0)$ is the empty set. The strategy s_i of player i must define a move $m_i(0)$ for the initially empty history, which is called the *initial move*. Note also that strategies must specify actions even for histories that may be unreachable given prior action choices. For a moment, suppose that the Repeated Forwarder's Dilemma has only two stages. Then one example strategy of each player is *FFDFD*, where the entries of the strategy define the following behavior: (i) Forward in the first stage, i.e. as an initial move, (ii) Forward if the history was $h(1) = \{(F, F)\}$, (iii) Drop if the history was $h(1) = \{(F, D)\}$, (iv) Forward if the history was $h(1) = \{(D, F)\}$, and (v) Drop if the history was $h(1) = \{(D, D)\}$. We notice that the strategy space grows very quickly as the number of stages increases: in the two-stage Repeated Forwarder's Dilemma, we have $|S| = 2^5 = 32$ strategies for each player. Hence in repeated games, it is typically infeasible to make an exhaustive search for the best strategy and hence for Nash equilibria.

The payoff in the repeated game might change as well. In repeated games, the users typically want to maximize their payoff for the whole duration T of the game. Hence, they maximize:

$$u_i = \sum_{t=0}^{T} u_i(t, s). \tag{B14}$$

In some cases, the objective of the players in the repeated game can be to maximize their payoffs only for the next stage (i.e., as if they played a static game). We refer to these games as *myopic games* as the players are short-sighted optimizers.

---

[18] Recall that in the static game, the strategy was a single move.

If the players maximize their total payoff during the game, we call it a *long-sighted game*.

Recall that we refer to a finite game if the number of stages T is finite. Otherwise, we call it an infinite game. We will see in Section B4.3 that we can also model finite games with an unknown number of stages T by means of infinite games.

B4.2 Nash equilibria in finite horizon games

Let us first solve the finite Repeated Forwarder's Dilemma using the concept of Nash equilibrium. Assume that each player is long-sighted and wants to maximize her total payoff. As we have seen, it is computationally infeasible to calculate the Nash equilibria based on strategies that are mutual best responses to each other as the number of stages increases. Nevertheless, we can apply the concept of backward induction that we have presented in Section B3.2. Because the game is of complete information, the players know the end of it (they know T). Now, in the last stage game, they both conclude that their dominant strategy is to drop the opponent's packet (i.e., to play D). Given this argument, their best strategy is to play D in the penultimate stage. Following the same argument, this technique of backward induction dictates that the players should choose a strategy that plays D in every stage.

As mentioned, the strategy space increases exponentially with the length of the game. Hence, one usually restricts the strategy space to a reasonable subset. A widely used family of strategies is the *strategies of history-1*. These strategies take only the moves of the opponents in the *previous stage* into account (meaning that they are "forgetful" strategies, because they "forget" the previous behavior of the opponents). In the games we have considered thus far, we have two players and hence the history-1 strategy of player i in the repeated game can be expressed by the *initial move* $m_i(0)$ and the following strategy function:

$$m_i(t + 1) = s_i(m_j(t)). \tag{B15}$$

Accordingly, we can define the strategies in the Repeated Forwarder's Dilemma as detailed in Table B5. Note that this definition of strategies can enable a feasible analysis, even in the presence of a large number of stages.

We can observe that in some strategies, such as All-D or All-C, the player does not condition her next move on the previous move of the opponent. We refer to these strategies as *non-reactive strategies*. Analogously, the strategies that take the opponents' behavior into account are called *reactive strategies* (for example, TFT or STFT).

Let us now analyze the Repeated Forwarder's Dilemma assuming that the players use the history-1 strategies. The conclusion is the same as the one derived for the single stage game.

Table B5. *History-1 strategies of player 1 in the Repeated Forwarder's Dilemma*

$m_1(0)$	$\begin{array}{c}m_1(t)\vert\\m_2(t)=F\end{array}$	$\begin{array}{c}m_1(t)\vert\\m_2(t)=D\end{array}$	Strategy function s_1	Name of the strategy
D	D	D	$m_1(t+1)=D$	Always Defect (All-D)
D	F	D	$m_1(t+1)=m_2(t)$	Suspicious Tit-For-Tat (STFT)
D	D	F	$m_1(t+1)=\overline{m_2(t)}$	Suspicious Anti Tit-For-Tat (SATFT)
D	F	F	$m_1(t+1)=F$	Suspicious Always Cooperate (S-All-C)
F	D	D	$m_1(t+1)=D$	Nice Always Defect (Nice-All-D)
F	F	D	$m_1(t+1)=m_2(t)$	Tit-For-Tat (TFT)
F	D	F	$m_1(t+1)=\overline{m_2(t)}$	Anti Tit-For-Tat (ATFT)
F	F	F	$m_1(t+1)=F$	Always Cooperate (All-C)

The entries in the first three columns represent: the initial move of player p_1, the move of player p_1 to a previous move $m_2(t) = F$ of player p_2, and the move of p_1 as a response to $m_2(t) = D$. The bar represents the alternative move (e.g., $\overline{F} = D$). As an example, the TFT strategy begins the game with forwarding (i.e., cooperation) and then copies the behavior of the opponent in the previous stage.

Theorem B3 *In the Repeated Forwarder's Dilemma, the strategy profile (All-D, All-D) is a Nash equilibrium.*

Although not proven formally, the justification of the above theorem is provided in [36].

B4.3 Infinite horizon repeated games with discounting

In the game theory literature, infinite games with discounting are frequently used to model a finite game in which the players are not aware of the duration of the game. Clearly, this is often the case in strategic interactions, in particular in networking operations. In order to model the unpredictable end of the game, one decreases the value of future stage payoffs. This technique is called *discounting*. In such a game, the players maximize their *discounted total payoff*:

$$u_i = \sum_{t=0}^{\infty} u_i(t, s) \cdot \delta^t, \tag{B16}$$

where δ denotes the *discounting factor*. The *discounting factor* δ determines the decrease of the value for future payoffs, where $0 < \delta < 1$ (in general, we can assume that δ is close to one). The discounted total payoff expressed in (B16) is often normalized, and thus we call it the *normalized payoff* for short:

$$u_i = (1 - \delta) \cdot \sum_{t=0}^{\infty} u_i(t, s)\delta^t. \tag{B17}$$

The role of the factor $1 - \delta$ is to let the stage payoff of the repeated game be expressed in the same unit as the static (stage) game. Indeed, with this definition, if the stage payoff $u_i(t, s) = 1$ for all $t = 0, 1, \ldots$, then the normalized payoff is equal to 1 as well, because $\sum_{t=0}^{\infty} \delta^t = \frac{1}{1-\delta}$.

We have seen that the Nash equilibrium in the finite Repeated Forwarder's Dilemma was a non-cooperative one. Yet, this rather negative conclusion should not affect our morale: in most networking problems, it is reasonable to assume that the number of iterations (e.g., of packet transmissions) is very large and a priori unknown to the players. Therefore, as discussed above, games are usually assumed to have an infinite number of repetitions. And, as we will see, infinitely repeated games can lead to more cooperative behavior.

Consider the history-1 strategies All-C and All-D for the players in the Repeated Forwarder's Dilemma. Thanks to the normalization in (B17), the corresponding normalized payoffs are exactly those presented in Table B1. A conclusion similar to the one we drew in Section B4.2 can be directly derived: the strategy profile (All-D, All-D) is a Nash equilibrium. Indeed, if the opponent always defects, the best response is All-D. A sketch of the proof is provided (for the Prisoner's Dilemma) in [144].

To identify other Nash equilibria, let us first define the *Trigger* strategy. If a player i plays Trigger, then she forwards in the first stage and continues to forward as long as the other player j does not drop. As soon as the opponent j drops her packet, player i drops all packets for the rest of the game. Note that Trigger is not a history-1 strategy. The Trigger strategy applies the general technique of *punishments*.

If no players drop a packet, the payoffs corresponds to (F, F) in Table B1, meaning that it is equal to $1 - c$ for each player. If a player i plays $m_i(t) = D$ at stage t, her payoff will be higher at this stage (because she will not have to face the cost of forwarding), but it will be zero for all the subsequent stages, as player j will then always drop. The normalized payoff of player i will be equal to:

$$(1 - \delta)[(1 + \delta + \cdots + \delta^{t-1})(1 - c) + \delta^t \cdot 1] = 1 - c + \delta^t(c - \delta). \tag{B18}$$

As $c < \delta$ (remember that, in general, c is very close to zero, whereas δ is very close to one), the last term is negative and the payoff is therefore smaller than $1 - c$. In other words, even a single defection leads to a payoff that is smaller than the one provided by All-C. Hence, a player is better off always forwarding in this infinite game, in spite of the fact that, as we have seen, the stage game only has (D, D) as an equilibrium point. It can be easily proven that (Trigger, Trigger) is a

Nash equilibrium and that it is also Pareto-optimal (the intuition for the latter is the following: there is no way for a player to go above her normalized payoff of $1 - c$ without hurting her opponent's payoff). Note that by similar arguments, one can show that (TFT, TFT) is also a Pareto-optimal Nash equilibrium, because it results in the payoff $1 - c$ for each of the players.

It is important to mention that the players cannot predict the end of the game and hence they cannot exploit this information. As mentioned in [144], reducing the information or the strategic options (i.e., decreasing her own payoff) of a player might paradoxically lead to a better outcome in the game. This uncertainty is the reason the cooperative equilibrium appears in the infinitely repeated version of the Forwarder's Dilemma game.

B4.4 The Folk theorem

We will now explore further the mutual influence of the players' strategies on their payoffs. We will begin by defining the notion of *minmax value* (sometimes called the *reservation utility*). The minmax value is the lowest stage payoff that the opponents of player i can force her to obtain with punishments, provided that i plays the best response against them; more formally, it is defined as follows:

$$\underline{u_i} = \min_{s_{-i}} \left[\max_{s_i} u_i(s_i, s_{-i}) \right]. \tag{B19}$$

This is the lowest stage payoff that the opponents can enforce on player i. Let us denote by $s_{\min} = (s_{i,\min}, s_{-i,\min})$ the strategy profile for which the minimum is reached in (B19). We call the $s_{-i,\min}$ the *minmax profile* against player i within the stage game.

It is easy to see that player p_i can obtain at least her minmax value $\underline{u_i}$ in any stage and hence we call *feasible payoffs* the payoffs higher than the minmax payoff. In the Repeated Forwarder's Dilemma, the feasible payoffs for any player p_1 are higher than 0. Indeed, by playing $s_1 = $ All-D, she is assured to obtain at least that value, no matter what the strategy of p_2 can be. A similar argument applies to player p_2. Let us graphically represent the feasible payoffs in Figure B10. We highlight the convex hull of payoffs that are strictly non-negative for both players as the set of feasible payoffs.

The notion of minmax that we have just defined refers to the stage game, but it has a very interesting application in the repeated game, as the following theorem shows.

Theorem B4 *Player i's normalized payoff is at least equal to $\underline{u_i}$ in any Nash equilibrium of the infinitely repeated game, regardless of the level of the discount factor.*

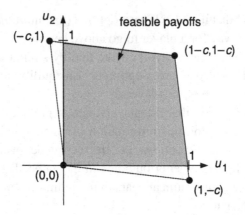

Figure B10. The feasible payoffs in the Repeated Forwarder's Dilemma.

The intuition can be obtained again from the Repeated Forwarder's Dilemma: a player playing All-D will obtain a (normalized) payoff of at least 0. The theorem is proven in [144].

We are now in a position to introduce a fundamental result that is of high relevance to our framework: the Folk theorem.[19]

Theorem B5 (Folk theorem) *For every feasible payoff vector $u = \{u_i\}_i$ with $u_i >$ $\underline{u_i}$, there exists a discounting factor $\underline{\delta} < 1$ such that for all $\delta \in (\underline{\delta}, 1)$ there is a Nash equilibrium with payoffs u.*

The intuition is that if the game is long enough (meaning that δ is sufficiently close to 1), the gain obtained by a player by deviating once is outweighed by the loss in every subsequent period, where the loss is caused by the punishment (minmax) strategy of the other players.

We have already seen the application of this theorem in the infinite Repeated Forwarder's Dilemma: a player is deterred from deviating, because the short term gain obtained by the deviation (1 instead of $1 - c$) is outweighed by the risk of being minmaxed (for example using the Trigger strategy) by the other player (provided that $c < \delta$).

The Repeated Forwarder's Dilemma is a two-player game, yet the Folk theorem applies to games with more players as well.

---

[19] This denomination of "folk" stems from the fact that this theorem was part of the oral tradition of game theorists, before it was formalized. Strictly speaking, we present the Folk theorem for the discounting criterion. There exist different versions of the Folk theorem, each being proved by different authors (as they are listed in [296] at end of Chapter 8).

B5 Discussion

One of the criticisms of game theory, as applied to the modeling of human decisions, is that human beings are rarely fully rational. Therefore, modeling the decision process by means of a few equations and parameters is questionable. In wireless networks, the (human) users do not interact with each other on such a fine-grained basis as forwarding one packet or accessing the channel once. Typically, they (or the device manufacturer, or the network operator, if any) program their devices to follow a protocol (i.e., a strategy) and it is reasonable to assume that they rarely reprogram their devices. Hence, such a device can be modeled as a rational decision maker. From this point of view, the application of game theory is easier to justify for (wireless) networking than, say, for economics, where the players are human beings.

Yet, there are several reasons for which the application of game theory to wireless networks can be criticized. We detail them here, as they are very rarely mentioned (for understandable reasons . . .) in research papers.

Benefit and cost

The first issue is the notion of benefit: for example, how important is it for a given user that a given packet is properly sent or received? This very much depends on the situation: the packet can be a crucial message, or could just convey a tiny portion of a figure appearing in a game. Likewise, the sensitivity to delay can also vary dramatically from situation to situation.

Similarly, the definition of cost might be a complex issue as well. In our examples (and in many examples of application of game theory to wireless networks), the cost usually represents the energy consumption of the devices. In some cases, however, a device can be power-plugged, thus its "cost" could be neglected. Likewise, a device whose battery is almost depleted should probably have a different evaluation of cost than when its battery is full. Furthermore, cost can include other considerations than energy, such as the previously mentioned delay or the consumed bandwidth.

Pricing and mechanism design

Mechanism design is concerned with the question of leading the players to a desirable equilibrium by changing (designing) some parameters of the game. In particular, pricing is considered to be a good technique to regulate the usage of a scarce resource by adjusting the costs of the players. Many network researchers have contributed to this field. These contributions provide a better understanding of specific networking mechanisms.

Yet it is not clear today, even for wired networks, how relevant these contributions are going to be in practice. Usually the pricing schemes used in reality by operators are very coarse-grained; operators tend to charge based on investment and personnel costs and on the pricing strategy of their competitors, and *not* on the instantaneous congestion of the network. If a part of the network is frequently congested, they will increase the capacity (deploy more base stations, more optical fibers, more switches) rather than throttle the user consumption by pricing. From this point of view, pricing is completely different from the cases in which it is very difficult to increase the available capacity, as is the case for example in a road network.

Hence, an area where pricing has more practical relevance is probably for service provisioning *among* operators (e.g., renting transmission capacity); but very little has been published so far on this topic.

Infinite games

As mentioned, games in networking are usually assumed to be infinite, in order to capture the idea that a given player does not know when the interaction with another player will stop. This is, however, not perfectly true: for example, a given player could "know" that it is about to be turned off and moved away (e.g., its owner is about to finish a given session for which the player has been attached at a given access point). Yet we believe this not to be a real problem: indeed, the required "knowledge" is clearly related to the application layer, whereas the games we are considering involve networking mechanisms (and thus are typically related to the MAC and network layers).

Discounting factor

As we have seen, in the case of infinitely repeated games, it is common practice to make use of the discounting factor. This notion is based on applications of game theory to economics: a given capital at time t_0 has "more value" than the same amount at a later time t_1 because, between t_0 and t_1, this capital can generate some (hopefully positive) interest. At first sight, transposing this notion into the realm of networking makes sense: A user wants to send (or to receive) information as soon as she expresses the wish to do so; in other words, now is better than later.

But this may be a very rough approximation, and the comment we made about the benefit can be applied here as well: the willingness to wait before transmitting a packet heavily depends on the current situation of the user and on the content of the packet. In addition, in some applications such as audio or video streaming, the network can *forecast* how the demand will evolve.

A more satisfactory interpretation of the discounting factor in our framework is related to the uncertainty that there will be a subsequent iteration of the stage game; for example, connectivity to an access point can be lost. With this interpretation in mind, the discounting factor represents the probability that the current round is not the last one.

It is important to notice that the average discounted payoff is not the only way to express the payoff in an infinitely repeated game. Osborne and Rubinstein [296] discuss other techniques, such as "Limit of means" and "Overtaking." None of them, however, captures the notion of users' impatience. In our opinion, they are therefore less appropriate for our purpose.

Reputation

In some cases, a player can include the reputation of another player in order to anticipate her moves; for example, a player observed to be non-cooperative frequently in the past is likely to continue to be so in the future. If the game models individual packet transmissions, this attitude would correspond to the suspicion that another player has been programmed in a highly "selfish" way. These issues go beyond the scope of this tutorial. For a discussion of these aspects, the reader is referred for example to Chapter 9 of the book by Fudenberg and Tirole [144].

Subgame perfection

The concept of subgame perfection has often been criticized with arguments based on equilibrium selection (recall the issue from Section B2.5). Many researchers pointed out that if several Nash equilibria exist in a given subgame, the players might not be able to determine how to play. As an example, they might both play T in the Multiple Access Game with Retransmissions in the second subgame as well. This disagreement can result in an outcome that is not an equilibrium.

Information

In this tutorial, we have studied games with *complete information*. This means that each player knows the identity of other players, their strategy functions and the resulting payoffs or outcomes. In addition, we consider games with *observable actions* and *perfect recall*. In wireless networking, these assumptions might not hold: For example, owing to the unexpected changes of the radio channel, a given player may erroneously reach the conclusion that another player is behaving selfishly. This can trigger the punishment (assuming there is one), leading to the risk of further retaliation, and so on. This means that, for any design of a self-enforcement protocol, special care must be devoted to the assessment of the amount and accuracy of the information that each player can obtain. The application of games with incomplete and imperfect information is an emerging field in wireless networking, with very few papers published so far.

Computational complexity

Even if all the necessary information is available, computing the Nash equilibria can be very complex and thus beyond the capabilities of low-tier devices.

Mobility

In highly mobile scenarios, the "players" have very short time for interaction. In a vehicular network, for example, two given vehicles can be in power range of each other just for a fraction of a second over their whole lifetime. In such a case, the very notion of game might not make sense.

Cooperative vs. non-cooperative games

In this tutorial, we assume that each player is a selfish individual, who is engaged in a non-cooperative game with other players. We do not cover the concept of cooperative games, where the players might have an agreement on how to play the game. Cooperative games include the issues of bargaining and coalition formation. These topics are very interesting and some of the problems found in wireless networks can be modeled using these concepts. The interested reader is referred to [296].

B6 Summary

In this tutorial, we showed how non-cooperative game theory can be applied to wireless networking. Using four simple examples, we explained how to capture wireless networking problems in a corresponding game, and we analyzed them to predict the behavior of players. We deliberately focused on the basic notions of non-cooperative game theory and studied games with complete information. We modeled devices as players, but there can be problems where the players are other participants, e.g. network operators, as explained in Chapter 11.

B7 To probe further

The first textbook in this area was written by von Neumann and Morgenstern, in 1944 [280]. A few years later, John Nash made a number of additional contributions [285, 286], the cornerstone of which is the famous *Nash equilibrium*. Since then, many other researchers have contributed to the field, and within a few decades, game theory has become a very active discipline; it is routinely taught in economics curricula. Game theory specialists (including Nash) have recently been awarded Nobel prizes in economics. An amazingly large number of game theory textbooks has been produced, but all of them consider economics as the premier application area (and all their concrete examples are inspired by that field).

This tutorial is based on three classic textbooks and we mention them here in ascending order of complexity. Gibbons [150] provides a very nice, easy-to-read introduction to non-cooperative game theory with many examples. Osborne and Rubinstein [296] introduce the game-theoretic concepts very precisely, yet this book is more difficult to read because of the more formal development. This is the only book out of these three that covers cooperative game theory as well. Finally, Fudenberg and Tirole's [144] book covers many advanced topics, in addition to the basic ones.

Additional mainstream textbooks are [43] by Basar and Olsder as well as [283] by Myerson.

Unsurprisingly, game theory has already been applied to networking, in most cases to solve routing and resource allocation problems in a competitive environment. We refer the reader to the "To probe further" sections of Chapters 9, 10, and 11 for references. A subset of these papers is included in [23]. For an analysis of the relationship between rate control and shadow prices in wired networks, the reader is referred to [220].

Recently, game theory was also applied to wireless communications (some of the papers are referenced in [355]). A recent tutorial of game theory for wireless engineers is [262]; it provides a synthesis of lectures on this topic.

B8 Questions

(1) Write the extensive form of the Joint Packet Forwarding Game.
(2) Let us consider a modified version of the Joint Packet Forwarding Game. Suppose that player p_1 can reach the destination at the cost of $2c$. In this case, she is the only one who receives the reward of 1. Hence, player p_1 has the choice to drop (D), forward to player p_2 (F'), or forward to the destination (F'').
 (a) Write the normal form of this modified game. Identify the Nash equilibria. Which equilibrium is Pareto-optimal?
 (a) Write the extensive form. Which equilibrium is subgame perfect?
(3) In this problem, we will model the connectivity between devices as a game. Consider the network presented in Figure B11.
 Suppose that the source S wants to send a packet to the destination D, but it needs the help of the forwarders to do this (one or both of them). Forwarder i has two possible moves: (a) set its power level to $P_i = 0$ or (b) set its power level to $P_i = P$. If either of the forwarders chooses P, then it connects S with D. If the connection is established, then both forwarder nodes get a reward of 1 (meaning a reward of 1 for *each* of them), no matter who established the connection. But forwarding has a cost: the forwarder who chooses the power level P has to pay the cost $2c$. We assume that $2c < 1$. Let us call this game the Connectivity Game 1.

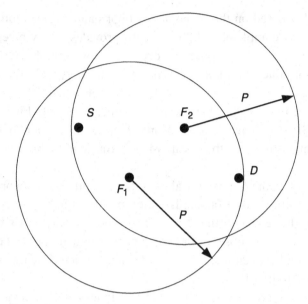

Figure B11. Connectivity Game 1 with two potential forwarder nodes.

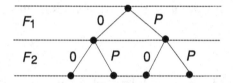

Figure B12. Extensive-form of the Connectivity Game 1.

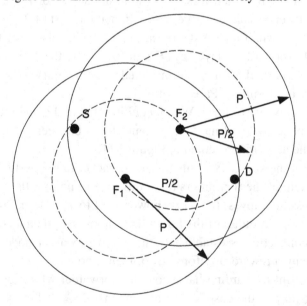

Figure B13. Connectivity Game 2 with two potential forwarder nodes.

(a) Suppose that F_1 chooses its power level first (F_2 decides after). The extensive-form of this game is provided in Figure B12. Write the payoffs on the leaves (bottom end dots) of the tree. Given that F_1 is the leader, show the Stackelberg equilibria in this game.

(b) Write the corresponding normal form of the same game (in a matrix). Identify the Nash equilibria and the Pareto-optimal strategy profiles.

(c) Let us now assume that the forwarders can use three power levels: $P_i = 0$, $P/2$, or P. With the new power level $P/2$ they can reach only their immediate neighbors, meaning that with $P_1 = P/2$ the node F_1 can reach only S and F_2. Similarly, if $P_2 = P/2$, the node F_2 reaches only F_1 and D. Clearly, both of them have to choose a power level $P_i \geq P/2$ to get the reward 1 each. Choosing $P/2$ costs only c. The new game (called Connectivity Game 2) is shown in Figure B13. Show the normal form of the Connectivity Game 2. Which are the Nash equilibria? Which are the Pareto-optimal strategy profiles (states)?

(d) Let us now define a variant of the Connectivity Game 2, in which the players have to share the reward of 1 proportionally to their contribution. For example, if $P_1 = P/2$ and $P_2 = P$, then F_1 gets $1/3$ and F_2 gets $2/3$. Let us call this variant the Connectivity Game 3. Write the matrix representation of the normal form of the Connectivity Game 3. Show the Nash equilibria in the Connectivity Game 3 assuming that the cost is very small (for example $c = 0.05$). Show the Pareto-optimal states.

(e) Assume now that in the Connectivity Game 3, $c = 1/4$. Fill the normal-form matrix numerically. Identify Nash equilibria and Pareto-optimal states. Compare the results with the results of the previous question.

(f) Assume now that in the Connectivity Game 3, $c = 1/6$. Fill the normal-form matrix numerically. Identify Nash equilibria and Pareto-optimal states. Compare the results with the results of the previous two questions.

(4) What is the difference between games of imperfect and incomplete information?

References

[1] PKCS #1. RSA cryptography standard. Public-Key Cryptography Standards #1, RSA Laboratories, 2002.

[2] FIPS 113. Computer Data Authentication. Federal Information Processing Standards Publication 113, US Department of Commerce, Bureau of Standards, National Technical Information Service (NIST), 1985.

[3] FIPS 180-1. Secure Hash Standard. Federal Information Processing Standards Publication 180-1, US Department of Commerce, Bureau of Standards, National Technical Information Service (NIST), 1995. supersedes FIPS PUB 180.

[4] FIPS 186. Digital Signature Standard. Federal Information Processing Standards Publication 186, US Department of Commerce, Bureau of Standards, National Technical Information Service (NIST), 1994.

[5] FIPS 197. Advanced Encryption Standard. Federal Information Processing Standards Publication 197, US Department of Commerce, Bureau of Standards, National Technical Information Service (NIST), 2001.

[6] FIPS 46. Data Encryption Standard. Federal Information Processing Standards Publication 46, US Department of Commerce, Bureau of Standards, National Technical Information Service (NIST), 1977. revised in 1988 and 1993.

[7] I. Aad, J.-P. Hubaux, and E. W. Knightly. Denial of service resilience in ad hoc networks. In *Proceedings of the ACM Conference on Mobile Computing and Networking (MobiCom)*, Philadelphia, Pennsylvania, USA, Sept. 2004.

[8] K. Aberer and M. Hauswirth. P2P Systems. In *Practical Handbook of Internet Computing*. CRC Press, 2004.

[9] B. Aboba, L. Blunk, J. Vollbrecht, J. Carlson, and H. Levkowetz. Extensible Authentication Protocol (EAP). RFC 3748, June 2004.

[10] B. Aboba and P. Calhoun. Remote Authentication Dial In User Service (RADIUS) support for Extensible Authentication Protocol (EAP). RFC 3579, September 2003.

[11] G. Ács, L. Buttyán, and I. Vajda. Provable security of on-demand distance vector routing in wireless ad hoc networks. In *Proceedings of the European Workshop on Security and Privacy in Ad Hoc and Sensor Networks (ESAS)*, Visegrad, Hungary, July 2005.

[12] G. Ács, L. Buttyán, and I. Vajda. Provably secure on-demand source routing in mobile ad hoc networks. *IEEE Transactions on Mobile Computing (TMC)*, 5(11), November 2006.

[13] C. Adams and S. Lloyd. *Understanding PKI: Concepts, Standards, and Deployment Considerations*. Addison-Wesley Professional, second edition, 2002.

[14] W. Aiello, S. M. Bellovin, M. Blaze, R. Canettia, J. Ioannidis, A. D. Keromytis, and O. Reingold. Efficient, DoS-resistant, secure key exchange for internet protocols. In *Proceedings of ACM Computer and Communications Security (CCS) Conference*, 2000.

[15] A. Akella, S. Seshan, R. Karp, and S. Shenker. Selfish behavior and stability of the internet: A game-theoretic analysis of TCP. In *Special Interest Group on Data Communications (SIGCOMM)*, Pittsburgh, PA, USA, 2002.

[16] K. Akkaya and M. Younis. A survey on routing protocols for wireless sensor networks. *Ad Hoc Networks*, **3**(3): 325–349, May 2005.

[17] I. Akyildiz, W.Y. Lee, M. Vuran, and S. Mohanty. Next generation/dynamic spectrum access/cognitive radio wireless networks: a survey. *Computer Networks*, **50**(13), 2006.

[18] I. Akyildiz, W. Su, Y. Sankarasubramaniam, and E. Cayirci. Wireless sensor networks: a survey. *Computer Networks*, **38**: 393–422, 2002.

[19] T. Alpcan and T. Basar. A game theoretic approach to decision and analysis in network intrusion detection. In *Proc. of 43rd IEEE Conference on Decision and Control (CDC)*, Paradise Island, Bahamas, 2006.

[20] T. Alpcan, T. Basar, R. Srikant, and E. Altman. CDMA uplink power control as a noncooperative game. In *Wireless Networks, The Journal of Mobile Communication, Computation and Information*. Kluwer Academic Publishing, November 2002.

[21] T. Alpcan, X. Fan, T. Basar, M. Arcak, and T. J. Wen. Power control for multicell CDMA wireless networks: A team optimization approach. In *Proc. of WiOpt'05*, Apr. 2005.

[22] E. Altman, R. El Azouzi, and T. Jiménez. Slotted Aloha as a stochastic game with partial information. In *Proceedings of the Symposium on Modeling and Optimization in Mobile, Ad Hoc, and Wireless Networks (WiOpt)*, Sophia-Antipolis, France, 2002.

[23] E. Altman, T. Boulogne, R. El Azouzi, T. Jimenez, and L. Wynter. A survey on networking games in telecommunications. *Computers and Operations Research*, **33**(2), Feb. 2006.

[24] L. Anderegg and S. Eidenbenz. Ad-hoc VCG: a truthful and cost-efficient routing protocol for mobile ad hoc networks with selfish agents. In *ACM-MobiCom*, San Diego, September 2003.

[25] R. Anderson, H. Chan, and A. Perrig. Key infection: Smart trust for smart dust. In *International Conference on Network Protocols*, Berlin, Germany, 2004.

[26] R. Anderson and M. Kuhn. Tamper resistance – a cautionary note. In *The Second USENIX Workshop on Electronic Commerce*, Oakland, CA, USA, 1996.

[27] W. Arbaugh, N. Shankar, J. Wan, and K. Zhang. Your 802.11 network has no clothes. *IEEE Wireless Communications Magazine*, **9**(4), 2002.

[28] J. Arkko, J. Kempf, B. Zill, and P. Nikander. SEcure Neighbor Discovery (SEND). In *RFC 3971*, 2005.

[29] M.G. Arranz, R. Aguero, L. Munoz, and P. Mahonen. Behavior of UDP-based applications over IEEE 802.11 wireless networks. In *Personal, Indoor and Mobile Radio Communications, 12th IEEE International Symposium on*, volume 2, San Diego, CA, USA, Sept.-Oct. 2001.

[30] N. Asokan and P. Ginzboorg. Key agreement in ad hoc networks. *Elsevier Computer Communications*, **23**, 2000.

[31] T. Aura. Cryptographically generated addresses (CGA). In *Information Security Conference*, Bristol, UK, 2003.

[32] T. Aura. Cryptographically generated addresses (CGA). In *RFC 3972*, 2005.

[33] G. Avoine. Fraud within asymmetric multi-hop cellular networks. In *Financial Crypto*, Roseau, The Commonwealth Of Dominica, February 2005.

[34] G. Avoine, E. Dysli, and P. Oechslin. Reducing time complexity in RFID systems. In *Proceedings of the 12th Annual Workshop on Selected Areas in Cryptography (SAC)*, 2005.

[35] B. Awerbuch, D. Holmer, C. Nita-Rotaru, and H. Rubens. An on-demand secure routing protocol resilient to Byzantine failures. In *Proceedings of the ACM Workshop on Wireless Security (WiSe)*, Atlanta, Georgia, USA, 2002.

[36] R. Axelrod. *The Complexity of Cooperation: Agent-Based Models of Competition and Collaboration*. Princeton University Press, 1997.

[37] F. Baccelli, B. Blaszczyszyn, and F. Tournois. Downlink admission/congestion control and maximal load in CDMA networks. In *Proceedings of the IEEE Conference on Computer Communications (INFOCOM '03)*, March 30 – April 3 2003.

[38] H. Balakrishnan, K. Lakshminarayanan, S. Ratnasamy, S. Shenker, I. Stoica, and M. Walfish. A layered naming architecture for the internet. In *Special Interest Group on Data Communications (SIGCOMM)*, Portland, OR, USA, 2004.

[39] M. Balazinska and P. Castro. Characterizing mobility and network usage in a corporate wireless local-area network. In *Proceedings of the ACM Conference on Mobile Systems, Applications and Services (MobiSys)*, San Francisco, CA, USA, May 2003.

[40] D. Balfanz, D. Smetters, P. Stewart, and H. Wong. Talking to strangers: authentication in ad-hoc wireless networks. In *Proceedings of the 9th Annual Network and Distributed System Security Symposium (NDSS)*, San Diego, California, USA, 2002.

[41] C. Barrett, M. Marathe, D. Engelhart, and A. Sivasubramaniam. Analyzing the short-term fairness of IEEE 802.11 in wireless multi-hop radio networks. In *Modeling, Analysis and Simulation of Computer and Telecommunications Systems, 10th IEEE International Symposium on*, Fort Worth, Texas, USA, 2002.

[42] S. Basagni, I. Chlamtac, V. Syrotiok, and B. Woodward. A distance routing effect algorithm for mobility (DREAM). In *Proceedings of the ACM Conference on Mobile Computing and Networking (MobiCom)*, 1998.

[43] T. Basar and G. J. Olsder. *Dynamic Noncooperative Game Theory*. Academic Press, London/New York, 1995.

[44] M. Bellare and P. Rogaway. Optimal asymmetric encryption – how to encrypt with RSA. In *Advances in Cryptology – Proceedings of the International Cryptology Conference (EUROCRYPT)*, Perugia, Italy, May 1994.

[45] R. E. Bellman. *Dynamic Programming*. Princeton University Press, Princeton, New Jersey, 1957.

[46] N. Ben Salem, L. Buttyán, J.-P. Hubaux, and M. Jakobsson. Node cooperation in hybrid ad hoc networks. *IEEE Transactions on Mobile Computing (TMC)*, **5**(4), April 2006.

[47] N. Ben Salem and J.-P. Hubaux. A fair scheduling for wireless mesh networks. In *Proceedings of the IEEE Workshop on Wireless Mesh Networks (WiMesh)*, Santa Clara, CA, USA, 2005.

[48] N. Ben Salem and J.-P. Hubaux. Securing Wireless Mesh Networks. *IEEE Wireless Communications*, **13**, 2006.

[49] N. Ben Salem, J.-P. Hubaux, and M. Jakobsson. Reputation-based Wi-Fi deployment. *Mobile Computing and Communications Review (MC2R)*, **9**(3), 2005.

[50] Y. Benkler. Some economics of wireless communication. *Harvard Journal of Law and Technology*, **16**(1), Fall 2004.

[51] A. Beresford and F. Stajano. Location privacy in pervasive computing. *IEEE Pervasive Computing Magazine*, **2**(1): 46–55, January–March 2003.

[52] A. Beresford and F. Stajano. Mix zones: User privacy in location-aware services. In *Proceedings of the IEEE Conference on Pervasive Computing and Communications – Workshops*, 2004.

[53] G. Bianchi. Performance analysis of the IEEE 802.11 distributed coordination function. *Selected Areas in Communications, IEEE Journal on*, **18**(3): 535–547, 2000.

[54] M. Blaze, J. Feigenbaum, J. Ioannidis, and A. Keromytis. The KeyNote trust-management system version 2. In *RFC 2704*, 1999.

[55] M. Blaze, J. Feigenbaum, and J. Lacy. Decentralized trust management. In *Proceedings of the IEEE Symposium on Research in Security and Privacy*, Oakland, CA, USA, 1996.

[56] L. Blazevic, L. Buttyán, S. Čapkun, S. Giordano, J.-P. Hubaux, and J.-Y. Le Boudec. Self-organization in mobile ad hoc networks: the approach of Terminodes. *IEEE Communications Magazine*, **39**(6): 307–321, June 2001.

[57] B. Blum, T. He, S. Son, and J. Stankovic. IGF: a state-free robust communication protocol for wireless sensor networks. Technical Report CS-2003-11, University of Virginia, Charlottesville, VA, USA, 2003.

[58] J. Blum and A. Eskandarian. The threat of intelligent collisions. *IT Professional*, **6**(1), January–February 2004.

[59] D. Boneh, C. Gentry, B. Lynn, and H. Shacham. A survey of two signature aggregation techniques. *CryptoBytes*, **6**(2), 2003.

[60] D. Boneh, C. Gentry, H. Shacham, and B. Lynn. Aggregate and verifiably encrypted signatures from bilinear maps. In *Advances in Cryptology – Eurocrypt 2003*. Springer, 2003.

[61] N. Borisov, I. Goldberg, and D. Wagner. Intercepting mobile communications: the insecurity of 802.11. In *Proceedings of the ACM Conference on Mobile Computing and Networking (MobiCom)*, Rome, Italy, 2001.

[62] P. Bose, P. Morin, I. Stojmenovic, and J. Urrutia. Routing with guaranteed delivery in ad hoc wireless networks. In *Proceedings of the ACM Workshop on Discrete Algorithms and Methods for Mobile Computing and Communications (DIAL M)*, 1999.

[63] P. Bose, P. Morin, I. Stojmenovic, and J. Urrutia. Routing with guaranteed delivery in ad hoc wireless networks. *Wireless Networks*, **7**(6): 609–616, 2001.

[64] J.-Y. Le Boudec and M. Vojnovic. Perfect simulation and stationarity of a class of mobility models. In *Proceedings of the IEEE Conference on Computer Communications (INFOCOM)*, Miami, FL, USA, 2005.

[65] C. Boyd and A. Mathuria. *Protocols for Authentication and Key Establishment*. Springer, 2003.

[66] S. Brands and D. Chaum. Distance-bounding protocols. In *Proceedings of the Workshop on Theory and Application of Cryptographic Techniques*, Lofthus, Norway, 1993.

[67] M. Briceno, I. Goldberg, and D. Wagner. GSM cloning. http://www.isaac.cs.berkeley.edu/isaac/gsm.htm, April 1998.

[68] A. Broder and M. Mitzenbacher. Network applications of bloom filters: A survey. *Internet Mathematics*, **4**, 2003.

[69] S. Buchegger and J.Y. Le Boudec. Performance analysis of the CONFIDANT Protocol (Cooperation Of Nodes–Fairness In Dynamic Ad-hoc NeTworks). In *Proceedings of the ACM Symposium on Mobile Ad Hoc Networking and Computing (MobiHoc)*, Lausanne, Switzerland, June 2002.

[70] M. Burrows, M. Abadi, and R. Needham. A logic of authentication. *ACM Transactions on Computer Systems*, **8**(1): 18–36, February 1990.

[71] L. Buttyán. Removing the financial incentive to cheat in micropayment schemes. *IEE Electronics Letters*, **36**(2), January 2000.

[72] L. Buttyán, L. Dóra, and I. Vajda. Statistical wormhole detection in sensor networks. In *Proceedings of the European Workshop on Security and Privacy in Ad Hoc and Sensor Networks (ESAS)*, Visegrad, Hungary, 2005. Springer.

[73] L. Buttyán, T. Holczer, and P. Schaffer. Spontaneous cooperation in multi-domain sensor networks. In *European Workshop on the Security of Ad Hoc and Sensor Networks (ESAS)*, Visegrad, Hungary, 2005.

[74] L. Buttyán, T. Holczer, and I. Vajda. Optimal key-trees for tree-based private authentication. In *Proceedings of the Privacy Enhancing Technologies Workshop (PET)*, 2006.

[75] L. Buttyán and J.-P. Hubaux. Enforcing service availability in mobile ad hoc WANs. In *Proceedings of the ACM Symposium on Mobile Ad Hoc Networking and Computing (MobiHoc)*, Boston, MA, USA, August 2000.

[76] L. Buttyán and J.-P. Hubaux. Rational Exchange – A Formal Model Based on Game Theory. In *2nd International Workshop on Electronic Commerce (WELCOM)*, Heidelberg, Germany, November 2001.

[77] L. Buttyán and J.-P. Hubaux. Stimulating cooperation in self-organizing mobile ad hoc networks. In *ACM/Kluwer Mobile Networks and Applications (MONET) Special Issue on Mobile Ad Hoc Networks*, volume 8, October 2003.

[78] L. Buttyán and I. Vajda. Towards provable security for ad hoc routing protocols. In *Proceedings of the ACM Workshop on Security in Ad Hoc and Sensor Networks (SASN)*, October 2004.

[79] M. Čagalj. *Thwarting selfish and malicious behavior in wireless networks*. EPFL Thesis Nr 3449, 2006.

[80] M. Čagalj, S. Ganeriwal, I. Aad, and J.-P. Hubaux. On selfish behavior in CSMA/CA networks. In *Proceedings of the IEEE Conference on Computer Communications (INFOCOM)*, Miami, Florida, USA, March 2005.

[81] M. Čagalj, S. Čapkun, and J.P Hubaux. Key agreement in peer-to-peer wireless networks. *Proceedings of the IEEE*, **94**(2), February 2006.

[82] S. Camtepe and B. Yener. Key distribution mechanisms for wireless sensor networks: a survey. Technical Report TR-05-07, Rensselaer Polytechnic Institute, Computer Science Department, March 2005.

[83] L. Cao and H. Zheng. Spectrum allocation in ad hoc networks via local bargaining. In *Proc. of the Second Annual IEEE Communications Society Conference on Sensor and Ad Hoc Communications and Networks (SECON)*, Santa Clara, CA, September 2005.

[84] S. Čapkun, L. Buttyán, and J.-P. Hubaux. SECTOR: Secure tracking of node encounters in multi-hop wireless networks. In *Proceedings of the ACM Workshop on Security in Ad Hoc and Sensor Networks (SASN)*, Fairfax, VA, USA, 2003.

[85] S. Čapkun, L. Buttyán, and J.-P. Hubaux. Self-organized public-key management for mobile ad hoc networks. *IEEE Transactions on Mobile Computing*, **2**(1), January–March 2003.

[86] S. Čapkun, M. Hamdi, and J.-P. Hubaux. GPS-free positioning in mobile ad-hoc networks. *Cluster Computing Journal*, **5**(2), April 2002.

[87] S. Čapkun and J.-P. Hubaux. BISS: Building secure routing out of an incomplete set of security associations. In *Proceedings of the ACM Workshop on Wireless Security (WiSe)*, 2003.

[88] S. Čapkun and J.-P. Hubaux. Secure positioning of wireless devices with application to sensor networks. In *Proceedings of the IEEE Conference on Computer Communications (INFOCOM)*, Miami, FL, USA, 2005.

[89] S. Čapkun and J.-P. Hubaux. Secure positioning in wireless networks. *IEEE Journal on Selected Areas in Communications (J-SAC)*, **24**(2), February 2006.

[90] S. Čapkun, J.-P. Hubaux, and L. Buttyán. Mobility helps peer-to-peer security. *IEEE Transactions on Mobile Computing (TMC)*, **5**(1), January 2006.

[91] B. Carbunar, I. Ioannidis, and C. Nita-Rotaru. Janus: Towards robust and malicious resilient routing in hybrid wireless networks. In *ACM-WiSe*, Philadelphia, October 2004.

[92] D. W. Carman, P. S. Kruus, and B. J. Matt. Constraints and approaches for distributed sensor network security. Technical Report 00-010, NAI Labs, 2000.

[93] C. Castelluccia, E. Mykletun, and G. Tsudik. Efficient aggregation of encrypted data in wireless sensor networks. In *International Conference on Mobile and Ubiquitous Systems (Mobiquitous)*, San Diego, CA, USA, 2005.

[94] D. Catrein, L. A. Imhof, and R. Mathar. Power control, capacity, and duality of uplink and downlink in cellular CDMA systems. *IEEE Transactions on Communications*, **52**(10): 1777–1785, Oct. 2004.

[95] R. Chaabouni. Break WEP faster with statistical analysis. Technical report, EPFL, LASEC, June 2006.

[96] H. Chan and A. Perrig. PIKE: Peer intermediaries for key establishment in sensor networks. In *Proceedings of the IEEE Symposium on Research in Security and Privacy*, Berkeley/Oakland, CA, USA, 2004.

[97] H. Chan, A. Perrig, and D. Song. Random key predistribution schemes for sensor networks. In *Proceedings of the IEEE Symposium on Research in Security and Privacy*, Berkeley/Oakland, CA, USA, 2003.

[98] D. Chaum. Untraceable electronic mail, return addresses, and digital pseudonyms. *Communications of the ACM*, **24**(2): 84–88, 1981.

[99] D. Chaum. The Dining Cryptographers Problem: Unconditional sender and recipient untraceability. *Journal of Cryptology*, **1**(1): 65–75, 1988.

[100] S. Cheung. An efficient message authentication scheme for link state routing. In *Proceedings of the 13th Annual Computer Security Applications Conference*, 1997.

[101] D. Choffnes and F. Bustamante. An integrated mobility and traffic model for vehicular wireless networks. In *The Second ACM International Workshop on Vehicular Ad Hoc Networks (VANET)*, Cologne, Germany, 2005.

[102] D. Clark, J. Wroclawski, K. Sollins, and R. Braden. Tussle in cyberspace: Defining tomorrow's internet. In *IEEE Transactions on Networking*, June 2005.

[103] T. Clausen and P. Jacquet. Optimized link state routing protocol (OLSR). Internet RFC 3626, October 2003.

[104] LAN/MAN Standards Committee. *ANSI/IEEE Std 802.11: Wireless LAN Medium Access Control (MAC) and Physical Layer (PHY) Specifications*. IEEE Computer Society, 1999.

[105] D. Coppersmith and M. Jakobsson. Almost optimal hash sequence traversal. In *Proceedings of the Financial Cryptography Conference (FC)*, Southampton, Bermuda, 2002.

[106] M. Corner and B. Noble. Protecting applications with transient authentication. In *Proceedings of the ACM Conference on Mobile Systems, Applications and Services (MobiSys)*, San Francisco, CA, May 2003.

[107] J.-S. Coron. Optimal security proofs for PSS and other signature schemes. In *Proceedings of the International Cryptology Conference (EUROCRYPT)*, Amsterdam, The Netherlands, April 2002.

[108] N. Cressie. *Statistics for Spatial Data*. John Wiley and Sons, 1993.

[109] J.-H. Cui, J. Kong, M. Gerla, and S. Zhou. Challenges: Building scalable mobile underwater wireless sensor networks for aquatic applications. *IEEE Network, Special issue on Wireless Sensor Networking*, **20**(3), 2006.

[110] C. Dellacrocas and P. Resnick. Online reputation mechanisms – a roadmap for future research. In *1st Interdisciplinary Symposium on Online Reputation Mechanism*, Cambridge, MA, USA, April 2003.

[111] J. Deng, R. Han, and S. Mishra. INSENS: intrusion-tolerant routing for wireless sensor networks. *Computer Communications*, **29**(2): 216–230, 2005.

[112] Z. Despotovic and K. Aberer. Trust and reputation in P2P networks. In *1st Interdisciplinary Symposium on Online Reputation Mechanism*, Cambridge, MA, USA, April 2003.

[113] C. Díaz, S. Seys, J. Claessens, and B. Preneel. Towards measuring anonymity. In *Proceedings of the Privacy Enhancing Technologies Workshop*, 2002.

[114] T. Dierks and C. Allen. The TLS protocol (version 1.0). Internet RFC 2246, 1999.

[115] W. Diffie and M. Hellman. New directions in cryptography. *IEEE Transactions on Information Theory*, **22**, 1976.

[116] T. Dimitriou. A lightweigth RFID protocol to protect against traceability and cloning attacks. In *Proceedings of the IEEE Conference on Security and Privacy in Communication Networks (SecureComm)*, 2005.

[117] S. Dohrmann and C. Ellison. Public-key Support for Collaborative Groups. In *Proceedings of the 1st Annual PKI Research Workshop*, Gaithersburg, MD, USA, 2002.

[118] D. Dolev, C. Dwork, and M. Naor. Nonmalleable cryptography. *SIAM Review*, **45**(4), 2003.

[119] D. Dolev and A. Yao. On the security of public-key protocols. In *IEEE Transactions on Information Theory*, 1983.

[120] J. R. Douceur. The Sybil attack. In *Proceedings of the International Workshop on Peer-to-Peer Systems (IPTPS)*, Cambridge, MA, USA, March 2002.

[121] M. Dramitinos, C. Courcoubetis, and G. D. Stamoulis. Auction-based resource reservation in 2.5/3G networks. *Kluwer/ACM Mobile Networks and Applications (MONET)*, 2003.

[122] W. Du, J. Deng, Y. S. Han, and P. K. Varshney. A pairwise key pre-distribution scheme for wireless sensor networks. In *Proceedings of the ACM Conference on Computer and Communications Security (CCS)*, Washington DC, USA, 2003.

[123] H. Dubois-Ferriere, M. Grossglauser, and M. Vetterli. Age matters: Efficient route discovery in mobile ad hoc networks using encounter ages. In *Proceedings of the ACM Symposium on Mobile Ad Hoc Networking and Computing (MobiHoc)*, Annapolis, MD, USA, 2003.

[124] J. Edney and W. Arbaugh. *Real 802.11 Security: WiFi Protected Access and 802.11i*. Addison-Wesley, 2004.

[125] E. Efstathiou and G. Polyzos. A peer-to-peer approach to wireless LAN roaming. In *Proceedings of ACM International Workshop on Wireless Mobile Applications and services on WLAN Hotspots (WMASH)*, San Diego, CA, USA, September 2003.

[126] S. Eidenbenz, G. Resta, and P. Santi. Commit: A sender-centric truthful and energy-efficient routing protocol for ad hoc networks with selfish nodes. *International Parallel and Distributed Processing Symposium (IPDPS)*, 13, 2005.

[127] T. ElGamal. A public key cryptosystem and a signature scheme based on discrete logarithms. *IEEE Transactions on Information Theory*, **31**(4): 469–472, July 1985.

[128] P. Erdős and A. Rényi. On the evolution of random graphs. *Publications of the Math. Institute of the Hungarian Academy of Sciences*, **5**, 1960.

[129] L. Eschenauer and V. Gligor. A key-management scheme for distributed sensor networks. In *Proceedings of the ACM Conference on Computer and Communications Security (CCS)*, Washington DC, USA, 2002.

[130] L. Eschenauer, V. Gligor, and J. Baras. On trust establishment in mobile ad-hoc networks. In *Proc. of the Security Protocols Workshop*, Cambridge, UK, April 2002.

[131] K. Fall and K. Varadhan. *ns notes and documentation*. UC Berkeley, LBL, USC/ISI, Xerox PARC, 2003.

[132] G. Faulhaber and D. Farber. Spectrum management: Property rights, markets, and the commons. In *Telecommunications Policy Research Conference Proceedings*, Arlington, Virginia, USA, 2003.

[133] M. Félegyházi, L. Buttyán, and J.-P. Hubaux. Equilibrium analysis of packet forwarding strategies in wireless ad hoc networks – the static case. In *Personal Wireless Communication (PWC)*, Venice, Italy, September 2003.

[134] M. Félegyházi and J.-P. Hubaux. Wireless operators in a shared spectrum. In *Proceedings of the IEEE Conference on Computer Communications (INFOCOM)*, Barcelona, Spain, April 2006.

[135] M. Félegyházi, J.-P. Hubaux, and L. Buttyán. Cooperative packet forwarding in multi-domain sensor networks. In *First International Workshop on Sensor Networks and Systems for Pervasive Computing (PerSeNS)*, Hawaii, USA, March 2005.

[136] M. Félegyházi, J.-P. Hubaux, and L. Buttyán. Nash equilibria of packet forwarding strategies in wireless ad hoc networks. In *IEEE Transactions on Mobile Computing (TMC)*, April 2006.

[137] M. Félegyházi, M. Čagalj, D. Dufour, and J.-P. Hubaux. Border games in cellular networks. In *Proceedings of the IEEE Conference on Computer Communications (INFOCOM)*, April 2007.

[138] B. A. Fette and B. Fette. *Cognitive Radio Technology (Communications Engineering)*. Elsevier Inc., 2006.

[139] D. Figueiredo, M. Garetto, and Don Towsley. Exploiting mobility in ad-hoc wireless networks with incentives. In *Technical Report. University of Massachussetts, Computer Science 04-66*, 2004.

[140] K. Finkenzeller. *RFID-Handbook: Fundamentals and Applications in Contactless Smart Cards and Identification*. John Wiley and Sons, 2003.

[141] S. Fluhrer, I. Mantin, and A. Shamir. Weaknesses in the key scheduling algorithm of RC4. In *Proceedings of the 8th Workshop on Selected Areas in Cryptography*, 2001.

[142] J. W. Friedman and C. Mezzetti. Learning in games by random sampling. *Journal of Economic Theory*, **98**: 55–84, 2001.

[143] D. Fudenberg and D. K. Levine. *The Theory of Learning in Games*. MIT Press, 1998.

[144] D. Fudenberg and J. Tirole. *Game Theory*. MIT Press, 1991.

[145] V. Gambiroza, B. Sadeghi, and E. Knightly. End-to-end performance and fairness in multihop wireless backhaul networks. In *Proceedings of the ACM Conference on Mobile Computing and Networking (MobiCom)*, Philadelphia, PA, USA, 2004.

[146] C. Gehrmann, C. J. Mitchell, and K. Nyberg. Manual Authentication for Wireless Devices. *RSA Cryptobytes*, 7(1), January 2004.

[147] C. Gehrmann and K. Nyberg. Enhancements to Bluetooth Baseband Security. In *Proceedings of Nordsec*, Copenhagen, Denmark, Nov. 2001.

[148] C. Gehrmann, J. Persson, and B. Smeets. *Bluetooth Security*. Artech House, June 2004.

[149] M. Gerlach. VaneSe – An approach to VANET security. In *Vehicle-to-Vehicle Communications (V2VCOM)*, San Diego, CA, USA, 2005.

[150] R. Gibbons. *A Primer in Game Theory*. Prentice Hall, 1992.

[151] S. Giordano and M. Hamdi. Mobility management: the Virtual Home Region. Technical report, EPFL, October 1999.

[152] V. Gligor, S. Luan, and J. Pato. On inter-realm authentication in large distributed systems. In *Proceedings of the IEEE Symposium on Research in Security and Privacy*, 1992.

[153] P. Golle, D. Greene, and J. Staddon. Detecting and correcting malicious data in VANETs. In *Workshop on Vehicular Ad hoc Networks (VANET)*, Philadelphia, PA, USA, 2004.

[154] P. Golle, M. Jakobsson, A. Juels, and P. Syverson. Universal re-encryption for mixnets. In *RSA Conference – Cryptographers' Track (CT-RSA)*, 2004.

[155] D. Goodman and N. Mandayam. Network assisted power control for wireless data. *Mobile Networks and Applications (MONET)*, 6: 409–415, 2001.

[156] M. Grossglauser and D. Tse. Mobility increases the capacity of ad hoc wireless networks. *IEEE Transactions on Networking*, **10**(4), August 2002.

[157] P. Gupta and P. R. Kumar. The capacity of wireless networks. *IEEE Transactions on Information Theory*, **46**: 388–404, 2000.

[158] Z. Haas and B. Liang. Ad hoc mobility management with uniform quorum systems. *IEEE/ACM Transactions on Networking*, **7**(2): 228–240, April 1999.

[159] J. Hall, M. Barbeau, and E. Kranakis. Enhancing intrusion detection in wireless networks using radio frequency fingerprinting. In *The IASTED Conference on Communications, Internet and Information Technology*, St. Thomas, US Virgin Islands, 2004.

[160] N. Haller, C. Metz, P. Nesser, and M. Straw. A one-time password system. Internet RFC 2289, 1998.

[161] S. V. Hanly and D. N. Tse. Power control and capacity of spread spectrum wireless networks. *Automatica*, **35**(12): 1987–2012, Dec. 1999.

[162] G. Hardin. The tragedy of the commons. *Science*, **162**(3859), 1968.

[163] D. Harkins and D. Carrel. The Internet Key Exchange (IKE). Internet RFC 2409, 1998.

[164] J. C. Harsanyi and R. Selten. *A General Theory of Equilibrium Selection in Games*. MIT Press, 1988.

[165] S. Haykin. Cognitive radio: Brain-empowered wireless communications. *IEEE Journal on Selected Areas in Communications (JSAC)*, **23**(2), Feb. 2005.

[166] C. He, M. Sundararajan, A. Datta, A. Derek, and J. C. Mitchell. A modular correctness proof of IEEE 802.11i and TLS. In *Proceedings of the ACM Conference on Computer and Communications Security (CCS)*, Alexandria, VA, USA, 2005.

[167] M. Hellman. A cryptanalytic time-memory tradeoff. *IEEE Transactions on Information Theory*, **IT-26**: 401–406, 1980.

[168] M. Hermelin and K. Nyberg. Correlation properties of the Bluetooth combiner. In *Proceedings of the International Conference on Information Security and Cryptology*, 1999.

[169] M. Heusse, F. Rousseau, G. Berger-Sabbatel, and A. Duda. Performance anomaly of 802.11b. In *Proceedings of the IEEE Conference on Computer Communications (INFOCOM)*, San Franciso, CA, USA, 2003.

[170] H. Holma and A. Toskala, editors. *WCDMA for UMTS*. John Wiley & Sons, Inc., New York, NY, USA, 2002.

[171] T.-C. Hou and V. O. K. Li. Transmission range control in multihop packet radio networks. *IEEE Transactions on Communications*, **34**(1): 38–44, January 1986.

[172] D. Houser and J. Wooders. Reputation in auctions: Theory, and evidence from ebay. Technical Report 00-01, University of Arizona, 2001.

[173] http://www.boingo.com/.

[174] http://www.isi.edu/nsnam/ns/.

[175] http://www.rsasecurity.com/products/securid/.

[176] L. Hu and D. Evans. Using directional antennas to prevent wormhole attacks. In *Proceedings of the Symposium on Network and Distributed Systems Security (NDSS)*, San Diego, CA, USA, 2004.

[177] Y.-C. Hu, D. Johnson, and A. Perrig. SEAD: Secure efficient distance vector routing for mobile wireless ad hoc networks. In *Proceedings of the IEEE Workshop on Mobile Computing Systems and Applications (WMCSA)*, 2002.

[178] Y.-C. Hu and A. Perrig. A survey of secure wireless ad hoc routing. *IEEE Security and Privacy Magazine*, 2(3): 28–39, May/June 2004.

[179] Y.-C. Hu, A. Perrig, and D. Johnson. Efficient security mechanisms for routing protocols. In *Proceedings of the Network and Distributed System Security Symposium (NDSS)*, San Diego, California, USA, 2003.

[180] Y.-C. Hu, A. Perrig, and D. Johnson. Packet leashes: a defense against wormhole attacks in wireless networks. In *Proceedings of the IEEE Conference on Computer Communications (INFOCOM)*, San Francisco, CA, USA, 2003.

[181] Y.-C. Hu, A. Perrig, and D. Johnson. Rushing attacks and defense in wireless ad hoc network routing protocols. In *Proceedings of the ACM Workshop on Wireless Security (WiSe)*, San Diego, CA, USA, 2003.

[182] Y.-C. Hu, A. Perrig, and D. Johnson. Ariadne: A secure on-demand routing protocol for ad hoc networks. *Wireless Networks*, 11(1–2): 21–38, January 2005.

[183] J. Huang, R. Berry, and M. Honig. Auction-based spectrum sharing. *ACM/Kluwer Journal of Mobile Networks and Applications (MONET) special issue on WiOpt'04*, 11: 405–418, 2006.

[184] J.-P. Hubaux, L. Buttyán, and S. Čapkun. The quest for security of mobile ad hoc networks. In *Mobihoc*, Long Beach, CA, USA, 2001.

[185] J.-P. Hubaux, T. Gross, J.-Y. Le Boudec, and M. Vetterli. Towards self-organizing mobile ad-hoc networks: the Terminodes project. *IEEE Communication Magazine*, 39(1), 2001.

[186] J.-P. Hubaux, S. Čapkun, and J. Luo. The security and privacy of smart vehicles. *IEEE Security and Privacy Magazine*, 2(3): 49–55, 2004.

[187] IEEE. Wireless LAN Medium Access Control (MAC) and Physical Layer (PHY) specifications. IEEE Standard 802.11, 1999.

[188] IEEE. Port-based network access control. IEEE Standard 802.1X, 2001.

[189] IEEE. Medium Access Control (MAC) security enhancements. IEEE Standard Amendment 802.11i, 2004.

[190] IEEE. Physical and medium access control layers for combined fixed and mobile operation in licensed bands. IEEE Standard Amendment 802.16e, December 2005.

[191] IEEE 802.11 WG. ANSI/IEEE Std 802.11: Wireless LAN Medium Access Control (MAC) and Physical Layer (PHY) Specifications: Medium Access Control (MAC) Enhancements for Quality of Service (QoS) IEEE 802.11/D2.0, 2001.

[192] IEEE 802.11 WG part 11a. Wireless LAN Medium Access Control (MAC) and Physical Layer (PHY) specifications, High-speed Physical Layer in the 5 GHz Band, 1999.

[193] IEEE 802.11 WG part 11b. Wireless LAN Medium Access Control (MAC) and Physical Layer (PHY) specifications, Higher Speed PHY Layer Extension in the 2.4 GHz Band, 1999.

[194] IEEE 802.11 WG part 11e/D4.3, Draft supplement to part 11: Wireless Medium Access Control (MAC) and Physical Layer (PHY) Specifications: Medium Access

Control (MAC) Enhancements for Quality of Service (QoS) IEEE Std. 802.11e/D4.3, 2003.

[195] IEEE 802.11 WG part 11g. Wireless LAN Medium Access Control (MAC) and Physical Layer (PHY) specifications, Further Higher Speed Physical Layer Extension in the 2.4 GHz Band, 2003.

[196] IEEE 802.11e WG. Amendment: Medium Access Control (MAC) Quality of Service (QoS) Enhancements, January 2005.

[197] IEEE P1609.2 Version 1 – Standard for Wireless Access in Vehicular Environments – Security Services for Applications and Management Messages (In development), 2006.

[198] INTUG. International mobile roaming. *An INTUG response to the DG Information Society Second Phase Consultation on Roaming Charges April 2006*, May 2006.

[199] R. Jain. *The art of computer systems performance analysis*. John Wiley and Sons, 1991.

[200] W. C. Jakes, editor. *Microwave Mobile Communications*. John Wiley & Sons, Inc. – IEEE Press, 1994.

[201] M. Jakobsson. Payments and Diffie-Hellman key exchange (presentation slides). Private communication with M. Jakobsson.

[202] M. Jakobsson. Ripping coins for a fair exchange. In *Advances in cryptology. EURO-CRYPT*, 1995.

[203] M. Jakobsson, J.-P. Hubaux, and L. Buttyán. A micropayment scheme encouraging collaboration in multi-hop cellular networks. In *Financial Crypto*, La Guadeloupe, France, January 2003.

[204] M. Jakobsson and S. Wetzel. Security weaknesses in Bluetooth. In *Progress in Cryptology – CT-RSA 2001: The Cryptographers' Track at RSA Conference 2001*, San Francisco, CA, USA, April 2001.

[205] H. Ji and C.-Y. Huang. Non-cooperative uplink power control in cellular radio systems. *Wireless Networks (WINET)*, **46**(3): 233–240, 1998.

[206] T. Jiang and J. S. Baras. Trust evaluation in anarchy: A case study on autonomous networks. In *Proceedings of the IEEE Conference on Computer Communications (INFOCOM)*, Barcelona, Spain, 2006.

[207] Y. Jin and G. Kesidis. Equilibria of a noncooperative game for heterogeneous users of an ALOHA network. *IEEE Comm. Letters*, 6, 2002.

[208] Y. Jin and G. Kesidis. Nash equilibria of a generic networking game with applications to circuit-switched networks. In *Proceedings of the IEEE Conference on Computer Communications (INFOCOM)*, San Francisco, USA, March-April 2003.

[209] R. Johari and J. N. Tsitsiklis. Routing and peering in a competitive Internet. In *Conference on Decision and Control*, December 2004.

[210] D. Johnson and D. Maltz. Dynamic source and routing in ad hoc wireless networks. In T. Imilienski and H. Korth, editors, *Mobile Computing*. Kluwer Academic Publishers, 1996.

[211] D. Johnston and J. Walker. Overview of IEEE 802.16 security. *IEEE Security and Privacy Magazine*, **2**(3): 40–48, May-June 2004.

[212] A. Juels. Minimalist cryptography for low-cost RFID tags. In *International Conference on Security in Communication Networks (SCN)*, 2004.

[213] A. Juels. RFID security and privacy: A research survey. Technical report, RSA Laboratories, September 2005.

[214] A. Juels and R. Pappu. Squealing Euros: Privacy protection in RFID-enabled banknotes. In Rebecca N. Wright, editor, *Proceedings of the Financial Cryptography Conference*, volume 2742 of *Lecture Notes in Computer Science*, Le Gosier, Guadeloupe, French West Indies, January 2003. IFCA, Springer-Verlag.

[215] A. Juels, R. Rivest, and M. Szydlo. The blocker tag: Selective blocking of RFID tags for consumer privacy. In *Proceedings of the ACM Conference on Computer and Communications Security (CCS)*, pages 103–111, Washington, DC, USA, October 2003.

[216] F. Kargl, A. Geiß, S. Schlott, and M. Weber. Secure dynamic source routing. In *Proceedings of the 38th Annual Hawaii International Conference on System Sciences (HICSS)*, Hawaii, USA, 2005.

[217] C. Karlof and D. Wagner. Secure routing in wireless sensor networks: attacks and countermeasures. *Elsevier's AdHoc Networks Journal, Special Issue on Sensor Network Applications and Protocols*, 1(2–3): 293–315, September 2003.

[218] B. Karp and H. T. Kung. Greedy perimeter stateless routing for wireless networks. In *Proceedings of the ACM Conference on Mobile Computing and Networking (MobiCom)*, Boston, Massachusetts, USA, August 2000.

[219] F. Kelly. Charging and rate control for elastic traffic. *European Transactions on Telecommunications*, 8, 1997.

[220] F. Kelly, A. Maulloo, and D. Tan. Rate control for communication networks: shadow prices, proportional fairness, and stability. *Journal of the Operational Research Society*, 49, 1998.

[221] Z. Kfir and A. Wool. Picking virtual pockets using relay attacks on contactless smart card systems. In *Proceedings of the IEEE Conference on Security and Privacy in Communication Networks (SecureComm)*, 2005.

[222] D. Kim, Y. Chang, and J. W. Lee. Pilot power control and service coverage support in CDMA mobile systems. In *Proceedings of IEEE Vehicular Technology Conference (VTC'99)*, 1999.

[223] Y.-B. Ko and N. H. Vaidya. Location-aided routing (LAR) in mobile ad hoc networks. *Wireless Networks*, 6(4): 307–321, 2000.

[224] M. Kodialam and T. Nandagopal. Characterizing the capacity region in multi-radio multi-channel wireless mesh networks. In *Proceedings of the ACM Conference on Mobile Computing and Networking (MobiCom)*, Cologne, Germany, 2005.

[225] R. Kohlas and U. Maurer. Confidence valuation in a public-key infrastructure based on uncertain evidence. In *Proceedings of PKC'00*, volume 1751 of *Lecture Notes in Computer Science*. Springer-Verlag, 2000.

[226] C. E. Koksal, H. Kassab, and H. Balakrishnan. An analysis of short-term fairness in wireless media access protocols. In *Proceedings of ACM Sigmetrics*, Santa Clara, CA, USA, 2000.

[227] J. Kong and X. Hong. ANODR: Anonymous on-demand routing with untraceable routes for mobile ad-hoc networks. In *Proceedings of the ACM Symposium on Mobile Ad Hoc Networking and Computing (MobiHoc)*, 2003.

[228] J. Kong, X. Hong, and M. Gerla. Modeling ad-hoc rushing attack in a negligibility-based security framework. In *Proceedings of the ACM Workshop on Wireless Security (WiSe)*, Los Angeles, CA, USA, 2006.

[229] J. Kong, Z. Ji, W. Wang, M. Gerla, and R. Bagrodia. On wormhole attacks in underwater sensor networks: A two-tier localization approach. Technical Report CSD-TR040051, UCLA Computer Science Department, December 2004.

[230] J. Konorski. Multiple access in ad hoc wireless LANs with noncooperative stations. In *NETWORKING*, volume 2345 of LNCS, Pisa, Italy, Springer, 2002.

[231] Y. Korilis, A. Lazar, and A. Orda. Architecting noncooperative networks. *IEEE Journal on Selected Areas in Communications (JSAC), Special Issue on Advances in the Fundamentals of Networking*, 13(7), Sep. 1995.

[232] Y. Korilis and A. Orda. Incentive compatible pricing strategies for QoS routing. In *Proceedings of the IEEE Conference on Computer Communications (INFOCOM)*, New York, NY, USA, March 1999.

[233] D. Kotz and K. Essien. Analysis of a campus-wide wireless network. In *Proceedings of the ACM Conference on Mobile Computing and Networking (MobiCom)*, Atlanta, Georgia, USA, Sept. 2002.

[234] E. Koutsoupias and C. Papadimitriou. Worst-case equilibria. In *Proceedings of the 16th Annual Symposium on Theoretical Aspects of Computer Science (STACS'99)*, March 1999.

[235] E. Kranakis, H. Singh, and J. Urrutia. Compass routing on geometric networks. In *Proceedings of the Canadian Conference on Computational Geometry*, August 1999.

[236] H. Krawczyk. SIGMA. http: //www.ee.technion.ac.il/ hugo/sigma.html.

[237] H. Krawczyk, M. Bellare, and R. Canetti. HMAC: Keyed hashing for message authentication. Internet RFC 2104, February 1997.

[238] P. Kruus, D. Sterne, R. Gopaul, M. Heyman, B. Rivera, P. Budulas, B. Luu, T. Johnson, N. Ivanic, and G. Lawler. In-band wormholes and countermeasures in OLSR networks. In *Proceedings of the IEEE Conference on Security and Privacy in Communication Networks (SecureComm)*, Baltimore, MD, USA, August 2006.

[239] D. Kügler. Man in the Middle Attacks on Bluetooth. In *Financial Cryptography*, Guadeloupe, French West Indies, 2003. Lecture Notes in Computer Science, Springer-Verlag.

[240] F. Kuhn, R. Wattenhofer, Y. Zhang, and A. Zollinger. Geometric ad-hoc routing: theory and practice. In *Proceedings of the ACM Symposium on Principles of Distributed Computing (PODC)*, Boston, MA, USA, 2003.

[241] F. Kuhn, R. Wattenhofer, and A. Zollinger. Worst-case optimal and average-case efficient geometric ad-hoc routing. In *Proceedings of the ACM Symposium on Mobile Ad Hoc Networking and Computing (MobiHoc)*, Annapolis, MD, USA, 2003.

[242] H. W. Kuhn. Extensive games and the problem of information. *Contributions to the Theory of Games II* (H. Kuhn and A. Tucker, Eds.), Princeton University Press, 1953.

[243] A. Kuzmanovic and E. Knightly. Low-rate TCP-targeted denial of service attacks (the shrew vs. the mice and elephants). In *Special Interest Group on Data Communications (SIGCOMM)*, Karlsruhe, Germany, Aug. 2003.

[244] P. Kyasanur and N. Vaidya. Detection and handling of MAC layer misbehavior in wireless networks. In *Dependable Systems and Networks*, San Francisco, CA, USA, June 2003.

[245] P. Kyasanur and N. Vaidya. Selfish MAC layer misbehavior in wireless networks. *IEEE-TMC*, 4(5), September/October 2005.

[246] T.V. Lakshman and M. Kodialam. Detecting network intrusions via sampling: A game theoretic approach. In *Proceedings of the IEEE Conference on Computer Communications (INFOCOM)*, Miami, FL, USA, 2005.

[247] B. Lamparter, K. Paul, and D. Westhoff. Charging support for ad hoc stub networks. *Computer Communications*, 26(13), August 2003.

[248] L. Lamport. Constructing digital signatures from a one-way function. Technical Report CSL-98, SRI International, Palo Alto, 1979.

[249] J.-O. Larsson and M. Jakobsson. SHAKE. Private communication with M. Jakobsson.

[250] L. Lazos and R. Poovendran. SeRLoc: Secure range-independent localization for wireless sensor networks. In *Proceedings of the ACM Workshop on Wireless Security (WiSe)*, Philadelphia, PA, USA, 2004.

[251] L. Lazos, R. Poovendran, C. Meadows, P. Syverson, and L. Chang. Preventing wormhole attacks on wireless ad hoc networks: a graph theoretic approach. In *Proceedings*

of the IEEE Wireless Communications and Networking Conference (WCNC), New Orleans, LA, USA, 2005.

[252] A. Lee. Optimizing traffic management. *Starhome GmbH*, (http://www.starhome. com), 2006.

[253] J. W. Lee, R. R. Mazumdar, and Ness B. Shroff. Downlink power allocation for multi-class CDMA wireless networks. In *Proceedings of the IEEE Conference on Computer Communications (INFOCOM'02)*, June 23–27 2002.

[254] T. Leinmuller, E. Schoch, F. Kargl, and C. Maihofer. Influence of falsified position data on geographic ad-hoc routing. In *Proceedings of the European Workshop on Security and Privacy in Ad Hoc and Sensor Networks (ESAS)*, Visegrad, Hungary, 2005.

[255] A. Lenstra and E. Verheul. Selecting cryptographic key sizes. *Journal of Cryptology: the journal of the International Association for Cryptologic Research*, **14**(4), December 2001.

[256] P. Levis, S. Madden, D. Gay, J. Polastre, R. Szewczyk, A. Woo, E. Brewer, and D. Culler. The emergence of networking abstractions and techniques in TinyOS. In *Proceedings of the First Symposium on Networked System Design and Implementation (NSDI)*, San Francisco, CA, USA, 2004.

[257] J. Li, C. Blake, D. De Couto, H. I. Lee, and R. Morris. Capacity of ad hoc wireless networks. In *Proceedings of the ACM Conference on Mobile Computing and Networking (MobiCom)*, Rome, Italy, 2001.

[258] J. Li, J. Jannotti, D. De Couto, D. Karger, and R. Morris. A scalable location service for geographic ad hoc routing. In *Proceedings of the ACM Conference on Mobile Computing and Networking (MobiCom)*, Boston, Massachusetts, USA, 2000.

[259] D. Liu and P. Ning. Establishing pairwise keys in distributed sensor networks. In *Proceedings of the ACM Conference on Computer and Communications Security (CCS)*, Washington DC, USA, 2003.

[260] P. Liu, P. Zhang, S. Jordan, and M. L. Honig. Single-cell forward link power allocation using pricing in wireless networks. *IEEE Transactions on Wireless Communications*, **3**(2): 533–543, March 2004.

[261] Y. Liu, C. Comaniciu, and H. Mang. A Bayesian game approach to intrusion detection in wireless ad hoc networks. In *GameNets '06: Proceeding from the 2006 Workshop on Game Theory for Communications and Networks*, Pisa, Italy, 2006.

[262] A. B. MacKenzie, L. Dasilva, and W. Tranter. *Game Theory for Wireless Engineers*. Morgan and Claypool Publishers, 2006.

[263] A. B. MacKenzie and S. B. Wicker. Game theory and the design of self-configuring, adaptive wireless networks. *IEEE Communications Magazine*, November 2001.

[264] A. B. MacKenzie and S. B. Wicker. Stability of multipacket slotted Aloha with selfish users and perfect information. In *Proceedings of the IEEE Conference on Computer Communications (INFOCOM)*, San Francisco, CA, USA, March–April 2003.

[265] D. Maher. United States Patent (No. 5,450,493): Secure communication method and apparatus. http: //www.uspto.gov, 1993.

[266] W. Mao. *Modern Cryptography, Theory & Practice*. Prentice Hall PTR, 2004.

[267] S. Marti, T. J. Giuli, K. Lai, and M. Baker. Mitigating routing misbehavior in mobile ad hoc networks. In *Proceedings of the ACM Conference on Mobile Computing and Networking (MobiCom)*, Boston, Massachusetts, USA, 2000.

[268] J. L. Massey, G. H. Khachatrian, and M. K. Kuregian. SAFER+. In *Proceedings of the First Advanced Encryption Standard Candidate Conference*. National Institute of Standards and Technology (NIST), 1998.

[269] M. Mauve, J. Widmer, and H. Hartenstein. A survey on position-based routing in mobile ad-hoc networks. *IEEE Network*, **15**(6): 30–39, November–December 2001.

[270] Jonathan M. McCune, Adrian Perrig, and Michael K. Reiter. Seeing-is-believing: Using camera phones for human-verifiable authentication. In *SP '05: Proceedings of the 2005 IEEE Symposium on Security and Privacy*, Oakland, CA, USA, 2005.

[271] A. Mehrotra and L. Golding. Mobility and security management in the GSM system and some proposed future improvements. *Proceedings of the IEEE*, **86**(7), July 1998.

[272] A. Menezes, P. van Oorschot, and S. Vanstone. *Handbook of Applied Cryptography*. CRC Press, 1997.

[273] F. Meshkati, M. Chiang, H. V. Poor, and S. Schwartz. A non-cooperative power control game for multi-carrier CDMA systems. In *Proc. IEEE Wireless Communications and Networking Conference (WCNC)*, March 2005.

[274] P. Michiardi and R. Molva. Core: A collaborative reputation mechanism to enforce node cooperation in mobile ad hoc networks. In *Proceedings of the 6th IFIP Communications and Multimedia Security Conference*, Portoroz, Slovenia, Sep. 2002.

[275] D. Mills. A computer-controlled LORAN-C receiver for precision timekeeping. Technical Report 92-3-1, Dept. of Electrical and Computer Engineering, University of Delaware, Newark, March 1992.

[276] D. Mills. A precision radio clock for WWV transmissions. Technical Report 97-8-1, Dept. of Electrical and Computer Engineering, University of Delaware, Newark, August 1997.

[277] Joseph III Mitola. *Cognitive Radio Architecture: The Engineering Foundations of Radio XML*. Wiley, 2006.

[278] D. Molnar and D. Wagner. Privacy and security in library RFID: issues, practices, and architectures. In *Proceedings of the ACM Conference on Computer and Communications Security (CCS)*, 2004.

[279] G. Montenegro and C. Castelluccia. Statistically Unique and Cryptographically Verifiable (SUCV) identifiers and addresses. In *Network and Distributed System Security Symposium (NDSS)*, San Diego, California, USA, 2002.

[280] Oskar Morgenstern and John Von Neumann. *Theory of Games and Economic Behavior*. Princeton University Press, May 1944.

[281] R. Moskowitz and P. Nikander. Host identity protocol architecture. In *Internet draft*, 2005.

[282] J. Mundinger and J.-Y. Le Boudec. Reputation in self-organized communication systems and beyond. In *Proceedings of Inter-Perf'06 (Invited Paper)*, 2006.

[283] R. Myerson. *Game Theory – Analysis of Conflict*. Harvard University Press, Cambridge/London, 1991.

[284] T. Nandagopal, T. Kim, X. Gao, and V. Bhargavan. Achieving MAC layer fairness in wireless packet networks. In *Proceedings of the ACM Conference on Mobile Computing and Networking (MobiCom)*, Boston, Massachusetts, USA, 2000.

[285] J. Nash. Equilibrium points in *n*-person games. *Proceedings of the National Academy of Sciences*, **36**, 1950.

[286] J. Nash. Non-cooperative games. *The Annals of Mathematics*, **54**(2): 286–295, 1951.

[287] R. Nelson and L. Kleinrock. The spatial capacity of a slotted Aloha multihop packet radio with capture. *IEEE Transactions on Communications*, **32**(6): 684–694, June 1984.

[288] B. C. Neuman and T. Ts'o. Kerberos: An authentication service for computer networks. *IEEE Communications*, **32**(9), 1994.

[289] J. Newsome, E. Shi, D. Song, and A. Perrig. The Sybil attack in sensor networks: analysis and defenses. In *Information Processing in Sensor Networks (IPSN)*, Berkeley, CA, USA, 2004.

[290] V. Niemi and K. Nyberg. *UMTS Security*. John Wiley & Sons, 2003.

[291] P. Nikander. A scalable architecture for IPv6 address ownership. In *Internet-draft*, 2001.

[292] P. Nikander, J. Kempf, and E. Nordman. IPv6 Neighbor Discovery (ND) trust models and threats. In *RFC 3756*, 2004.

[293] Y. Nohara, S. Inoue, K. Baba, and H. Yasuura. Quantitative evaluation of unlinkable ID matching schemes. In *Proceedings of the Workshop on Privacy in the Electronic Society (WPES)*, 2005.

[294] M. Ohkubo, K. Suzuki, and S. Kinoshita. Efficient hash-chain based RFID privacy protection scheme. In *Proceedings of the Interantional Conference on Ubiquitous Computing (UbiComp)*, 2004.

[295] G. Orwell. *Nineteen Eightyfour*. Penguin Books, 1983.

[296] M. J. Osborne and A. Rubinstein. *A Course in Game Theory*. The MIT Press, Cambridge, MA, 1994.

[297] G. O'Shea and M. Roe. Child-proof Authentication for MIPv6 (CAM). In *ACM Computer Communication Review*, San Diego, CA, USA, 2001.

[298] P. Kyasanur and N. Vaidya. Detection and handling of MAC layer misbehavior in wireless networks. In *Dependable Systems and Networks*, San Francisco, CA, USA, June 2003.

[299] P. Papadimitratos and Z. Haas. Secure routing for mobile ad hoc networks. In *Proceedings of SCS Communication Networks and Distributed Systems Modelling Simulation Conference (CNDS)*, 2002.

[300] P. Papadimitratos and Z. Haas. Secure data transmission in mobile ad hoc networks. In *Proceedings of the ACM Workshop on Wireless Security (WiSe)*, pages 41–50, 2003.

[301] P. Papadimitratos and Z. Haas. Secure link state routing for mobile ad hoc networks. In *IEEE Workshop on Security and Assurance in Ad hoc Networks*, Orlando, FL, USA, January 2003.

[302] P. Papadimitratos and Z. Haas. Secure message transmission in mobile ad hoc networks. *Elsevier Ad Hoc Networks Journal*, 1(1), July 2003.

[303] P. Papadimitratos and Z. Haas. Secure on-demand distance vector route discovery in ad hoc networks. In *IEEE Sarnoff Symposium*, Prinction, NJ, USA, April 2005.

[304] P. Papadimitratos and Z. Haas. Secure data communication in mobile ad hoc networks. *IEEE Journal on Selected Areas in Communications*, 24(2), February 2006.

[305] P. Papadimitratos, Z.J. Haas, and J.-P. Hubaux. How to specify and how to prove correctness of secure routing protocols for MANET. In *Proceedings of the IEEE CS BroadNets Conference*, San Jose, CA, USA, 2006.

[306] Boingo Wi-Fi Industry White Paper. Towards ubiquitous wireless broadband. Technical report, http: //www.boingo.com/wi-fi_industry_basics.pdf, 2003.

[307] B. Parno and A. Perrig. Challenges in securing vehicular networks. In *Workshop on Hot Topics in Networks (HotNets-IV)*, College Park, Maryland, USA, 2005.

[308] B. Parno, A. Perrig, and V. Gligor. Distributed detection of node replication attacks in sensor networks. In *IEEE Symposium on Security and Privacy*, Oakland, CA, USA, 2005.

[309] B. Patel and J. Crowcroft. Ticket based service access for the mobile user. In *Proceedings of the ACM Conference on Mobile Computing and Networking (MobiCom)*, Budapest, Hungary, September 1997.

[310] C. Peng, H. Zheng, and B. Y. Zhao. Utilization and fairness in spectrum assignment for opportunistic spectrum access. *Mobile Networks and Applications (MONET)*, **11**: 555–576, May 2006.

[311] C. Perkins. *Ad Hoc Networking*. Addison Wesley, 2001.

[312] C. Perkins, E. Belding-Royer, and S. Das. Ad-hoc on-demand distance vector (AODV) routing. Internet RFC 3561, July 2003.

[313] C. Perkins and P. Bhagwat. Highly dynamic destination-sequenced distance-vector routing (DSDV) for mobile computers. In *Proceedings of the SIGCOMM'94 Conference on Communications Architectures, Protocols, and Applications*, August 1994.

[314] A. Perrig, R. Canetti, J. D. Tygar, and D. Song. Efficient authentication and signing of multicast streams over lossy channels. In *Proceedings of the IEEE Symposium on Research in Security and Privacy*, May 2000.

[315] A. Perrig and D. Song. Hash visualization: A new technique to improve real-world security. In *Proceedings of the International Workshop on Cryptographic Techniques and E-Commerce (CrypTEC)*, Hong Kong, July 1999.

[316] A. Perrig, R. Szewczyk, V. Wen, D. Culler, and J. D. Tygar. Spins: Security protocols for sensor networks. In *Proceedings of the ACM Conference on Mobile Computing and Networking (MobiCom)*, Rome, Italy, 2001.

[317] A. Pfitzmann and M. Koehntopp. Anonymity, unobservability and pseudonymity – a proposal for terminology. In *Proceedings of the Workshop on Design Issues in Anonymity and Unobservability*, 2001.

[318] D. Plummer. An Ethernet Address Resolution Protocol, November 1982. IETF Standards Track RFC 826.

[319] Y. Qiu and P. Marbach. Bandwidth allocation in wireless ad hoc networks: A price-based approach. In *Proceedings of the IEEE Conference on Computer Communications (INFOCOM)*, San Francisco, CA, USA, March–April 2003.

[320] M.O. Rabin. Efficient Dispersal of Information for Security, Load Balancing, and Fault Tolerance. *Journal of ACM*, **36**(2): 335–348, April 1989.

[321] T. S. Rappaport. *Wireless Communications: Principles and Practice (2nd Edition)*. Prentice Hall, 2002.

[322] M. Raya, I. Aad, J.-P. Hubaux, and A. Elfawal. DOMINO: Detecting MAC layer greedy behavior in IEEE 802.11 hotspots. *IEEE Transactions on Mobile Computing (TMC)*, **5**(12), 2006.

[323] M. Raya and J.-P. Hubaux. The security of vehicular ad hoc networks. In *Workshop on Security in Ad hoc and Sensor Networks (SASN)*, Alexandria, VA, USA, 2005.

[324] M. Raya, J.-P. Hubaux, and I. Aad. DOMINO: A system to detect greedy behavior in IEEE 802.11 hotspots. In *Proceedings of the ACM Conference on Mobile Systems, Applications and Services (MobiSys)*, Boston, Massachusetts, USA, June 2004.

[325] M. Raya, P. Papadimitratos, and J.-P. Hubaux. Securing vehicular communications. *IEEE Wireless Communications*, 13(5), 2006.

[326] M. Reiter and A. Rubin. Crowds: Anonymity for web transactions. *ACM Transactions on Information and System Security*, **1**(1): 66–92, November 1998.

[327] P. Resnick and R. Zeckhauser. Trust among strangers in Internet transactions: Empirical analysis of eBay's reputation system. In *NBER workshop on empirical studies of electronic commerce*, January 2001.

[328] P. Resnick, R. Zeckhauser, J. Swanson, and K. Lockwood. The value of reputation on eBay: A controlled experiment. In *ESA Conference*, June 2002.

[329] M. Rieback, B. Crispo, and A. Tanenbaum. The evolution of RFID security. *IEEE Pervasive Computing Magazine*, **5**(1), January–March 2006.

[330] R. Rivest and A. Shamir. PayWord and MicroMint: Two simple micro-payment schemes. Technical report, MIT Laboratory for Computer Science, 1996.

[331] R. Rivest, A. Shamir, and L. Adleman. A method for obtaining digital signatures and public-key cryptosystems. *Communications of the ACM*, **21**, 1978.

[332] Tim Roughgarden. *Selfish Routing and the Price of Anarchy*. The MIT Press, 2005.

[333] E. M. Royer and C.K. Toh. A review of current routing protocols for ad hoc mobile wireless networks. *IEEE Personal Communications*, **6**(2): 46–55, April 1999.

[334] J. Saltzer. On the naming and binding of network destinations. In P. Ravision *et al.*, editor, *Local computer networks*, 1982.

[335] K. Sampigethaya, L. Huang, M. Li, R. Poovendran, K. Matsuura, and K. Sezaki. CARAVAN: providing location privacy for VANET. In *Workshop on Embedded Security in Cars (ESCAR)*, Cologne, Germany, 2005.

[336] L. Samuelson. *Evolutionary Games and Equilibrium Selection*. MIT Press, 1997.

[337] K. Sanzgiri, B. Dahill, B. Levine, C. Shields, and E. Belding-Royer. A secure routing protocol for ad hoc networks. In *Proceedings of the International Conference on Network Protocols (ICNP)*, 2002.

[338] S. Sarma, S. Weis, and D. Engels. Radio-frequency identification: Security risks and challenges. *CryptoBytes*, **6**(1), 2003.

[339] B. Schneier. *Applied Cryptography – Protocols, Algorithms, and Source Code in C*. John Wiley and Sons, 2nd edition edition, 1996.

[340] M. Schwartz. *Mobile Wireless Communications*. Cambridge University Press, 2005.

[341] R. Selten. Spieltheoretische Behandlung eines Oligopolmodells mit Nachfrageträgheit. *Zeitschrift für die gesamte Staatswissenschaft*, **12**, 1965.

[342] A. Serjantov and G. Danezis. Towards an information theoretic metric for anonymity. In *Proceedings of the Privacy Enhancing Technologies Workshop (PET)*, 2002.

[343] A. Seshadri, A. Perrig, L. van Doorn, and P. Khosla. SWATT: SoftWare-based ATTestation for embedded devices. In *IEEE Symposium on Security and Privacy*, 2004.

[344] S. Seys and B. Preneel. ARM: Anonymous routing protocol for mobile ad hoc networks. In *Proceedings of the IEEE Workshop on Pervasive Computing and Ad Hoc Communications (PCAC)*, Vienna, AU, 2006.

[345] R. C. Shah and J. M. Rabaey. Energy aware routing for low energy ad hoc sensor networks. In *IEEE Wireless Communications and Networking Conference (WCNC)*, Orlando, FL, USA, 2002.

[346] Y. Shaked and A. Wool. Cracking the Bluetooth PIN. In *Proceedings of the ACM Conference on Mobile Systems, Applications, and Services (MobiSys)*, Seattle, WA, USA, 2005.

[347] S. Shakkottai, E. Altman, and A. Kumar. The case for non-cooperative multi-homing of users to access points in IEEE 802.11 WLANs. In *Proceedings of the IEEE Conference on Computer Communications (INFOCOM '06)*, April 23–29, 2006.

[348] S. Shakkottai and R. Srikant. Economics of network pricing with multiple ISPs. In *Proceedings of the IEEE Conference on Computer Communications (INFOCOM '05)*, March 13–17 2005.

[349] Y. Shang, W. Ruml, Y. Zhang, and M. Fromherz. Localization from mere connectivity. In *Proceedings of the ACM Symposium on Mobile Ad Hoc Networking and Computing (MobiHoc)*, Annapolis, MD, USA, 2003.

[350] C. Shannon. Communication theory of secrecy systems. *Bell Systems Technical Journal*, **28**: 656–715, 1949.

[351] D. Singelée and B. Preneel. Security overview of Bluetooth. Technical report, COSIC, June 2004.

[352] D. Song. dsniff. http: //naughty.monkey.org/~dugsong/dsniff/.

[353] Specification of the Bluetooth System (Core). Version 1.1. http://www. bluetooth.org, 2001.

[354] V. Srinivasan, P. Nuggehalli, C. Chiasserini, and R. Rao. Cooperation in wireless ad hoc networks. In *Proceedings of the IEEE Conference on Computer Communications (INFOCOM)*, 2003.

[355] V. Srivastava, J. Neel, A. B. MacKenzie, R. Menon, L. A. DaSilva, J. Hicks, J. H. Reed, and R. Gilles. Using game theory to analyze wireless ad hoc networks. *IEEE Communications Surveys and Tutorials*, 7(4), Fourth Quarter 2005.

[356] F. Stajano. *Security for Ubiquitous Computing*. John Wiley and Sons, 2002.

[357] F. Stajano and R. Anderson. The resurrecting duckling: security issues for ad hoc wireless networks. In *Security protocols – 7th international workshop on security protocols, Volume 1796 of LNCS*. Springer-Verlag, 2000.

[358] F. Stajano and F.-L. Wong. Location privacy in Bluetooth. In *Proceedings of the 2nd European Workshop on Security and Privacy in Ad Hoc and Sensor Networks (ESAS)*, Visegrad, Hungary, June 2005.

[359] S. Steinbrecher and S. Koepsell. Modelling unlinkability. In *Proceedings of the Privacy Enhancing Technologies Workshop (PET)*, 2003.

[360] D. Stinson. *Cryptography – Theory and Practice*. CRC Press, 1995.

[361] Y. Sun, Z. Han, W. Yu, and K. Liu. A trust evaluation framework in distributed networks: Vulnerability analysis and defense against attacks. In *Proceedings of the IEEE Conference on Computer Communications (INFOCOM)*, Barcelona, Catalunya, Spain, 2006.

[362] H. Takagi and L. Kleinrock. Optimal transmission ranges for randomly distributed packet radio terminals. *IEEE Transactions on Communications*, 32(3): 246–257, March 1984.

[363] G. Theodorakopoulos and J.S. Baras. On trust models and trust evaluation metrics for ad hoc networks. *IEEE Journal on Selected Areas in Communications*, 24(2): 318–328, Feb. 2006.

[364] P. Värbrand and D. Yuan. A mathematical programming approach for pilot power optimization in WCDMA networks. In *Proceedings of the Australian Telecommunications, Networks and Applications Conference (ATNAC '03)*, December 2003.

[365] Vehicle Safety Communications Project. Task 3 Final Report – Intelligent vehicle safety applications enabled by DSRC. Technical Report DOT HS 809 859, US Department of Transportation, National Highway Traffic Safety Administration, March 2005.

[366] W. Diffie and M. Hellman. New directions in cryptography. *IEEE Transactions on Information Theory*, November 1976.

[367] L. M. Wahl and M. A. Nowak. The continuous prisoner's dilemma: I. linear reactive strategies. In *Journal of Theoretical Biology vol. 200*, 1999.

[368] J. Walker. Unsafe at any key size: An analysis of the WEP encapsulation. IEEE 802.11-00/362, October 2000.

[369] W. Wang and B. Bhargava. Visualization of wormholes in sensor networks. In *Proceedings of the ACM Workshop on Wireless Security (WiSe)*, Philadelphia, PA, USA, 2004.

[370] W. Wang, S. Eidenbenz, Y. Wang, and X.-Y. Li. OURS: Optimal unicast routing systems in non-cooperative wireless networks. In *ACM-MOBICOM 2006*, Cologne, Germany, 2006.

[371] W. Wang, X. Liu, and Hong Xiao. Exploring opportunistic spectrum availability in wireless communication networks. In *IEEE Vehicular Technology Conference (VTC)*, Dallas, TX, USA, Sep. 2003.

[372] X. Wang, A. Yao, and F. Yao. New collision search for SHA-1. Rump Session at Crypto'05, 2005.

[373] X. Wang, Y.L. Yin, and H. Yu. Finding collisions in the full SHA-1. In *Advances in Cryptology – Crypto'05*. Springer-Verlag, 1985.

[374] C. Ware, J. Judge, J. Chicharo, and E. Dutkiewicz. Unfairness and capture behaviour in 802.11 ad hoc networks. *IEEE International Conference on Communications (ICC)*, 1, 2000.

[375] S. Weis, S. Sarma, R. Rivest, and D. Engels. Security and privacy aspects of low-cost radio frequency identification systems. In *Proceedings of the First International Conference on Security in Pervasive Computing*, 2003.

[376] D. Westhoff, J. Girao, and M. Acharya. Concealed data aggregation for reverse multicast traffic in sensor networks: Encryption, key distribution, and routing adaptation. *IEEE Transactions on Mobile Computing (TMC)*, 5(10): 1417–1431, 2006.

[377] A. Wood, L. Fang, J. Stankovic, and T. He. SIGF: A family of configurable, secure routing protocols for wireless sensor networks. In *Proceedings of the ACM Workshop on Security in Ad Hoc and Sensor Networks (SASN)*, Alexandria, Virginia, USA, 2006.

[378] H. Wu, Y. Peng, K. Long, and J. Ma. Performance of Reliable Transport Protocol over IEEE 802.11 Wireless LAN: Analysis and Enhancement. In *Proceeding of the IEEE Conference on Computer Communications (INFOCOM)*, New York, NY, June 2002.

[379] X9.17. Financial institution key management (wholesale) – Appendix C. American National Standard X9.17-1995, American National Standards Institute (ANSI), 1995.

[380] X9.62. The Elliptic Curve Digital Signature Algorithm (ECDSA). American National Standard X9.62-2005, American National Standards Institute (ANSI), 2005.

[381] M. Xiao, N. B. Schroff, and E. K. P. Chong. A utility-based power control scheme in wireless cellular systems. *IEEE/ACM Trans. on Networking*, 11(10): 210–221, March 2003.

[382] W. Xu, W. Trappe, Y. Zhang, and T. Wood. The feasibility of launching and detecting jamming attacks in wireless networks. In *Proceedings of the ACM Symposium on Mobile Ad Hoc Networking and Computing (MobiHoc)*, Urbana-Champaign, IL, USA, 2005.

[383] G. Xylomenos and G. Polyzos. TCP and UDP performance over a wireless LAN. In *Proceedings of the IEEE Conference on Computer Communications (INFOCOM)*, volume 2, New York, NY, USA, March 1999.

[384] H. Yaiche, R. R. Mazumdar, and C. Rosenberg. A game theoretical framework for bandwidth allocation and pricing in broadband networks. In *IEEE/ACM Transactions on Networking*, Oct. 2000.

[385] S. Yang and J. Baras. Modeling vulnerabilities of ad hoc routing protocols. In *Proceedings of the ACM Workshop on Security in Ad Hoc and Sensor Networks (SASN)*, Fairfax, VA, USA, 2003.

[386] F. Ye, A. Chen, S. Lu, and L. Zhang. A scalable solution to minimum cost forwarding in large sensor networks. In *Proceedings of the Tenth International Conference on Computer Communications and Networks*, 2001.

[387] T. Ylonen and C. Lonvick. The secure shell (SSH) protocol architecture. Internet RFC 4251, 2006.

[388] J. Yoon, M. Liu, and B. Noble. Random waypoint considered harmful. In *Proceedings of the IEEE Conference on Computer Communications (INFOCOM)*, San Francisco,CA, USA, 2003.

[389] T. Yu, M. Winslett, and K. Seamons. Interoperable strategies in automated trust negotiation. In *Proceedings of the ACM Conference on Computer and Communications Security (CCS)*, Philadelphia, PA, USA, 2001.

[390] J. Zander. Jamming games in slotted Aloha packet radio networks. In *Military Communications Conference (MILCOM)*, Monterey, CA, USA, 1990.

[391] M. G. Zapata and N. Asokan. Securing ad hoc routing protocols. In *Proceedings of the ACM Workshop on Wireless Security (WiSe)*, Atlanta, GA, USA, 2002.

[392] M. El Zarki, S. Mehrotra, G. Tsudik, and N. Venkatasubramanian. Security issues in a future vehicular network. In *European Wireless*, Florence, Italy, 2002.

[393] A. Zemlianov and G. de Veciana. Cooperation and decision making in wireless multi-provider setting. In *Proceedings of the IEEE Conference on Computer Communications (INFOCOM '05)*, March 13–17 2005.

[394] K. Zetter. Feds rethinking RFID passport. *Wired News*, April 26 2005.

[395] J. Zhang, J. Li, S. Weinstein, and N. Tu. Virtual operator based aaa in wireless LAN hot spots with ad hoc networking support. *ACM-MC2R*, 6(3), July 2002.

[396] M. Zhang and Y. Fang. Security analysis and enhancements of 3GPP Authentication and Key Agreement protocol. *IEEE Transactions on Wireless Communications*, 4(2), March 2005.

[397] Y. Zhang, W. Liu, and W. Lou. Anonymous communications in mobile ad hoc networks. In *Proceedings of the IEEE Conference on Computer Communications (INFOCOM)*, Miami, FL, USA, 2005.

[398] J. Zhao, H. Zheng, and G. H. Yang. Distributed coordination in dynamic spectrum allocation networks. In *Proc. of IEEE DySPAN*, Baltimore, MD, November 2005.

[399] H. Zheng and L. Cao. Device-centric spectrum management. In *Proc. of IEEE DySPAN*, Baltimore, MD, November 2005.

[400] S. Zhong, L. E. Li, Y. G. Liu, and Y. R. Yang. On designing incentive-compatible routing and forwarding protocols in wireless ad-hoc networks. *ACM Springer Wireless Networks (WINET), Special Issue of Selected Papers of Mobicom 2005*, 2007.

[401] S. Zhong, Y. R. Yang, and J. Chen. Sprite: A simple, cheat-proof, credit-based system for mobile ad hoc networks. In *Proceedings of the IEEE Conference on Computer Communications (INFOCOM)*, San Francisco, CA, USA, March–April 2003.

[402] C. Zhou, P. Zhang, M. L. Honig, and S. Jordan. Two-cell power allocation for downlink CDMA. *IEEE Transactions on Wireless Communications*, 3(6), Nov. 2004.

[403] L. Zhou and Z. Haas. Securing ad hoc networks. *IEEE Network*, 13(6), November–December 1999.

[404] S. Zhu, S. Setia, and S. Jajodia. LEAP: Efficient security mechanisms for large-scale distributed sensor networks. In *Proceedings of the ACM Conference on Computer and Communications Security (CCS)*, Washington DC, USA, 2003.

[405] C. Zouridaki, B. L. Mark, M. Hejmo, and R. K. Thomas. Robust cooperative trust establishment for MANETs. In *Proceedings of SASN '06*, Alexandria, VA, USA, 2006.

Index